Building in Value

Building in Value

Pre-design Issues

Edited by
Rick Best and Gerard de Valence
University of Technology, Sydney, Australia

A member of the Hodder Headline Group
LONDON • SYDNEY • AUCKLAND

Copublished in North, Central and South America by
John Wiley & Sons Inc., New York • Toronto

First published in Great Britain in 1999 by
Arnold, a member of the Hodder Headline Group,
338 Euston Road, London NW1 3BH

http://www.arnoldpublishers.com

Copublished in North, Central and South America by
John Wiley & Sons Inc.,
605 Third Avenue,
New York, NY 10158-0012

British Library Cataloguing in Publication Data
A catalogue record for this book is available from the British Library

Library of Congress Cataloging-in-Publication Data
A catalog record for this book is available from the Library of Congress

ISBN 0 340 74160 0
ISBN 0 470 35566 2 (Wiley)

1 2 3 4 5 6 7 8 9 10

Commissioning Editor: Eliane Wigzell
Production Editor: Rada Radojicic
Production Controller: Priya Gohil

Typeset in 10/12 pt Times by AFS Image Setters Ltd, Glasgow
Printed and bound in Great Britain by JW Arrowsmith Ltd, Bristol

What do you think about this book? Or any other Arnold title?
Please send your comments to feedback.arnold@hodder.co.uk

Contents

PART 3: IMPLEMENTATION AND MANAGEMENT

To our wives and families

Acknowledgements

We would like to thank a number of people who have helped in one way or another in the production of this book.

Steve Harfield, Associate Dean, in the Faculty of Design Architecture and Building at the University of Technology, Sydney, for his valuable support; Craig Langston for his support and editorial assistance; Eliane Wigzell, for her patience and help with all manner of queries; Valli Moffitt and her team at the DAB Design Studio for their work on the illustrations; and all the contributing authors for their hard work and good humour throughout the lengthy process of preparing the book.

Above all, we thank our wives, Davidia and Natasha, for their patience and tolerance during those long days and nights which we spent chained to our desks doing what editors do.

List of Contributors

Editors

Rick Best – Director of the Construction Economics department at the University of Technology, Sydney (UTS). He has degrees in architecture and quantity surveying and has research interests in information technology in construction, energy in buildings and low energy design. He is currently enrolled in a PhD by thesis in which he is investigating the costs and benefits associated with the use of district energy systems in Australia. Rick worked in both architectural and quantity surveying practices before joining UTS where he lectures in computing, measurement, estimating and design-related subjects.

Gerard de Valence – Lecturer in Construction Economics at the University of Technology, Sydney. He has an honours degree in Economics from the University of Sydney. He has worked in industry as an analyst and economist, and as an economist and task leader in the Policy and Research Division of the Royal Commission into Productivity in the Building Industry in NSW. His principal areas of research activity and interest include the measurement of project performance, the study of economic factors relevant to the construction industry, the analysis of the construction industry's role in the national and international economy, the study of interrelationships between construction project participants, and the impact of emerging technologies.

Contributors

Allan Ashworth – formerly Course Director in Quantity Surveying at the University of Salford and then HMI (Her Majesty's Inspector) with a particular responsibility for surveying education with the Department for Education (UK). He is the visiting Professor in Construction Education at Liverpool John Moores University and adjunct Professor in Quantity Surveying at UNITEC in New Zealand and a consultant with the Quality Assurance Agency for

Higher Education (UK) and the Royal Institution of Chartered Surveyors. His main research interests are in cost modelling and life cycle costing. He is the author of several standard texts in quantity surveying and construction economics.

Stephen Ballesty – General Manager of Rider Hunt Terotech, Sydney, Property Asset Management Consultants. He holds an honours degree in Construction Science and an MBA in Technology Management. In addition, he is professionally qualified as both a quantity surveyor and a cost engineer. He has over 17 years of extensive experience in both property and construction consulting, predominantly in Australia and the USA. He is currently the NSW Chairman of the Facilities Management Association of Australia.

Roy Barton – Associate Professor in the Faculty of Education at the University of Canberra with a background in Construction Management and Economics and Adjunct Professor in Strategic Asset Management at Queensland University of Technology. He has extensive experience in the application of value management and partnering to major projects both in Australia and internationally and was one of the authors of both the *NSW Total Asset Management Manual* and the *Australian/New Zealand Standard for Value Management*, AS/NZS 4183:1994.

Tony Collins – currently Property Development and Investment Consultant for Commercial Property Finance at ABSA Bank, South Africa, following 16 years as an academic. From 1990 to 1998 he was Head of the Construction Management and Quantity Surveying Department at the Cape Technikon in Cape Town, teaching and researching Property/Real Estate Economics and Development. He holds an honours degree in Real Estate Economics and a masters degree in Quantity Surveying.

Drury Crawley – Program Manager for Building Energy Tools at the US Department of Energy in Washington DC. He manages a variety of government-sponsored building energy simulation software projects including EnergyPlus, DOE-2, Building Design Advisor and SPARK. He graduated with a Bachelor of Architecture from the University of Tennessee and is a registered architect.

Grace Ding – Lecturer in Construction Economics at UTS. She has a Diploma from Hong Kong Polytechnic, a Bachelors degree in QS from the University of Ulster and a Masters degree by thesis from the University of Salford. She has practised as a Quantity Surveyor in Hong Kong, England and Australia. Grace is enrolled as a PhD student at UTS and is involved in research and teaching in the area of environmental economics and general practice.

Margaret Durham – holds a Bachelors degree in Economics, and Masters degrees in Law and Politics. She has had a long interest in local government, including serving for two years as Mayoress of her municipality. Her professional career has been in tertiary education, and she has been lecturing at the University of Technology, Sydney, since 1991. She also runs a private consultancy, assisting people through the development process.

Alan Gilpin – Honorary Visiting Fellow in the School of Civil and Environmental Engineering, University of New South Wales; formerly Commissioner of Inquiry (Environment and Planning) for the New South Wales Government, and Chairman of the Environment Protection Authority in Victoria. He has published widely and is recognized internationally as an expert in the field of environmental impact assessment.

Jon Hand – Architect and Senior Research Fellow in the Energy Systems Research Unit at the University of Strathclyde. He has been involved in solar architecture, energy conservation and building thermal assessments in the United States, Africa, Europe and Australia for almost two decades. In terms of building thermal assessments he has worked with and been part of the development efforts on several simulation tools, the latest being ESP-r. Particular areas of interest are the training of simulation users and the evolution of design decision support tools.

Nathan Huon – holds a degree in quantity surveying and an associate diploma in civil engineering. He is employed as a project engineer and contract administrator with a large municipality in Sydney, where he is actively engaged in the procurement of a broad range of construction, refurbishment and maintenance works.

John Knott — currently visiting Scholar to the Faculty of Education, University of Canberra, Associate Director, Centre for Professional and Vocational Education, University of Canberra, a Director of Knott and Associates and an Associate Director of The Australian Centre for Value Management. He has designed courses, established accreditation, mentoring, academic and competency performance requirements and played a key role in the delivery of programmes in Strategic Asset Management and Value Management.

Craig Langston – Associate Professor in Construction Economics at the University of Technology, Sydney. Before commencing at UTS, he worked for nine years in a professional quantity surveying office in Sydney. His PhD thesis was concerned with discounting and life-cost studies. He developed two cost planning software packages (PROPHET and LIFECOST) which are sold internationally, and is the author of two textbooks concerning sustainable practices in the construction industry.

Martin Loosemore – a Senior Lecturer in project management at the University of New South Wales, Sydney. He received his PhD from the University of Reading in the UK and his research interests are in the area of crisis management but particularly in the interface between construction law and management.

Marton Marosszeky – an Associate Professor in the Faculty of the Built Environment at the University of New South Wales and Director of the Australian Centre for Construction Innovation. He is also Chair of the NSW Construction Industry Consultative Committee, the industry's peak body for industrial relations. He has a background in civil engineering in road and building construction, and early in his career he specialized in the structural design of tall buildings. His primary research interest is into innovation, especially in relation to construction project and process management. He has a particular interest in performance measurement and benchmarking as well as safety risk management, quality management and waste management.

Roger Miller – Associate Professor at the University of NSW where he teaches High Rise Building Construction and Information Technology Applications in Building. Upon completing university studies in building and civil engineering, he was employed in contracting in Australia, and engineering in the USA, Sweden and Switzerland. Prior to joining the University, he was employed as a consultant for a major computer company in the application of computer-based systems to building and structural engineering.

John Mitchell – Group IT Manager for Woods Bagot, an international architectural practice with offices in Australia, and South-east Asia. He has focused on the use of computing in the AEC sector, complex building design and health planning since 1975. He holds a degree in architecture and a postgraduate Diploma in Architectural Computing. His current focus is on developing techniques for the modelling of buildings, and the effective transfer and integration of data in CAD building projects. He is a member of several international committees working on solutions for the integrated electronic exchange of information in the construction industry.

Mark Neasbey – a founding director of The Australian Centre for Value Management, with wide experience in workshop facilitation, benchmarking and strategic asset management. He also delivers postgraduate courses accredited through the University of Canberra and the Australian Graduate School of Engineering Innovation. He holds a Bachelor of Arts degree in Human Geography and South East Asian Studies and a Graduate Certificate in Human Resources Development.

L.Y. Shen – Associate Professor in project management and economics in the Department of Building and Real Estate at Hong Kong Polytechnic University. Before becoming an academic in Hong Kong, he worked as a post-doctoral researcher on information management systems within construction companies at Reading University, UK, where he also completed his PhD degree. His major research interests include risk management and sustainable development in construction and real estate.

Peter Smith – Senior Lecturer in Construction Economics at the University of Technology, Sydney. Prior to his current appointment Peter worked for a large professional quantity surveying practice – an international construction and property development company – and has run his own quantity surveying consultancy practice. He has Bachelors and Masters degrees in Quantity Surveying and is currently completing his PhD. He is involved in a wide range of research and consulting activities with a specialty in consumer investment advice in the residential property industry.

Paul Strachan – Lecturer at the University of Strathclyde, Glasgow, and a member of the Energy Systems Research Unit (ESRU). His primary research interests involve the development, validation and application of simulation tools for use in the performance evaluation of building energy systems. He has participated in a number of European and other research projects involving building-integrated photovoltaics, advanced glazings, passive cooling systems, model validation and design tool integration.

John Twyford Lecturer in Construction Economics at the University of Technology, Sydney. He was admitted as a solicitor in 1965 after completing a Diploma of Law. He holds a Masters degree in Law and is in the final stages of an SJD. John spent five years in private practice before becoming the Legal Executive Officer of the Master Builders Association of New South Wales. Whilst at the MBA John helped develop the standard building contracts presently in use in Australia. He is the author of *The Layman and the Law in Australia* and is recognized as a leading contracts expert. His special legal interests are dispute resolution and the theoretical underpinning of contractual relations.

Foreword

Two things are especially striking about the last 800 years of building history. The first is that at no time has there been widespread use of a single or a standard method of procuring buildings; time and again people have tried new ways as a result of their dissatisfaction with previous methods. The second is how little the problems have changed, and how regularly the same ones crop up – in Britain in 1256 Henry III complained he had 'suffered much (financial) damage' at the hands of builders. It is clear from many historical records that, from at least the time of Henry III, the quest for most people commissioning a building has been to get a clear idea, before construction starts, of what the building will be like, when it will be completed, how much it will cost, and that the building will represent good value for the money expended. The records also tell us that, throughout this period, clients have sometimes been satisfied in some or all of these respects, and sometimes not.

Technology and advances in materials science has made buildings more complex and efficient; it would have been an impossible dream to build a 100-storey project in 1256 or to build 30-kilometre-long tunnels under the sea or through rock. The industry is safer, more efficient and better at project delivery. Customers today are not always looking for a bargain (although they're happy when one comes along!), instead they want a building that as well as offering value for money fulfils their cost, time, and performance requirements. Most importantly, they want certainty, not surprises.

This book is all about describing and understanding value in projects. The topics are wide-ranging all related to value for clients and constructors. A project does not start with procurement and end with hand-over – the pre-design stage is increasingly important, as is facilities management and disposal. Many clients not only face the cost of construction, but also the cost of ownership. Concern for the global environment is driving change in the industry, setting new targets in efficiency of design and use.

'Getting it right at the start', managing conflict, risk and information and knowing the way forward will provide an essential guide to students who will be a part of tomorrow's companies.

The editors and contributors have a clear understanding of the issues facing construction now and in the future. This book is a distillation of that knowledge.

In a fast-changing world of construction this book will be an invaluable reference to the many players – it recognises value in building and building in value.

Roger Flanagan
University of Reading, UK

Preface

This book is intended, primarily, to serve as a first reader for students undertaking tertiary-level courses related to property and construction – construction managers, project managers, architects, services engineers, quantity surveyors, cost engineers, facilities managers, to name a few. It should also be a very valuable book for people who are already in the property industry as it not only provides a great deal of solid background material but also seeks to look ahead at some of the ways in which the industry may change in coming years.

As the title suggests, the aim of the book is to explore the relationships which exist between value in buildings and the earliest parts of the process of construction. In order to do this effectively, a wide range of topics have been included, all of which should be addressed at the beginning of the design and construction cycle, and which have some impact on the value of the buildings which are produced.

The individual chapters are not intended to cover their respective topics in such depth that the students need never investigate the subjects further, as each of the topics is worthy of its own book. Instead, the fundamentals of each aspect of pre-design activity are presented and suggestions for further reading are included in the bibliographies. The value of this book is that it brings a range of topics together in one volume, thus providing students with an excellent starting point as they begin to explore the complexities of building procurement and value in building.

The idea for such a book came during a meeting of the staff of the Construction Economics Program at the University of Technology, Sydney. It was noticed that the research interests of the individual staff members, when viewed as a block, were all related to various aspects of promoting better value for money for clients who procure buildings. Those interests ranged from life cycle costing, through total facility management to the intricacies of procurement law. The feeling was that, if all these strands, together with contributions from others who had a special interest in other aspects of value, appraisal, investment and construction, could be consolidated into one book, it would be very useful for students and professionals alike.

The contributing authors are recognized experts in their respective fields, and all have published extensively. While the majority of the authors are from Australia, contributions have also come from England, Scotland, the USA, Hong Kong and South Africa. Many of the authors are academics, while some are from industry, and a significant number are involved in both industry and education.

The concepts and techniques described in the book are not specific to any one country, in fact every effort has been made to make the content relevant to construction in all parts of the world. Inevitably, however, some examples and case studies have been included that are specific to a particular area. These should be viewed merely as examples and readers should make their own comparisons with their local situation.

We hope that those who read this book will gain an understanding of the concept of value in building, and how they, as practitioners in their chosen field, can help their clients to maximize the value they get when they invest in buildings.

Rick Best
Gerard de Valence
November 1998

1

Getting it right at the start

Rick Best and Gerard de Valence†

1.1 Introduction

Everyone likes to think they get value for their money whether they are buying a hamburger, a new car or a holiday. Clients who commission the design and construction of buildings hope to maximize the value that they obtain for the large sums of money they invest in building procurement.

Modern buildings are complex artefacts, and the design, construction and commissioning of a new building is a long and complicated process that involves input from a great many people. These contributors, or participants, include a wide range of design professionals (e.g., architects, engineers, landscape architects), lawyers, financiers, building contractors and subcontractors, project managers and an array of specialist consultants concerned with such factors as environmental impact, acoustics, fire safety and structural integrity.

Some of the issues that influence building projects and their construction are the 'one-off' nature of many projects, the boom–bust cycle characteristic of the industry, short-term thinking and opportunism, the dominance of workplace issues in the minds of many of the participants, and the wide variation in skill levels across the industry which limits the use of new techniques and precludes the use of others.

Analysis of the construction process is commonly expressed in terms of establishing an equilibrium between the three primary concerns of time, cost and quality. The concept is not new but it remains valid: any client would like to construct a facility of the highest quality at minimum cost in the shortest possible time. Therefore, it is the goal of the project team to maximize quality while minimizing cost and time, with various members of the team being responsible, to a greater or lesser degree, for one or more of the three components.

Any change in one component is reflected in changes in the other two, e.g. higher quality may be achieved by the injection of more money and/or an increase in the time allocated to the process; alternatively, an increase in the resources (personnel, plant) applied to a project may reduce the time but increase the cost, and so on.

† University of Technology, Sydney, Australia

Many projects are sufficiently unique that they have many of the characteristics of a prototype. With the exception of projects involving repetitive processes (e.g. freeways or pipelines) construction projects are not much like manufactured items, and the owners and users do not have the benefit of extensive testing and refinement of the product. Nevertheless, users and owners are increasingly applying the same standards of performance and reliability, which they apply to manufactured products, to buildings. The individuality of many projects explains, at least partially, the reluctance of clients to expend the same sort of money and effort that would be applied to more repetitive works, because all of the upfront costs have to be allocated to just the one project.

In order to appreciate the role of the client in the building and construction industry it is important firstly to understand the reasons why varying types of clients undertake building and construction work and also to establish the motivations behind their priorities. The reasons for each client embarking on a building or construction project will vary extensively and it is therefore not surprising that the objectives of the client on any particular project may be significantly different from those of other clients on other projects. These motivations will range from the need for shelter to facilities for the production of complex goods and services.

Many clients have become dissatisfied with the performance and level of service they are receiving from the construction industry and, as such, clients are increasingly demanding better value from the industry. This client dissatisfaction has resulted from poor performance in terms of time, cost and quality, adversarial attitudes resulting in claims and disputes, poor constructability, and low levels of innovation combined with a perceived lack of customer focus. It should also be noted that these concerns are not only directed at constructors within the industry but also towards industry professionals acting as consultants and advisers during the construction process. Cox and Townsend (1998, p. 21) identify a number of barriers to achieving value for money, including: low and discontinuous demand, frequent changes in specification, inappropriate (contractor or consultant) selection criteria, inappropriate allocation of risk, poor quality, inefficient methods of construction, poor management, inadequate investment, an adversarial culture, and a fragmented industry structure.

In general, all clients need to find a balance between time, cost and quality. It has been common in the past for clients to place most weight on cost within this balance. However, with the increasing understanding of the importance of the concept of value adding, many clients are now placing significantly less emphasis on cost in terms of price and greater emphasis on cost in terms of value for money.

1.2 Phases of procurement

De Camillis (1988, p. 64) has suggested that '. . . the designing phase is nothing but the correct premise for the correct management of a project . . .', however it may be more the case that it is, in fact, the *pre-design* activities that set this premise for management, as it is during this phase that the real basis for the design process is set, and it is these activities that determine the 'life' of the design process and of the building which is the end product of that process.

Building projects are dynamic and complex, passing through several discrete phases of initiation, documentation and delivery. Obviously the life of a building does not cease when it is handed over to the client for occupation; however, once the building has been commissioned it becomes the concern of the client and his or her facility managers to run and maintain the building throughout its working life.

With building projects there is often a tendency for clients to rush the front end and to eliminate or diminish the benefits that could result from the exercise of greater care during the pre-design phase. Factors such as the boom–bust nature of the industry help to reinforce short-term thinking and promote this truncation of the pre-planning phase. Other factors, such as the lack of available skills in option evaluation and feasibility analysis and assessment, together with the tendency for site-based issues to dominate the cost equation and the thinking of those involved in building procurement, have added to the relatively low emphasis on the pre-design phase.

1.3 The pre-design phase

The aim of this book is to examine only the *pre-design* stage of the process, and to discuss those factors that affect the client's aim of maximizing the value obtained for money invested in a building and which are part of that pre-design stage. The three underlying ideas on which this book are based are:

1. the need for rigorous early testing of ideas or concepts for function, economic viability and physical suitability
2. the need for a procurement strategy, cost plan, time plan, and project action plan in the pre-design phase
3. the need for clear briefing, particularly in the area of building services, to be provided in the pre-design phase.

Figure 1.1 shows the benefits of good concept development, planning and brief preparation and the decreasing ability to influence final cost as construction proceeds. The aim is to improve the quality of solutions at the earliest stage and thus have the greatest impact on outcomes.

The book is divided into three sections: the first section looks at the concept of value in building and why clients choose to build, and goes on to examine various aspects of the regulatory and statutory controls that must be satisfied before design work can commence.

Section 2 outlines a range of analytical techniques that can be employed and which will assist clients in making informed decisions about questions such as project selection, project feasibility, and the use of cost modelling techniques.

Section 3 deals with the mechanics of building procurement – contracts, dispute management, traditional and emerging procurement systems – and discusses a number of management and implementation mechanisms, such as value

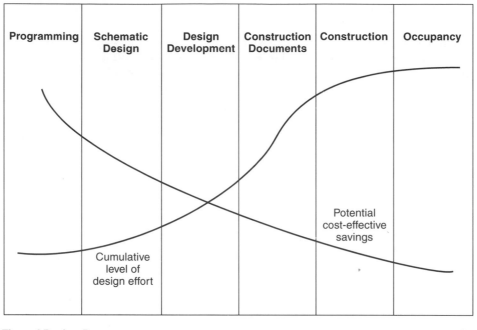

| Programming | Schematic Design | Design Development | Construction Documents | Construction | Occupancy |

Potential cost-effective savings

Cumulative level of design effort

Time of Design Process

Fig. 1.1 Cost/time influence curves.

management and risk management, which can be employed, or at least established, within the project framework at the start of procurement but which flow on into the later stages of the process.

This book, then, is concerned with those things that are done, or can be done, or should be done, before any design work, in the popular sense, is commenced – before even a single sketch for a proposed building is committed to paper or computer screen. Many decisions made during the pre-design phase determine the fundamental nature of many of the final product's attributes, and in some cases such decisions are made by default, occurring only because important questions *are not even addressed* during the early stages of the procurement process.

A common example is the decision fully to aircondition a proposed building – building orientation, or window type and area may receive scant attention if it is simply assumed that air conditioning is to be installed. Even though air conditioning may be an appropriate strategy, the size of the plant required will still be heavily dependent on factors such as orientation, fenestration and structure. Many buildings have sub-optimal internal environmental conditions due to inadequate mechanical systems – inadequate due to budgetary constraints, perhaps, but only inadequate because fundamental decisions regarding siting and envelope were made without clear overall design goals being established. The resultant buildings then are not as valuable as they might be; the potential quality of the final product has not been realized.

1.3.1 Project inception

The decision to build may be based on any of several premises – investment, speculative profit-making, and owner occupation are primary drivers for the building process. Whatever the driving force there must be some perception of a need for a facility, whether it is to house a process (such as manufacturing) or an activity, and the decision must be made that a 'building' is the best solution for the client's needs.

It may be the case that an analysis of the needs or functions that the client seeks to satisfy will suggest alternative solutions, and it may become clear that commissioning a new building project may not be the optimum choice. Alternatives such as a BOOT (build, own, operate and transfer) arrangement, the refurbishment of an existing building or facility, or contracting with an independent service provider (outsourcing) may be seen as more cost effective. Such departures from the traditional solution, where the client commissions the design and construction of a new building, can have major impacts on the quality, and consequently on the value, of the final product as different stakeholders have different expectations.

At the outset, some careful analysis of the client's needs is required to establish the best method by which those needs can be satisfied. Techniques such as cost–benefit analysis, feasibility studies, and multicriteria analysis can aid those involved in reaching a balanced decision to build or to pursue other options.

1.3.2 Impact of pre-planning

The success of any venture generally depends on the quality of the planning that occurs before one embarks on the actual project, whether it be organizing a barbecue or putting a man on Mars. Modern buildings are highly complex and sophisticated mechanisms. Some would be better described as organisms, so complex are their sensory and control systems. They are subject to many constraints and characterized by an ever-increasing range of parameters, all of which must addressed, considered, balanced, and ultimately brought into equilibrium in the final built product.

Adequate pre-planning is essential if the project team is to achieve this equilibrium while balancing time, cost and quality against client requirements, economic, legal and statutory constraints and so on. It requires careful analysis of all constraints and the establishment of a robust framework for decision making and communication.

Particular items that need to be addressed at the outset include planning controls, environmental impacts, the possible effects of future controls such as carbon taxes or emission permits, or changes in fuel prices.

1.3.3 Briefing

Once the decision to build has been made, then regardless of the procurement system adopted, the design brief is the key to arriving at a solution that best satisfies the client's expectations. Inadequate briefing can leave consultants acting in isolation without clearly defined goals and targets. Designers often have difficulty in arriving at successful design solutions as there is insufficient pre-design analysis, i.e. a clear underlying purpose with specific targets is necessary if the final solution is properly to satisfy the client and/or the building occupants (Bordass and Leaman, 1996).

Fundamental to the preparation of a complete and comprehensive brief is the identification of all those constraints that cannot be controlled by the designers. These include:

- physical factors such as the size, shape and location of the site
- financial constraints, usually in the form of a budget or cost limit
- legislative restrictions including zoning and planning constraints, height limits, and floor space ratios
- operational factors such as availability of reticulated services, and hours of operation (which may affect flows of traffic, people or materials into a completed building)
- social concerns including the impacts on others in the neighbourhood, or possible concerns with vandalism
- environmental concerns such as energy use, harmful emissions, and waste disposal
- technological constraints including availability of skilled labour and specialized components and materials.

Functional analysis and needs analysis are techniques that aim to facilitate the briefing process and these techniques should be applied at the outset of the procurement process in order that there is a comprehensive understanding of the needs and constraints that will shape both the design process and the final product of that process.

1.3.4 Assembling the project team

In the same way that a group of individual champions may not make the best sporting team, so an uncoordinated group of brilliant designers may not produce the best buildings. Successful building design requires a great deal of careful management and it is fundamental to the success of the project that the design team be assembled with a view to establishing a cooperative environment based on mutual respect among the various consultants.

Given the complexity of modern buildings, it is important that all members of the team are included in the process from the earliest possible time. Input from all team members must be actively sought, and their participation encouraged and facilitated from the very beginning of the process. The benefits of an integrated, holistic design approach are discussed in detail in a later chapter.

Similarly, the establishment of pre-qualification criteria for consultants and other team members, including contractors and sub-contractors, may provide substantial benefits to the project as the best equipped and qualified people can be added to the project.

1.3.5 Procurement methods

In recent years a number of innovative methods or systems of building procurement have emerged, generally aimed at improving, for a variety of reasons, on the traditional 'design, tender, construct' method. These include design and construct (D&C; also called design and build), fast-track (usually under a construction management arrangement), BOT (build, own, transfer) and BOOT (build, own, operate and transfer).

Selection of a method which best suits any individual project must be based on a careful analysis of a range of project parameters, including risk management and allocation, provision for dispute resolution, fee structures, best practice, performance-based contracts and a host of others.

In a rapidly changing world, with many new construction techniques and materials becoming available, almost on a daily basis, prudent clients will employ procurement methods that encourage and facilitate innovative problem-solving by individual members of the project team and by the team as a whole. This is now more true than ever as the complexity of buildings and the range of available solutions continues to expand. Important considerations include allowing adequate time for pre-design research, design development and documentation, encouraging adoption of best practice, and making provision for dispute avoidance and/or resolution.

1.3.6 Cost

Building cost depends on much more than mere construction cost; it includes related costs such as design fees, interest and holding charges, legal costs, costs of repairs and maintenance, and occupancy and operating costs.

Many studies in recent years (e.g. Peck, 1993) have pointed out that the initial cost of construction represents only a small part of the cost of a building over its lifetime. The cost of design is so small when considered against the life cycle cost of a building that it is almost insignificant – somewhere between 0.1 and 1%. It is in the early stages of the design and documentation process that many of the most important decisions that determine the ultimate value of the final product are made, yet clients regularly seek to cut corners during this period of procurement. It is equally apparent that for many clients capital cost is the dominant factor, particularly in areas, such as Australia, where commercial buildings are more often constructed as marketable commodities rather than as facilities to be occupied by those who initiate their construction.

Peck goes on to say that '. . . many client departments, developers and project managers [make] a virtue out of achieving pre-stated costs and time deadlines without any consideration of the total cost to the community that results from inadequate research, hasty design development, ill considered technical evaluation and poor documentation.'

Cost management is usually associated with monitoring predicted costs during the design process, followed by monitoring and reporting on expenditure during construction including valuation of variations, extensions of time and so on.

During the pre-design phase, however, which is the focus of this book, there are techniques available that can provide valuable assistance to the project team as they set the scene for the start of the design phase: various cost modelling methods, traditional cost planning, functional cost analysis, the establishment of value management procedures, and a commitment to a life-cost approach to the design problem can all lead to increased value for the client.

Similarly, if designers and owners are provided with appropriate advice on possible taxation benefits which could be attached to some aspects of their buildings there can be considerable savings for owners in the longer term.

1.3.7 Time

There has been increasing emphasis on reducing the time taken to design and construct buildings in recent years. This has led to the adoption of procurement methods such as fast-tracking, and design and construct. Many practitioners feel that this has led to a decline in the quality of the buildings produced, yet these methods remain popular as savings in holding charges and earlier returns on investments in construction provide sound financial reasons for their adoption. The question which must be asked is whether the short-term gains are sufficient to outweigh the longer-term inadequacies of buildings that are constructed without the total design being completed before construction commences.

It can be argued that questions of planning of activities (on and off site), constructability and so on should be addressed as part of the pre-design process, thus setting the scene for adequate consideration to be given to these questions during the actual design phase. If these questions are not addressed until the design is largely or even totally complete, construction times may well be increased to a point where any cost benefit arising from fast-tracking is reduced and ultimately exceeded by increased construction costs arising from inadequate pre-planning. This, in turn, reduces the value of the final product.

1.3.8 Information management

Cornick (1996) describes the design process as a series of information inputs and outputs, with value-adding occurring, ideally, during each phase of the process. Efficient and appropriate management of the information flow is necessary if time and cost are to be minimized and quality maximized.

Practical information management system design depends on a robust conceptual model of the design process and a clear understanding of the roles and requirements of all participants. An efficient information management system will ensure that accurate project information will reach those who need it when they need it, while establishing a clear audit trail so that information not only reaches intended recipients but its path to those recipients can be easily traced.

Effective information management can minimize abortive work and re-work, avoid disputes during design and construction, reduce clashes between services, systems and structure, and assist in the speedy resolution of disputes if they do arise.

1.4 Conclusions

The central concern in this book is the value that is embodied in buildings, and how that value can be increased through the application of appropriate measures before any design work actually begins.

In some cases the steps which are suggested may be little more than an agreement between participants in the procurement process that certain procedures or approaches will be part of the overall process, e.g. a commitment to a life-cost approach during design, or the setting up of value management protocols which will be applied at predetermined times during the process.

It is generally accepted that around 80% of the cost of any building is generated by only about 20% of the work items required to construct it. Much of that 80% is related to design decisions made in the earliest stages of the design process. This emphasizes the importance of having a suitable management framework in place before design work commences, based on the application of the sort of techniques discussed here. The alternative is that decisions are made that have far-reaching effects on the ultimate value of the building without a full appraisal or understanding of client requirements, cost implications, constructability problems and so on.

The techniques and methods introduced in the following chapters offer a range of potential benefits to building owners, occupants and designers, as they are intended to assist in the realization of buildings which cost less to build and operate, perform better for their owners and users, and generally provide improved amenity for occupants and the public, and therefore allow designers to serve their clients better – in short, to create better buildings.

References and bibliography

Bordass, B. and Leaman, A. (1996) Future buildings and their services: strategic considerations for designers and their clients. In: *Proceedings of CIBSE/ASHRAE Joint National Conference*, 88–92.

Cornick, T. (1996) *Computer-integrated Building Design* (London: E. and F.N. Spon).

Cox, A. and Townsend, M. (1998) *Strategic Procurement in Construction* (London: Thomas Telford).

De Camillis, G. (1988) Integrated design. *L'Arca*, **15**, 64–5.

Peck, M. (1993) Role of design professionals. *BOMA Magazine*, February, 54–7.

PART 1

Strategic and statutory issues

2

Value in building

Rick Best and Gerard de Valence†

Construction is often discussed in terms of the relationships between time, cost and quality. This has typically been illustrated as a simple triangle (Fig. 2.1), with the three parameters balanced at the corners. Atkin (1990) places feasibility or value for money at what he calls the 'pivotal point' in the centre of the triangle (Fig. 2.2). While it may be the client's objective to have a project that perfectly balances the competing criteria of time, cost and quality, thereby maximizing quality whilst minimizing cost and time, in practice there must be compromise between the three. For example, an increase in quality may require either an increase in cost and/or time. It is important, however, to realize that such an increase in time may be an increase in *design* time rather an increase in *construction*

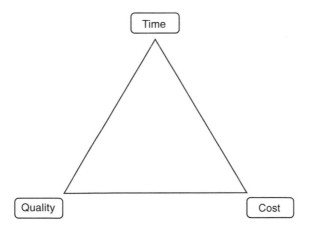

Fig. 2.1 The time/cost/quality triangle (after Atkin, 1990).

† University of Technology, Sydney, Australia

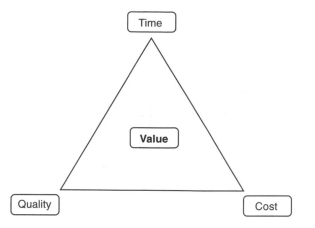

Fig. 2.2 The time/cost/quality/value triangle (after Atkin, 1990).

time, or that more money invested in design time, with a consequent increase in cost (design fees) may shorten construction time and so on. The interdependencies which exist between the three criteria are complex and require careful analysis during decision-making procedures.

Design costs represent as little as 0.5% of the cost of a building over its life yet, when as little as 1% of a project's up-front costs have been spent, 70% of the project's life costs have been committed (Romm, 1994). During construction the bulk of variations and cost extras are the result of inadequate research or poor documentation during the design phase of the project (Newton and Hedges, 1996). RMI (1998) discusses 'front-loaded design', which sees money spent in the initial stages of design that is repaid many times over during the course of construction and operation of the building.

2.2 Value and quality

It is pertinent at this point to clarify the meaning of the terms, 'value' and 'quality' in the context of this discussion.

In classical economics the *value* of an item is related to its utility, and this is subject to the law of diminishing returns (the more of an item one has the lower the utility or satisfaction gained from acquiring another one). The price of something is determined by demand and its relative scarcity; when there is plenty of some commodity available or there is little demand for it then the item has a low price, which expresses the relationship between supply and demand for the item. When discussing the value of a building there are usually other factors that must be considered apart from the basic supply and demand question, and there are many aspects of satisfaction or degree of utility derived from completing a building project successfully that are not captured by price alone.

In an important book, Price (1993) defined value as 'the amount of desirability obtainable from . . . a product consumed'. In the present context desirability may exist in one of two main forms: a building may be desirable because it can be used ('consumed') to accommodate a set of activities that are important to the user (housing, manufacturing, storage etc.), or it may be desirable as a producer of revenue (selling, renting). Additionally, it may be desirable through some association (history, sentimentality) or as a symbol (corporate, political, nationalistic). The varying expectations of building owners means that value has different meanings for different people: in considering the differing systems of procurement for office buildings which prevail in the UK, compared with that in other parts of Europe, Duffy (1988) suggested that 'the exchange value of office property [in the UK] matters more than use value'. This, he suggests, follows from the fact that 'most creative energy in current British office design stems from developers who build for profit first and consumers second.'

The cost of a building may be viewed as its *exchange value* – the amount of money that one party is willing to pay in exchange for ownership of an asset. For a developer the value of a building may be seen as the difference between total capital cost and the maximum possible selling price. Keeping capital cost to a minimum while achieving the highest quality possible should maximize the value of the final product and thereby maximize the return on the developer's investment.

If a developer plans to build, then retain and lease the building which he constructs, then the equation changes; value now depends on capital cost plus maintenance costs compared with rental income, while disposal costs or sale value at some time in the future may also be significant. Simply keeping capital cost at a minimum is no longer sufficient, and the question of quality in terms of attracting and retaining tenants, takes on greater importance.

The scene changes again when the developer is in fact an owner/occupier, as the costs to be considered and minimized are now not only capital cost, but must also include occupancy costs and maintenance costs. The value to the owner is the *use value* and, apart from avoided rent and tenancy costs, this may include corporate symbolism, an image of permanence, or an impression of the progressive nature of a firm, as well as the obvious value of providing a place for the owner to conduct business and generate profits.

In calculating the use value to an owner some other factors may be significant, such as reduced mobility (it is a greater undertaking to sell and relocate than to finalize a tenancy and relocate), and reduced corporate flexibility (changes in the size of the company, or the nature of its business may be more easily accommodated by changing premises, but owner/occupiers are more constrained than leaseholders).

A third type of value can be termed *esteem value*; this represents the 'attractiveness' or 'desirability' of an object.

In practical terms, however, the value of an article, whether a building or some other object, is actually a combination of the various types of value outlined above. In assessing whether an outcome represents value for money there must be some balancing of both the subjective and objective considerations that are of importance to the person investing:

> Value includes subjective considerations that highlight the relationship between what someone wants and what he/she is willing to give up in order to get it. Value is thus relative and not an inherent feature of any object. Value

is commonly applied to assets, is measured by comparison with other assets of similar function, attractiveness, cost and/or exchange worth and cannot be assessed in isolation. (UTS, 1996)

Any attempt to quantify value in monetary terms must take account of the varying worth that different individuals will attach to the subjective factors related to value. These include intangibles such as prestige, views and historical significance. When market value, that is, the most probable selling price a property would attract if offered for sale, is assessed, due allowance must be made for these factors as well as the more readily quantified component such as the cost of physical replacement of the actual building.

The Building Research Establishment (1976) defined value as 'quality in relation to cost' and suggested that 'maximum value is then in theory obtained from a required level of quality at least cost, the highest level of quality for a given cost, or from an optimum compromise between the two.'

When we speak of the quality of an article, such as a car or a piece of furniture, we tend to think in terms of how well the item is made: whether the components fit together well, whether the finished surfaces are smooth, or how long we expect the item to last.

Quality in a building depends on a range of variables, and it involves much more than the simple parameters such as the visible standard of finishes, structural soundness, or making components fit within close tolerances. Brandon (1984) suggests that quality in building design embraces 'all the aspects by which a building is judged including spatial arrangement, circulation, efficiency, aesthetic[s], flexibility as well as its functional ability as a climate modifier and as a suitable structure.'

Quality in this context, then, may include the following:

- level and type of services provided – air conditioning, communications, lighting and the like
- performance of services – how well the services fulfil their intended functions
- flexibility – the capacity for re-use or change of use
- fitness for purpose – how well the final product serves the intended function
- uniqueness – this may have value if the building is a corporate headquarters or has some other symbolic role, e.g. as a model of environmentally sensitive design
- natural site attributes – availability/utilization of a view, or access/proximity to other localities, installations
- minimized costs-in-use – resilience, durability
- minimized occupancy costs – low operating costs
- extended useful life – durability, flexibility
- capacity to produce a financial return – sale or lease
- capacity to provide a productive working environment – comfortable, stimulating
- provision of optimum indoor environment – thermal comfort, indoor air quality, absence of sick building symptoms
- security
- exclusion of external climatic factors – wind, rain, temperature extremes
- minimized environmental impact – of increasing importance as public concern for 'green' issues grows.

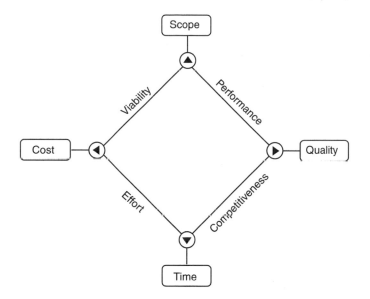

Fig. 2.3 The scope/time/cost/quality diamond.

Value then may be maximized, at least in part, by maximizing the various quality parameters applicable to any given project. These are all in relation to the specific characteristics of a given project. As Fig. 2.3 shows, one can go beyond the time, cost, quality triangle to include the scope of a project (the project objectives and their relationship to the client's strategic plan) and create a diamond.

2.3 Value, quality and better buildings

If increasing the quality of a building in turn increases the value of that building, will the result be a 'better' building? The answer to this question depends greatly on the perspective from which the building is evaluated, but in general terms the result should be better buildings – inasmuch as improved quality and value should lead to greater client satisfaction, provided, of course, that these goals are achieved without excessive cost.

Ferry and Brandon (1984) illustrate the search for value in building design by considering a theoretical minimum expenditure that would be required to provide the client with the most basic structure that would accommodate his or her needs. The process of design then searches for improvements to this basic facility which will eventually raise the quality of the design to a level the client finds acceptable in terms of quality and cost. There is a point beyond which additional expenditure may not provide more value to the client: extra money spent on further improvements would actually produce greater *utility* for the client if it were spent on other projects or simply invested with a financial institution.

A common example of this situation occurs when someone spends a great deal of money on making improvements to their home that are specific to their lifestyle but which are not attractive to future potential buyers of the property. Often, money spent in this way does not translate into a corresponding increase in market value, yet the increase in utility for the homeowner who makes the improvements may have been increased substantially. The value of those improvements therefore varies with the needs and expectations of the owner or buyer.

A similar situation arises when improvements are made to a property that is located in a less desirable area: a mansion in an industrial area will not be worth as much as the same house located in a leafy suburb where it stands among other comparable properties regardless of the fact that the cost to construct the two properties is the same.

2.4 Cost and quality

A common approach in the past has seen clients aim to build their buildings to an 'acceptable' standard, with such standards determined by a combination of statutory requirements and market forces, at minimum cost and in minimum time, with cost in this case being determined, at least in part, by the quality required by the target market.

In recent times the attributes of minimum cost and acceptable quality have undergone some searching analysis, with the result that for many theorists, and more recently for some enlightened clients, perceptions of these attributes have changed:

- Building cost is now generally acknowledged to include more than capital cost alone, with clients and owners becoming more aware of life-cost considerations.
- Capital cost, as well as life cost, may be reduced by reapportioning budgets, with savings in overall initial cost being made by judicious extra outlay devoted to certain building components, e.g. 'superwindows' that have a higher initial cost may allow the elimination of expensive items of mechanical plant, thus producing a double benefit in the form of avoided capital cost and reduced running costs. As Lovins (1995, p. 80) suggests: '. . . big savings are cheaper than small savings' based on the notion of '. . . single expenditures with multiple benefits'.
- 'Acceptable quality' may now include positive public perception of buildings ('green is good') which may even provide a competitive advantage for owners and occupiers of buildings, while informed clients are looking for buildings that provide a healthy, stimulating and more productive working environment for their staff, and are becoming increasingly aware of the possibilities that exist for managing, and thus minimizing, the energy and other running costs associated with their buildings.

Lovins and Browning (1992) have suggested that the barriers which presently stand in the way of achieving 'better' buildings – i.e. higher quality – and therefore more valuable buildings, are not technological but institutional. They define these barriers as 'design process, fee structure, and communication among disciplines'. These areas are discussed in detail in Chapter 19.

2.5 Procurement and value

The history of procurement over the last few decades shows a strong trend toward greater professionalism and specialization by industry organizations, with the growth of project management, structured financing and increased subcontracting of site work being good examples. The main trends are shown in Table 2.1, which suggests that this will be a major area of innovation and evolving ideas in the near future.

The actual definition of procurement that applies here is from Hawk's study of the 60 largest global contractors, from which Table 2.1 was adapted. He describes procurement as a strategic activity:

> Procurement . . . includes the R&D for new and traditional materials for new and old uses. It includes development of new production methods of materials and new means of producing products with these materials. It can involve issues of automation, efficiency, product life cycle studies and responses to problems of environmental deterioration and pollution. Linked to an improved model of design, procurement can be extremely important to transforming the industry. (Hawk, 1992, p. 47)

Procurement has become a more central issue to the industry for several reasons. Clients have been demanding 'better value for money' since the early 1970s, driven partly by the industry's own poor productivity performance and partly by comparisons with manufacturing productivity growth. Studies such as Naoum (1994), Walker (1996) and lean construction reports (e.g. Miles and Ballard, 1997) show clearly that non-traditional procurement systems have better time performance and cost benefits. Following Hawk's argument above, non-traditional procurement systems would also offer better opportunities for 'development of new production methods of materials and new means of producing products with these materials'.

This is the driving force behind the current thinking about procurement and innovation. As better benchmarks and performance measures evolve across a range of building features the idea of value is beginning to lose its fuzzy 'eye of the

Table 2.1 Trends in procurement

1970s	Emergence of formal construction management systems as clients expand their methods and scope. Professionals trained in architecture and engineering assume a greater role in development process.
1980s	Real estate, development, financial arrangements and cash management become critical. Alternative forms of ownership emerge, e.g. property trusts, syndicates, corporatization and privatization.
1990s	Total design solutions linked to new forms of company and project organization become critical to value adding. Clients seek single-point responsibility with the objectives of improved quality and efficiency of product.
2000	New procurement and organizational forms, ideas and methods emerge to deliver further improvement in product quality and efficiency.

beholder' quality and become focused on the key performance indicators that are relevant to that building. Measures of building performance may be based on earnings per square metre of office space, or revenue per square metre of retail space, and so on. In all cases the better the building's internal environment and the better the systems that support it, the better the performance. The benefits include lower health costs and sick days in offices, fewer mistakes and improved productivity in the workplace, and higher sales in well designed retail, and these can often be delivered through the combination of good design and quality services.

One of the issues that value in buildings raises is liability for post-completion defects. Latham (1994) argued that clients are entitled to satisfaction at the end of the defects liability period, and that defects should not be inevitable in a building or other construction. He suggested a better option than retention mechanisms, which build up funds to offset defects during the course of the project, could be retention bonds, which reduce in value as each milestone of the project is completed. Problems arise if the defects remain hidden (latent) during the liability period and only emerge afterwards. Latham recommended provision for compulsory latent defects insurance for 10 years from practical completion for all future new commercial, retail and industrial building work (including any public sector schemes), with a minimum value cut-off point subject to periodic review in the light of current building costs.

The method of project procurement and delivery chosen by the client determines the nature of competition in the industry. If those in the industry are seen as players of a competitive game, then the rules of that game are set by clients. The evolution of procurement and delivery strategies and policies developed by clients has changed the rules of the game over time, and clients are seeking ways to raise the standard of performance. At the end of the day, the industry competes on a playing field defined by clients under rules set by clients. Thus, the Latham Report (1994) urged the UK public sector to play a leading role in restructuring the industry, as did Gyles in Australia (RCBI, 1992). In the US the CII has promoted best practice to both clients and contractors with significant project time and cost results. Singapore's CIDB is the archetypal leading edge client (CIDB, 1994), and there are many other government ministries and agencies who are influencing construction industry performance around the world. Examples from the private sector include British Airways (BA), Toyota, Coca-Cola, Shell, Hewlett-Packard, Intel and Ciba-Geigy.

As a result of the push for better value, there is a renewed effort in investigating tendering and procurement practices around the world. The factors behind this interest are diverse, but greater efficiency and lower costs are universal goals, and a significant opportunity for the industry comes from the fact that improved environmental performance comes from better designed, more efficient buildings, which deliver lower through-life costs. In addition, a major impetus has come from the increasingly widespread use of IT and electronic procurement by large public and private clients.

In Australia, governments are experimenting with electronic procurement and alternative delivery methods. The NSW Government (1997) argues that innovative ways of procuring goods, services and infrastructure will increasingly distinguish one organization from another. Clients are raising their expectations and setting new standards and, in determining the standards of performance, they expect they will increasingly monitor performance and reward the better performers. The NSW Government is challenging local construction industry firms to show that they can produce client-driven, high quality, value-for-money outcomes, and sustain long-term relationships:

> In future, the services packaged for clients will extend beyond the familiar range of construction services associated with a single project. For example, to gain competitive advantage and improve profitability a client may seek the means by which purchase and ownership costs can be reduced (including investment in capital infrastructure) and the introduction instead of management systems that transform purchasing, inventory, and materials management activities into strategic, value-added, business functions/ outcomes. (NSW Government, 1997, p. 15)

The argument is that owners and clients are increasingly using a variety of alternative procurement methods aimed at reducing cost, achieving time schedules and milestones, shortening duration, reducing claims, and improving constructability and innovation. The overall trend is toward versions of design-build and turnkey construction because of the advantages of a project delivery system that combines designers, builders and sometimes suppliers into a single entity, to solve the problems inherent with traditional low-bid procurement (de Valence, 1997, 1998). The effects of alternative procurement methods on the structure and performance of the construction industry will be profound, and will be a major driver of industry development.

2.6 Conclusions

Cost, quality and value in buildings are closely interrelated. The concepts of quality and value cannot be simply defined, and cost may be viewed in many ways across a broad spectrum of possibilities, from simple capital cost to a complex array of cost components which include design fees and even demolition costs.

Any attempt to increase value for money in building must be based on an understanding of the various factors that influence the ultimate value of a construction project, and requires the selection and application of a range of appropriate techniques. The remainder of this book is devoted to the introduction and explanation of many of the available tools that can be used early in the procurement process, and which will assist clients in improving the value of the building projects that they initiate.

References and bibliography

Atkin, B. (1990) *Information Management of Construction Projects* (Sydney: T.W. Crow Associates and Crow Maunsell).

Bordass, W. and Leaman, A. (1996) Future buildings and their services: strategic considerations for designers and their clients. In *Proceedings of CIBSE/ASHRAE Joint National Conference*. Harrowgate.

Building Research Establishment (1976) *A Survey of Quality and Value in Building*, BRE Report, UK.

Brandon, P.S. (1984) Cost versus quality: a zero sum game? *Construction Management and Economics*, **2**, 111–26.

CIDB (1994) *Ten Year Commemorative Publication: Building on Quality* (Singapore: Construction Industry Development Board).

Cornick, T. (1996) *Computer Integrated Design* (E. & F.N. Spon).

de Valence, G. (1997) Endogenous growth theory, construction productivity and industry development. In *Proceedings: First International Conference on Construction Industry Development*, National University of Singapore.

de Valence, G. (1998) Trends in buildings and procurement: implications for construction economists in the twentyfirst century. In *Proceedings: PAQS Conference, Construction Economics in the Twentyfirst Century*, Queenstown, NZ, June.

Duffy, F. (1988) The European challenge. *Architects Journal*, 17 August, 30–40.

Ferry, D. and Brandon, P. (1984) *Cost Planning of Buildings*. 5th Edition (Granada).

Hawk, D.L. (1992) *Forming a New Industry: International Building Construction*. (Stockholm International Business School).

Latham Report (1994) *Constructing the Team: Joint Review of Procurement and Contractual Arrangements in the U.K. Construction Industry* (London: HMSO).

Lovins, A. (1995) The super-efficient passive building frontier. *ASHRAE Journal*, June, 79–81.

Lovins, A. and Browning, W. (1992) Green architecture: vaulting the barriers. *Architectural Record*, December.

Miles, R.M. and Ballard, G. (1997) Lean construction: application in high technology facility construction. In *Proceedings: First International Conference on Construction Industry Development*, National University of Singapore.

Naoum, S.G. (1994) Critical analysis of time and cost of management and traditional contracts. *Journal of Construction Engineering & Management-ASCE*, **120**, 687–705.

Newton, A. and Hedges, I. (1996) The improved planning and management of multi-disciplinary building design. *CIBSE/ASHRAE Joint National Conference*, pp. 120–30.

NSW Government (1996) *The Construction Industry in NSW: Opportunities and Challenges* (Sydney).

NSW Government (1997) *A Perspective of the Construction Industry in NSW in 2005*, Discussion Paper. Sydney.

Peck, M. (1993) Role of design professionals. *Building Owner and Manager*, February, 54–7.

Price, C. (1993) *Time, Discounting & Value* (Blackwell).

RCBI (1992) *Final Report* (Sydney: Royal Commission into Productivity in the Building Industry in NSW).

RMI (1998) *Green Development: Integrating Ecology and Real Estate* (Rocky Mountain Institute, John Wiley & Sons).

Romm, J. (1994) Lean and clean management. In RMI (1998) *Green Development: Integrating Ecology and Real Estate* (Rocky Mountain Institute, John Wiley & Sons).

UTS (1996) Value in Building, Data Sheet 13, *Owners and Builders Information Series*, Construction Economics Unit (University of Technology, Sydney), Building Services Corporation.

Walker, D.W. (1996) Construction time performance and traditional versus non-traditional procurement methods. *Journal of Construction Procurement*, **3**(1), 42–55.

3

Project initiation

Gerard de Valence†

Editorial comment

There are many ways that the earliest stages of a building project are described, and there is probably no single 'right' way to go about the many tasks involved. However, when faced with a complex, interrelated set of tasks and requirements it is often better to have a roadmap, or equivalent guide to the process, at hand. Many project management texts cover this topic in various ways, and there is surprisingly little agreement on what project initiation actually involves. The view taken here is a broad one.

Project initiation as described here is a sequence of activities that have the property of being logically structured in such a way so that the chances of making a bad decision are minimized. At each step in the sequence there is a feedback and checking loop that gives a 'reject/proceed' result, and puts a premium on the rigorous analysis of the project and options available. The various external and internal factors are accounted for in the analysis, and the regulatory and institutional requirements are considered.

This chapter outlines one approach to the beginnings of a building project, and sets out a sequence of activities that could or should be carried out in order to get value for the money spent on the project. Many of these activities are discussed in more detail in later chapters, feasibility studies and value management for example, and it is not the intention here to detail the particular techniques used in them.

What this chapter does is provide a context for many of the tools and techniques introduced in this book. The book itself can, in a sense, be seen as a toolkit that provides a range of tools to attack specific problems that can prevent achieving value in building. This chapter is, therefore, a guide to that toolkit, and it identifies the role and purpose of many of these tools.

By focusing on the importance of using the range of tools covered in this book as early as possible in the project's life, and showing how many of these tools can support each other by bringing more analysis and information to the project team, this chapter gives an insight into the process of project initiation. By emphasizing the importance of completing each stage of the initiation process satisfactorily, the chapter highlights how easy it can be for a project to get carried away by its own momentum.

† University of Technology, Sydney, Australia

This book essentially argues that it is through applying all the analytical tools available to a project that value is optimized, and that not doing the analysis leads to value destruction. Getting project initiation right is a complex and multi-faceted task, as this chapter shows.

3.1 Introduction

Before any of the detailed planning activities commonly associated with constructing a building begin, such as design or scheduling, there is a phase where the idea of the building is taken from a general set of ideas and requirements to a specific and well-defined project. This process of project definition can be described as project initiation. Although various authors have their own terminology and descriptions, the idea that the first stage of a building project should lead to a clear definition of the project is the common theme. Getting to a good project definition is the subject of this chapter.

For example, Turner (1992) describes projects as having four stages: germination, growth, maturity and death. In his terms, germination relates to what is generally known as the initiation stage. Here, a broad feasibility is produced on the basis of an initial project definition and, after a review of the conceptual estimates against the corporate objectives, the go or no-go decision is made.

The growth stage relates to the design appraisal process, where design alternatives are generated from the project objectives and are evaluated. The end result should be the selection of design solutions that best meet the project objectives, and cost estimates for these solutions. Maturity relates to the execution phase where the detailed planning of the selected design solution is implemented and monitored. Project estimates continue to be refined and the development of detailed forecasting and monitoring systems that are to be put in place continues. Death describes project finalization and covers debriefing of the design team, identification of achievements and problems, the establishment of historical data, and audit and review of the process.

By contrast Healy (1998) takes a different approach to this early stage in a project. He views the early stages of a project as essentially political in nature, with a need for a project champion to emerge and shepherd the development of the project idea through the concept stage. In his book this is described as the planning stage, where the outline of the project and the client requirements are refined before a design brief is prepared.

This chapter draws on the Best Practice *Project Initiation Guide* produced by the Construction Industry Development Agency (CIDA, 1993), an Australian Commonwealth Government agency that ran from 1992 to 1995. CIDA's mission was to be a catalyst for change in the Australian building and construction industry through leadership, motivation and the development of a culture of learning and continual improvement. By setting up performance standards, and undertaking consultation with stakeholders to identify factors for success and remove barriers to change, CIDA aimed to make the dynamic change process in the industry self-sustaining. CIDA worked with industry to produce a number of major policy initiatives in pre-qualification criteria, Codes of Conduct and Tendering, and to develop best practice and continuous improvement tools. The Pre-Qualification Criteria (PQC) were the major outcome from CIDA, and were designed to lift the performance of the whole construction industry for the benefit of clients. That said, the original *Guide* has been extensively revised and rewritten for this chapter.

3.2 The project initiation process

The project initiation process can be generalized into four steps. These steps and the important elements in each one are shown in Fig. 3.1. The first starts with the

Methods		Outcomes
• strategic planning • opportunity identification • intuition	**Project Idea Sources** ▼ **Idea**	• ideas
• strategic planning check • develop a project strategy • market research • strategic value management	**Concept Development** ▼ **Evaluation Brief**	• outline description • basic criteria • objectives • functional analysis • evaluation brief and actions
• identity constraints • describe option range • select short list • test function and objectives • cash flows • risk and NPV • special risk	**Evaluation** ▼ **Definition Brief**	• identification range of options • preferred option described: physical; functional; financial • definition brief and actions
• identify trade-offs • project value management • document • define procurement strategy • revise feasibility • cost and time planning	**Definition** ▼ **Delivery Brief**	• descriptive and illustrative definition of approved option • feasibility statement • procurement strategy • time and cost plan • delivery brief

Fig. 3.1 The project initiation process.

strategic planning process to identify opportunities and generate ideas on what is required, the second leads into development of the concept and the development of a project strategy, in the third there is evaluation of the options and analysis of project financials, and finally the definition of the project and preparation of the design brief.

3.2.1 Project requirements

The crucial ways that working through these steps adds value to a building are, firstly, by avoiding costly mistakes, and secondly, by making sure that the project will meet the objectives of the client and users. An example of the first is environmental planning and approval requirements, which can cause significant project delays and increased costs. Awareness of environmental planning and approval policies and regulation helps clients to choose options and solutions that reduce these impacts and provide greater certainty for project outcomes. Aligning business objectives with projects is a major topic in its own right, Turner devotes the first third of his (widely used) book to this, and it is the focus of the emerging field known as corporate real estate.

Rushing the concept and evaluation phase misses the best and lowest cost opportunity to get the project right. Changes made at the concept stage cost little but can have a major impact. In contrast, late changes are expensive or difficult, and disrupt and dislocate project delivery, leading to cost and time overruns.

In the first stage of developing the idea of the project the methods used are the same as those found in most strategic planning exercises. This is because clients need to develop a project concept to a stage that would allow evaluation against their Business Plan: a project is either needed to further business goals or it is unnecessary.

3.3 Concept development

Concept development should include a range of reasonable alternative options, including a no-build option. There are, however, numerous examples of projects proceeding to detailed design stage without this evaluation because there are common failures in the current practice of concept development. These failures include poor definition of the concept behind the project and project concepts that are not supported by the market or business needs of the client. A project needs to have clearly stated objectives and functional requirements, and actively to involve relevant stakeholders in concept development and analysis, if there is to be adequate briefing for the evaluation phase.

The lack of agreed standards and competing methodologies for business planning have made it difficult to build business planning into the project initiation process. Many projects still tend to be initiated on the basis of a good idea, an urgent need or circumstances, rather than flowing from a properly developed business plan. Figure 3.2 shows that the first step in concept development is to develop a project strategy that accords with the client's strategic business planning.

Methods		Outcomes
• strategic planning check • develop a project strategy	**Concept Identification**	• concepts
• market research • strategic value management	**Needs Analysis**	• outline description • basic criteria
• identify constraints • describe range of options • select short list	**Evaluation Brief Preparation**	• refined description • ranked set of objectives • establish analysis by: function/use; cost/benefit
• identify trade-offs • project value management • document • define procurement strategy • revise feasibility • cost and time planning	**Evaluation Brief**	• detailed description of action • required scope of work for next phase

Fig. 3.2 Concept development.

The concept development phase of project initiation has three stages: concept identification, needs analysis, and development of the evaluation brief.

3.3.1 Concept identification

Concept identification should ensure that the project concept is consistent with the client's strategic business planning. To satisfy the requirements of this stage, the client's strategic direction and strategic planning (which should use measurable performance indicators) will identify construction solutions that support these quantified objectives.

Concept identification also tests alternative concepts and the impacts of environmental and planning approval. It assesses whether the concept conforms with planning policies and regulation and asks if the project requires utility or infrastructure adjustments.

The outcomes from this stage include detailed results from conceptual analysis, identification of the catalyst for the project and the expected results. At this point a recommendation whether to proceed or not to proceed to the preliminary phases of the project can be made. Based on the developed concept a cost plan framework can be put into place.

3.3.2 Needs analysis

A comprehensively compiled and studied needs analysis will avoid poor outcomes. Without clearly identifying the concept and ideas driving the project, design development can result in significant rework of design or the risk of poor functional or service outcomes. In order to avoid this a needs analysis has to be undertaken. Needs analysis involves asking questions like 'What are the project's objectives?', 'What is the state of the market (in terms of demand etc.)?', 'What are the expectations and plans of likely users and customers?' and 'Which business or service needs are met by the project and how well?'

Inclusion of a needs analysis in the initiation process ensures rational analysis and avoids decisions that are driven by, for instance, budget allocation, availability of finance, or are an outcome of some political process.

Market research and analysis of historical trends (for example rates of absorption of office space) and comparable rates of return (for example income per square metre for retail centres) is the starting point for this stage. Understanding the market in which the building will exist and compete is essential if it is to be a successful investment.

There is considerable research that suggests that successful outcomes require all parties clearly to understand project objectives. Strategic Value Management (SVM) is one technique that can be used as a key to presenting a clear set of client objectives/needs for the project and to answer the key functional questions up front. By establishing this at the earliest stage of the project and tying it to the business plan for the organization, a coherent framework is established that enables the project to proceed to the next stage. This applies also to the users and workers who will be occupying the building and who have the clearest ideas about the operational requirements that the building must meet.

Options should be built into the Strategic Value Management process and treated by broadly describing and pricing each option, testing each option against function and objectives, and subjecting a smaller subset of options to more detailed description and pricing.

The tasks in this stage, then, consist of market research and analysis, identification of the functional needs and outcomes of the project, prioritization of the objectives and functions, and an evaluation of the project objectives against strategic business or service plans.

The outcomes expected from the needs analysis are a refined description of the project and what it must do with a ranked set of objectives. These will also be an analysis of the project by function or use, the costs and benefits of these functions, and a statutory planning check.

3.3.3 The evaluation brief

The main elements of this stage are the identification of trade-offs within the project that the following evaluation stage will consider. The beginnings of a procurement strategy will emerge from the concept development and SVM, and the documentation required for a feasibility study and cost and time planning will be prepared.

The evaluation phase requires analysis of the project concept to test for viability and robustness of assumptions. The evaluation brief ensures the concept phase is recorded clearly as input to the evaluation process. The evaluation brief will define the actions required and scope of work for the evaluation phase and contain the outcomes from the concept phase, which include an outline description of the concept, the project strategy and objectives, and a functional analysis based on SVM.

3.3.4 Summary

A best practice approach to concept development will ensure:

- A project strategy is produced by reducing the concept to writing and making comparisons with the broad objectives of the Strategic or Business plan of the sponsoring organization. This process includes ensuring the project strategy is aligned with the cost plan developed for the project.
- Research into market or service need is undertaken to determine the priority of objectives rather than opportunity.
- Strategic value management processes are established to define functions and objectives by an analysis of Function/Use and Cost/Benefit.
- An action statement detailing procedures for meeting time objectives is prepared.
- A written evaluation brief containing the outcome of the concept phase is completed.
- A statutory planning check of the concept and scale of development will determine if the project is permitted under existing guidelines or will require changes to existing statutory planning controls.

3.4 The evaluation stage

The evaluation stage of project initiation has three phases: options generation, testing and feedback, and development of the definition brief. Figure 3.3 shows how these stages relate and the methods and outcomes for each stage.

3.4.1 Options generation

With manufactured products, the range of options available to the purchaser is taken for granted. There is a market need for a range of forms, qualities and prices. Likewise, for most projects there is always more than one solution to a given problem and, in most cases, a range of options within a solution. However, the one-off nature of projects too often results in the 'one right answer' being locked in too early in the process. This approach can be expedient or it may be a result of the reluctance to spend money on the adequate development of meaningful options.

Provided the function or uses are clear and the project objectives are well defined there is a firm basis for generating a range of meaningful design options for analysis and consideration by the project team. This will include early design

Methods		Outcomes
• prepare options: flow charts; diagrams; block plans	**Options Generation**	• identification of range of possible alternatives each described and priced
• test functional utility and objectives • prepare cash flows • sensitivities preparation • define risk of achieving desired outcome • project risks	**Testing and Feedback**	• preferred option selected and tested • preliminary feasibility study with outcomes designed for each acceptance
• describe in writing preferred option: cost; time; quality • redefine functional, physical and financial constraints	**Definition Brief Preparation**	• detailed description of action required and scope of work for next phase
	⊙ **Definition Brief**	

Fig. 3.3 Evaluation.

studies by the project consultants. This does not imply the examination of all possible options, a path of action that would be most likely to delay or stop a project, but instead the careful selection of a meaningful set of options. A glance at the range of solutions which usually emerge in a design competition reveals the spectrum of ways in which the form of some projects can be expressed. Similarly, the range of forms for consumer goods and the way they have evolved over time gives some idea of the power of options.

The adequate examination of options also provides a review of risk for the project and may help to add flexibility for evolving choices or changed circumstances.

The methods used in the options generation phase are mainly descriptive. Group processes, expert input and the project consultant team are used to identify a range of options. Flow charts, outline cost plans, diagrams and simple block plans are then used to describe the options. This step in the process is very important and requires the full input of the client.

The outcomes from this stage include a range of identified possible alternatives. Each alternative is fully described qualitatively and quantitatively. The options considered will determine the length of the planning path. Choosing an option that complies with existing guidelines may significantly reduce the time and complexity of the planning and approval process.

The impact on the project programme of alternative designs statutory approval requirements must be carefully considered at the evaluation stage. This is particularly relevant where it has been identified (at concept stage) that the project critical path is highly sensitive to time delay. The complete evaluation of the complexity of the approval process at this stage is likely to add greater certainty to the timing of the project, which will subsequently minimize the associated cost risk. Alternative design concepts result in different 'approval paths' for a project.

3.4.2 Testing and feedback

This is the analytical step of project evaluation and the point at which a preferred option is selected, based on the criteria established for the project from the needs analysis at the concept stage. It involves taking the various options and applying a range of tests and techniques to determine fully the range of likely outcomes and risks for the project. This is the essential groundwork for the decision in the definition phase to proceed with the project.

Project feasibility studies deal with the decision to invest (the investment decision), and the break-up of capital into equity and debt (the financing decision). Cash flows for feasibility analysis should identify these decisions and be discounted at a net present value (NPV) hurdle rate, commensurate with the risk of the project's cash flows. The analysis in the testing and feedback phase is essentially specialist and highly skilled analysis.

The other focus of the testing process is to determine how well the options fit the stated functional and business objectives for the project set in the concept phase. This process will involve feedback into some of the outcomes and processes of the concept phase in order to refine and modify the objectives of the project and to incorporate the better state of knowledge which is building on the project. In addition, the new players on the project required for this specialist phase will be able to assist the refinement of the concepts for the project.

The methods used in this phase will most likely include:

- quantitative analysis of project functional and business objectives
- preparation of project option cash flows
- preparation of indicative project management information including project time schedules
- determination of the range of quantitative parameters which influence the project viability
- preparation of NPV analysis of the project options including sensitivity analysis
- completion of a project and business risk assessment
- undertaking probability of outcome assessment and
- consideration of special project risks.

The two outcomes from this study should be, first, a preliminary feasibility study which describes and analyses a range of options for the project, and defines a range of likely outcomes for the project as a whole, and second, a preferred option that has been selected and tested physically, functionally and financially.

3.4.3 The definition brief

The definition brief again signals a shift in approach to the project. The emphasis in this phase is getting to the point of making a decision to commit to the project. The preferred option is refined, the definition developed and expanded and the project management plan is developed in more detail. It is a transition phase between project evaluation and full project implementation. Again it is essential to capture and effectively communicate the results of the evaluation phase in the definition phase brief.

The definition phase brief may include a written description of the preferred option with cost targets, time requirements, and quality considerations. It may involve a redefinition of the functional, physical and financial constraints and objectives for the project.

3.4.4 Summary

A best practice approach to evaluation will produce:

- a range of meaningful options (including a no-build option), described: qualitatively, quantitatively and priced
- a Project Option Evaluation, comprising a comparison of estimates for meaningful options against the adopted cost plan, and a preliminary feasibility study which analyses options and defines a range of likely outcomes for the project as a whole
- further refinement and testing of the project functional and business objectives developed by the SVM processes in the Concept Development stage
- planning/approval constraints for the project
- an 'in principle' approval for the project, based on the quantified analysis
- the written Definition Brief based on the preferred option, consisting of cost targets, time requirements, quality considerations and the functional, physical constraints and objectives of the project.

3.5 The definition stage

The project definition stage of project initiation has three phases: schematic design, testing and feedback, and development of the delivery brief. Figure 3.4 shows how these stages relate and the methods and outcomes for each stage. The definition stage is where the preferred option is developed and documented as a design to bring the project plan, cost plan and time schedule to a point where the client can make a decision to proceed. It is important to have achieved regulatory planning approval by the end of this stage.

Methods		Outcomes
• identify trade-offs • rerun: value MST; strategic and functional objectives • document, drawings/diagrams, single line systems	**Schematic Design**	• descriptive and illustrative definition of preferred option
• schedules define procurement strategy and risk • rerun all quantitive analysis • cost plans • life cycle costing	**Testing and Feedback**	• revised feasibility • time schedules in place • financial constraints set • go/no go decision
	Delivery Brief Preparation	• identification of scope of works and procurement strategy • prepare report containing defined and tested project
	⊙ **Delivery Brief**	

Fig. 3.4 Definition.

3.5.1 Schematic design

The description of the design is, for the first time, now handed to the design professionals to document the project. At the same time, a fully developed project plan is prepared for the project management team for implementation of the project. This developed documentation for the project is then used for a value management study that confirms the SVM but also tackles the specifics of value for money with the chosen design. This process is used to ensure the money is being spent to deliver the function required and to eliminate high cost/low function elements of the design. The developed cost plan and the SVM study will result in ranked trade-offs in the design process and make the choices on these visible to the owner or sponsor.

The methods used in this phase include documentation and drawing of the project to a schematic or single line design and refinement of the project plan to develop a project time schedule, and cost plan. The outcomes from this phase include:

- a descriptive and illustrative definition of the preferred option including:
  time schedule
  cost plan
  schematic drawings and documentation
- value management report confirming value for money and
- ranked trade-offs identified.

3.5.2 Testing and feedback

This phase is the final confirmation of the project prior to a decision to proceed. The analysis developed and refined in the evaluation stage is rerun and further developed to confirm that the project, which is now defined in some detail, still meets the needs of the owner or sponsor and will still perform and deliver outcomes that are beneficial to the enterprise.

Also included in this stage is an evaluation of the costs of ownership and operation which flow from the chosen design. Techniques like life cycle cost analysis are used at this point. The methods used in the testing and feedback step for project definition include a rerun of all quantitative analysis and a full costing of ownership and operation of the building.

The outcomes from this stage include:

- revised feasibility analysis
- time schedules in place
- financial constraints set
- planning approval achieved
- cost of ownership and operation defined and
- a go/no-go decision in principle from the owner or sponsor.

3.5.3 The delivery brief

The delivery brief consolidates and documents all of the project analysis, description and planning, which has led to the approval of the owner, into a brief for the implementation of the project. The delivery brief is expected to contain:

- the enterprise objectives for the project
- the functional objectives for the project – what it must do
- the financial constraints and objectives
- a summary of the conclusions from the feasibility and risk analysis
- details of planning approvals
- the project implementation plan, actions and schedules
- a procurement plan
- a cost plan and
- the project documentation, description and illustrative definition.

3.5.4 Summary

A best practice approach to project definition will produce:

- a described and illustrated definition of the preferred option, which will include documentation and drawings of the project to a schematic or single line stage
- a rerun of the value management and project evaluation techniques applied to the alternative choices within the preferred outcome
- a refinement of the project plan including a specific project time schedule and a limit-of-cost plan
- a rerun and refinement of the feasibility of the chosen option
- information for the project sponsor to make a clear Go/No-Go decision for the chosen option
- the written Delivery Brief which fully details:
  project objectives
  functional objectives
  financial constraints and objectives
  conclusions of the feasibility and Risk Analysis
  planning approvals
  project implementation plans, actions and schedules
  a procurement plan
  a project budget
  the project documentation, description and illustrative definition
- an information memorandum containing the Delivery Brief, Financing Options and all relevant material.

3.6 Conclusions

The project initiation phase gathers a large and diverse amount of information together for building project clients. The purpose of this chapter is to outline a best practice approach to the project initiation process, and to locate many of the individual tasks undertaken in this phase that are discussed in greater detail in other chapters.

Project implementation is improved by the introduction of quantified objectives which define the success criteria for the project. This results in a more focused project and higher chances of achieving the goals that are set. The client achieves a better business outcome simply because the business goals are clearly linked to the project and properly thought out.

Financial modelling of a project's options is an important part of the initiation process, because these techniques are the basis for key decisions, and result in an analytical understanding of the project. An economic evaluation consisting of extensive sensitivity analysis applied to building and construction projects must be supplied to clients who make decisions about projects. Given the long-term nature of buildings and structures and the relatively high costs of ownership and operation, clients need careful evaluation of each project.

Historically, there has been very uneven use of quantitative techniques for evaluation of the feasibility of building projects. In addition, the assumptions on which they are based and the completeness of many evaluations is open to question. Nevertheless, adequate testing of needs, functions and objectives is crucial to determine the best value for money.

If the concept and evaluation phases are completed as recommended in this chapter the design process will not include decisions that should have been made earlier, avoiding lack of owner involvement in decision making, delays in design while options studies are completed, and projects that are poorly conceived or not viable.

Often the project initiation process is project delivery focused and does not generate innovative alternative solutions. This can lead to projects that are poorly conceived, and delays in design while options studies are completed. By following the concept, evaluation and definition stages outlined in this chapter, a project will have criteria or objectives for testing the success of the project; and the project will deliver value for money.

References and bibliography

CIDA (1993) *Project Initiation Guide* (Sydney: Construction Industry Development Agency).

Healy, P. (1998) *Project Management: Getting the Job Done on Time and in Budget* (Sydney: Butterworth Heinemann).

PMI (1987) *The Revised Project Management Body of Knowledge* (Project Management Institute).

Turner, J.R. (1992) *The Handbook of Project-Based Management: Improving the Processes for Achieving Strategic Objectives* (London: McGraw-Hill).

4

Procurement strategies

Gerard de Valence and Nathan Huon†

Editorial comment

Tendering and procurement in the building and construction industry is undergoing a general move away from traditional methods towards single point project delivery systems. The increase in popularity of these alternative procurement systems shows clients are attempting to improve time and cost performance and reduce the number of disputes on their projects.

Contributing to this trend are factors such as: responses to changes in national and state regulations and government policies; the effects of international trade agreements; new financial structures; innovation in the industry; the importance of constructability; use of IT; the selection of the most appropriate strategies; pre-qualification of contractors; and the role of strategic alliances and partnering.

Procurement can be broadly defined as the process that deals with project definition and delivery and the technical capabilities of the industry. Traditional procurement methods such as lump sum or cost plus are giving way to design and construct and other types of turnkey contracts. Turnkey places all design, construction and building, and performance responsibilities under a single entity, and the primary advantage to turnkey work is this single point of responsibility.

Clients are increasingly seeing the procurement method used as a key element in getting value from their projects. Clients have been demanding 'better value for money' since the early 1970s, and the evolution of procurement and delivery strategies and policies used by clients shows them seeking ways to raise the standard of performance. There is renewed effort in investigating tendering and procurement practices around the world, with greater efficiency and lower costs as universal goals.

Clients are raising their expectations and setting new standards and, in determining the standards of performance they expect, they will increasingly monitor performance and reward the better performers. The challenge to construction industry firms is to show they are genuinely competitive, that they can produce client-driven, high quality and value-for-money outcomes.

In future, the services packaged for clients will extend beyond the familiar range of construction services associated with a single project. Clients are turning to single point responsibility for project delivery, with expectations of cutting cost and

† University of Technology, Sydney, Australia

completion time, eliminating claims and disputes, and promoting constructability and innovation on large and small projects. These are fundamental to achieving value in building.

4.1 Overview

The term procurement relates to the strategic organizational management of resources in a logical sequence in order to meet project objectives. Resources in this sense can mean time, money, equipment, technology, people and materials. The activities of procurement relate to the set of actions required to design, manage and deliver the project objectives.

Fundamental to all procurement systems is the development of a framework that clearly establishes the boundaries of roles, responsibilities and relationships between the parties to a construction project. The parties may include a client, management consultants, financial institutions, design consultants, contractors, subcontractors and suppliers. Because of their multidisciplinary arrangement, due the nature of the participant's background, the procurement system selected will determine the allocation risks inherent in a project, including those relating to time, cost and quality.

A broad definition of procurement comes from Hawk's study of the international building and construction industry:

> Procurement in the opportunity sense is not just the selection and purchase of materials for construction but includes the R&D for new and traditional materials for new and old uses. It includes development of new production methods of materials and new means of producing products with these materials. It can involve issues of automation, efficiency, product life cycle studies and responses to problems of environmental deterioration and pollution. Linked to an improved model of design, procurement can be extremely important to transforming the industry. (Hawk, 1992, p. 47)

Procurement has become a more central issue to the industry for several reasons. First, clients have been demanding 'better value for money' since the early 1970s, driven partly by the industry's own poor productivity performance and partly by comparisons with manufacturing productivity growth. Clearly, non-traditional procurement systems have better time performance and cost benefits in many cases. Following Hawk's argument above, non-traditional procurement systems would also offer better opportunities for 'development of new production methods of materials and new means of producing products with these materials'.

Secondly, the method of project procurement and delivery used by the client determines the nature of competition in the industry. If the industry is seen as players of a competitive game, the rules of that game are set by clients. The evolution of procurement and delivery strategies and policies developed by clients has changed the rules of the game over time, and clients are seeking ways to raise the standard of performance. At the end of the day, the industry competes on a playing field defined by clients under rules set by clients. Thus, the Latham Report

(1994) urged the UK public sector to play a leading role in restructuring the industry, as did Gyles (1992) in Australia. In the US the CII has promoted best practice to both clients and contractors with significant project time and cost results.

Thirdly, there is a renewed effort in investigating tendering and procurement practices around the world. The factors behind this interest are diverse, but greater efficiency and lower costs are universal goals, and a major impetus has come from the increasingly widespread use of IT and electronic procurement by large public and private clients.

The main types of procurement systems used in the construction industry can be categorized into four groups; these are: Traditional, Non-Traditional, Single Source and Collaborative systems. Within each category are a number of recognized procurement forms (nomenclature may vary depending on the country of use), and Table 4.1 summarizes this.

Strategies are differentiated by management structure, liability and risk sharing, design management, tendering procedures and determination of the contract price. The purpose of this chapter is to discuss the various methods and identify their respective advantages and disadvantages.

In recent years the clear trend is towards the use of alternative non-traditional procurement methods. This has been driven by industry's clients' drive to improve time and cost performances and also to reduce the incidence of disputes.

4.2 Procurement strategies

4.2.1 Traditional lump sum

The traditional procurement strategy typically involves the client appointing an architect to procure the project design and documentation prior to any contract being entered with the building contractor. The architect in this case takes the lead role in the project. The architect's scope of work will also typically include the appointment of other consultants required for the design delivery, and may include engineers (civil, structural, services), planners, interior designer, quantity surveyor, environmental and specialist equipment procurers. All of the consultants are required to report directly to and act under the direct instructions of the architect. The architect then directly communicates with the client. The client enters into contracts directly with all consultants as advised by the architect as well as with the building contractor. Figure 4.1 shows these relationships.

The architect has a direct relationship with the client and may continue to act as agent for the client in administering the head contract with the builder. It is important to note that the architect has no contractual relationship with either the consultants or building contractor.

The architect typically provides advice on the most appropriate contract form, specific contract conditions and tendering method. In addition, the architect coordinates the tendering process and evaluation of the tenders on submission on behalf of the client.

Table 4.1 Summary of procurement strategies

| | Traditional | Non-Traditional | | Single Source | | | Collaborative | | |
	Traditional Lump Sum	Construction Management	Project Management	Design & Construct (D&C)	Turnkey	BOOT (& Variations)	Partnering	Joint Venture	Alliancing
Management Structure	Traditional Architect as project team leader.	Construction Manager (Builder) initially joint team leader with Architect then as lead in construction phase.	PM lead role appoints and manages all consultants and contractor required to deliver project within objectives set.	Builder is lead for project delivery. Design team typically part of the Builder's organization.	As for D&C.	As for Turnkey but with some partnership agreement with Client.	As for BOOT but could involve more than one lead 'builder'.	JV partners manage through cooperative process.	Business relationship between parties necessary for the delivery of the project.
Liability	Client directly contracted to Builder.	Construction Manager acts as agent for Client. Client contracts direct with sub-contractors & suppliers.	Client contracts directly with all consultants and builder. PM responsible for performance of same.	Client and builder with direct contractual relationship. All risk for delivery on Contractor.	Client contracts builder for funding, site procurement and construction to meet project objectives.	As for Turnkey but Builder will undertake to operate facility for a contracted period of time on completion.	As for Traditional but Client and Builder have a Charter of cooperation which empowers but is not legally binding.	The Board of JV has liability as an organization for non-performance or non-delivery.	As for JV with equitable balance of risk and reward.
Design Management	Architect as lead. Full documentation prior to construction.	Architect and CM facilitate design process. Client has direct contract with consultants.	PM appoints and manages the design process within the project objectives.	Builder appoints and manages all consultants to deliver within terms of project brief.	Builder appoints and manages all consultants to deliver within terms of project brief.	As for Turnkey.	As for Turnkey.	As for Turnkey.	Design consultants part of the Alliance organization.
Tendering Process	Open, Selective or Negotiated on the basis of contract documentation.	Open, Selective or Negotiated. Typically no detailed documentation at tender.	Selective or Negotiated. Either single or two-stage tendering.	Selective or Negotiated on the basis of performance brief or DA/BA. Single stage process.	Selective or Negotiated on the basis of performance brief or DA/BA. Single stage process.	Selective in single or two stages or value assessment method.	Selective in single or two stages or value assessment method or by negotiation.	By negotiation.	By negotiation.
Contract Price	Lump sum. Schedule of Quantities or cost plus.	Fee + Costs.	Lump sum. Schedule of Quantities or cost plus.	Typically lump sum.	Typically lump sum.	By negotiation.	By negotiation.	By negotiation.	Cost plus margin.

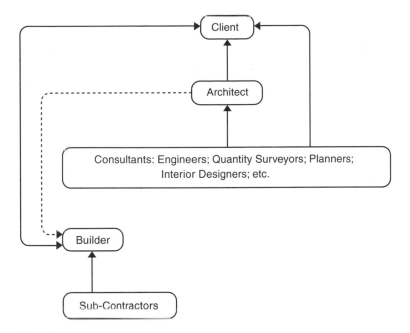

Fig. 4.1 Traditional management structure.

A fully detailed package of design documents will be issued for tender and may include all detailed designs, specifications, conditions of contract (between the client and the head contractor) together with other relevant documentation such as a bill of quantities (where used) for pricing by the building contractor. The conditions of contract may be those of a standard nature or may have been formulated specifically for the project with the assistance of the client's legal council.

The appointed building contractor will usually break up the project documentation into trade packages and subcontract these components of the work through individual contractual arrangements with specialist subcontractors. The builder is contracted to provide the management services necessary to plan, coordinate and build the works within an agreed time and budget as required by the contract documents.

Benefits of the traditional strategy

The benefits associated with this form of procurement include:

1. Client involvement in the design and construction process is limited (Barrie and Paulson, 1992).
2. Because this form has existed for so long, the control systems, documentation and organization of a large number of industry players are set up to manage in this way (Rougvie, 1987).
3. There are a number of developed standard forms of contract that are continually updated as they are tested in arbitration and court.

4. Where the design is properly completed prior to tenders being called the client can have a high degree of confidence in the contract price. Gould (1997) notes however, that the final fixed price is only as good as the contract documents.
5. The provision of detailed drawings, specifications and quantities provides a basis for obtaining competitive tenders for the work and facilitates the comparison of tenders by the client and architect (Moxley, 1993).
6. Typically, the contract provides for changes and variations to the design and/or specification by the client during the works. The contract also provides a means for valuing these variations through comparison with the bill of quantities.
7. As a result of sufficient time being provided for the detailed design and documentation, the risk associated with the number of design changes required during the works is relatively low. However, all too often this is not the case, with many tender packages being rushed out at the last minute to satisfy the client's already inadequate timeframe.
8. Where a quantity surveyor is engaged on the project, a high level of cost control and monitoring, using the priced bill of quantities, can be achieved.

Disadvantages of the traditional system

There are a number of significant disadvantages to the traditional system of procurement; these include:

1. The significant risk taken on by the client, particularly in the case where the client lacks experience and the contract forms are not wholly appropriate for the project.
2. The traditional system fundamentally ignores the benefits of integration between the design and construction processes, and this leads to a number of problems. First, as the design system is more complex, effective management and communication is difficult (Franks, 1987). Second, the exclusion of the contractor from the design phase eliminates the scope to improve the design and/or buildability using the contractor's vast knowledge and experience in the construction process. This in turn can result in a design that is not easily built, or exceeds the client's budget, or one that could have been done better and cheaper.
3. Due to the fact that the design and construction phases of the project are 'end on', with no parallel working, the overall speed of the system is adversely affected. The additional time required in procuring the works impacts upon the client's finances, holding costs and consultant's fees. Further, during periods of rapid inflation the client's objectives to have the project completed in the shortest possible timeframe are unlikely to be achieved.
4. Unless the client is relatively informed or experienced the architect must take on the conflicting roles of the principal designer and project manager. A significant conflict of interest may exist in the case of the architect acting as project superintendent, particularly where errors or omissions in the design plans are found during the course of the works (Uher et al., 1993).
5. The client's liability and risk is significant, particularly where the client is inexperienced. The limited involvement in the design and construction process can also be a significant disadvantage.

6. In many cases, due to time pressures, the documentation may have errors or omissions, which may result in extensive price and time variations.
7. Unless the lead consultant has significant project management skills it can be argued that the traditional system is not appropriate for large/complex projects that require substantial coordination and complex management procedures.
8. The traditional method almost precludes innovation on the part of the contractor with the threat that the contractor's tenders be determined as non-conforming should they decide to use their initiative and suggest better ways to achieve the client's objectives.
9. The tendency under traditional procurement strategies to accept the lowest bidding contractor has led to sub-optimal management of the construction process resulting in poor project speed and quality. Low-bid practices by the client may also result in the head contractor being forced into using marginal subcontractors.

4.2.2 Alternative procurement – background

The limitations of the traditional procurement method have contributed to the poor performance of the construction industry and have prompted the development of alternative procurement strategies designed to facilitate improvements in the way buildings and structures are delivered (Cox and Townsend, 1998, p. 32). Client dissatisfaction with the industry's failure to provide high value solutions to their building needs is the major reason for the development of alternative procurement strategies.

It has, however, been suggested that the discontentment of clients is not necessarily the only reason for the development of alternative procurement strategies. Venmore-Rowland *et al.* (1991) suggested that some contractors, quantity surveyors and architects have seen alternative strategies as a means of promoting their role in the procurement process, or an opportunity to establish themselves as key players in the industry.

The term 'alternative' can suggest a new approach, but many of the emerging 'alternative' procurement strategies such as 'design and construct' have been around for years. Each strategy has developed a certain degree of flexibility so that, in reality, many of the alternatives tend to overlap one another. For the purpose of this discussion the various alternative strategies available to clients for the procurement of their building and construction projects are classified under three broad classifications: non-traditional method; single source strategies; and collaborative strategies, as in Table 4.1. A number of variations within each of these classifications have emerged as specific project requirements and these have needed to be addressed.

Non-traditional procurement strategy
These strategies are sometimes referred to as managed contracts; they generally follow the principles of the traditional system with subtle variations to the management structure and to that of the architect-dominated traditional strategy.

Managed contracts provide for a more active role of the client as a member of the project team in the management of the project. The degree of involvement by the client will, however, depend upon their experience and the availability of resources. Uher *et al.* (1993) suggest that managed contracts provide the lowest level of risk to the building contractor, who acts as a construction or project manager; the client, as a result of the increased involvement on the management team, accepts a greater share of the total risk. The two main types of managed contracts – Construction Management and Project Management – are discussed below.

Construction management

Under a construction management contract the client will engage an architect and a construction manager with the aim of introducing a team approach to the design, documentation and construction phases of the project. This team approach helps ensure decisions are made in the best interests of the client. These decisions include evaluation of cost and time schedules and overall construction performance during the project. The 'interactions between construction cost, quality and completion schedules are carefully examined by the team so that a project of maximum value to the [client] is realised in the most economic time frame' (Uher *et al.*, 1993, p. 60).

Under this arrangement the client is contracted to the architect, construction manager and the individual trade contractors through separate contracts. The construction manager will usually be engaged as an agent to the client on a fee for service basis. Under this agency arrangement the construction manager, who has not directly contracted the trade contractors, will not accept a great deal of the risk inherent in the project. Where an agency arrangement is adopted, the construction manager's services may include arranging the individual contracts between the client and the trade contractors and being responsible for the administration of the same. During the project the construction manager will be required to coordinate the work of the trade contractors and to ensure that they adhere to the construction programme and budget. In some cases, however, the construction manager may be required to contract directly with the individual trade contractors and as such this non-agency construction management arrangement may shift significant risk away from the client and to the construction manager who would also act as the general contractor. Figure 4.2 represents the arrangement.

Advantages of construction management

1. Promotes a team approach to design and construction issues.
2. The builder is part of the design team, which should lead to efficiencies that may not necessarily flow from the traditional approach (Gould, 1997, p. 62).
3. It is an 'open-book' approach, which ensures that the builder's costs are reimbursed and will reduce the risk of litigation and spurious claims for additional costs.
4. The client gains the financial benefit of the builder's buying and negotiating power with subcontractors and suppliers.
5. Time savings are possible through the ability to 'fast track' the design.

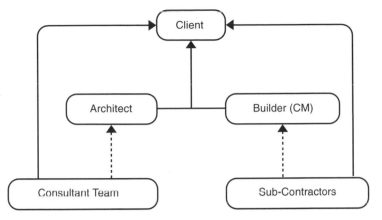

Fig. 4.2 Construction management structure.

Disadvantages of construction management

1. There is a high degree of risk to the client associated with the direct contracts with the subcontractors, particularly where the builder is responsible for the tendering and letting of trade packages. In the event that the builder fails to assess subcontracts properly the direct cost of any failure is borne by the client.
2. This strategy demands the continuous involvement of the client as part of the three-party team and, as such, inexperienced clients may be exposed to considerable risks that they are not in a position to manage.
3. As the builder has a guarantee that costs will be reimbursed there may be less incentive to pursue competitive bids for stages of work.

4.2.3 Project management

Project management has been defined as 'the overall planning, control and coordination of a project from inception to completion aimed at meeting a client's requirements in order that the project will be completed on time within authorised cost to the required quality standards' (CIB, 1992, p. 6).

A project manager is appointed by the client as the first member of the project team and adopts the broad function of delivering the project, within the project time, cost and quality objectives. Typically, external project managers would consist of a contracting organization or consultancy with experience in delivering project management services. In any case, the project manager must have both professional and commercial expertise in the building and construction industry.

In many cases the project manager may be required to work directly with representatives of the client's organization or with other individuals nominated by the representatives to complement the project management team. This gives the client a great deal of flexibility, particularly where they have experienced in-house staff but do not have the resources to commit them to the entire project management function. The significant advantage that this presents to the client is the degree of transparency in the procurement process.

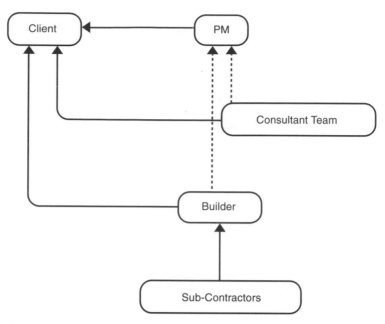

Fig. 4.3 Project management structure.

Figure 4.3 shows the typical management arrangement. The client enters separate contracts with the project manager, design consultant(s) and builder. The project manager must coordinate the activities of each of the team members. In this regard a greater degree of singular point responsibility is achieved with this strategy than under the construction management arrangement, due to the singular communication link between the client and the project manager.

While no direct contract exists between the project manager and the design consultant(s), the project manager will usually be required, under the terms of the management agreement, to take the lead in the overall management of the design process.

Uher *et al.* (1993, pp. 72–3) suggested the key functions of the project manager are to: advise the principal as to finance, land acquisition and other planning issues; prepare feasibility studies and advise on optimum design solutions; select the other members of the project team; prepare tender documentation and administer the contracts as the superintendent; coordinate all members of the team throughout the project; administrate the project budget and programme; report to the principal the status of the project's time, cost and quality progress; lease areas of the building where appropriate; and to commission the finished facility.

Advantages of project management
1. Single point of responsibility in the project manager. The client's risks to various aspects of the project are reduced.
2. Because the project manager is responsible for the delivery of the project, it enables the client to focus on his or her usual business operations rather than expend time and effort on the procurement of the facility.

3. This strategy lends itself to large and complex projects that require greater overall management skills than offered by the traditional method.

Disadvantages of project management
The nature of the contractual arrangements can lead to a highly adversarial state that is more likely to lead to lengthy and expensive litigation.

4.2.4 Single source procurement

Many clients have become increasingly dissatisfied with the difficulties and ambiguities presented by traditionally based procurement strategies. With fragmentation in the structure, procedures and the relationships between the contracting parties, clients are now looking towards new procurement strategies that simplify the formal procedures of construction and concentrate resources and effort to improve performance. 'Each link between the client and the groups involved on a building or construction project is a potential break in the chain of communication and responsibility' (Moxley, 1993: 136).

These strategies are also known as 'performance contracts', owing to the single point of responsibility for producing an end product that meets all of the client's performance requirements.

Single source strategies can be clearly identified by their predominant allocation of risk to the contractor. This is not to say that the client avoids all of the project risks; however, where a comprehensive brief is prepared, the client's risks can be minimized compared with traditional strategies. Thus, in the event of a project failure the contractor will be solely responsible and the client's interests will be safeguarded in this respect (Franks, 1984). Essentially, from the perspective of the client, there can be no 'buck-passing' between the architect and the builder as is often the case with traditional-type strategies.

Moves towards a centralized point of responsibility have seen increased interest from clients in strategies such as 'design and construct' and the development of other new and innovative strategies such as 'build own operate transfer' (BOOT).

Design and construct
Design and construction and turnkey procurement methods are closely aligned; the essential difference between the strategies is the degree to which the builder is involved in the pre-planning and post-construction activities of the project. The common theme, however, is that in each the contractor is singularly responsible for both the design and construction of the building or facility. Figure 4.4 shows the management structure.

A client must take care to ensure that the concept design and/or performance specifications for the project are thoroughly prepared (Collins, 1991). This work will usually be undertaken by an independent adviser, often an architect, engaged by the client for the duration of the project to provide impartial advice on the suitability of the detailed design as developed by the design and construct contractors.

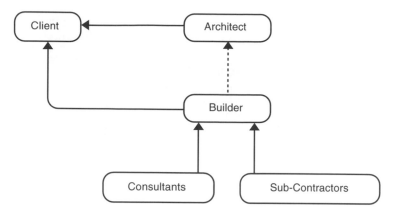

Fig. 4.4 Typical design and construct management structure.

Variations to the standard design and construct procurement strategy may include the 'design, novate and construct' strategy where the client provides a performance specification, includes a preliminary design and also requires its consultants who prepared the concept design to be novated across to the contractor. The contractor subsequently assumes responsibility for those consultants to undertake the detailed design.

The successful contracting organization will often utilize its own design staff to undertake the design component of the project. Alternatively, a joint venture arrangement between the contractor and an architectural firm may be necessary depending upon the size and complexity of the project. In other cases the contractor may decide to subcontract the design work under a standard subcontract agreement.

The contractor's role in design development will vary from one contract to the next depending upon the degree to which preliminary design work is undertaken by the client. The client may attempt to retain some design control through the inclusion of certain contract conditions to protect his or her interests.

Advantages of design and construct
1. Single point of responsibility from the client's perspective.
2. Builder adopts a major portion of the project risk through the design development of the initial concept.
3. Builder can retain control of the design process and has the ability to 'fast track' in order to be more cost effective.

Disadvantages of design and construct
1. Cost pressures on builder's side can lead to short cuts in the design, and designing only to meet the minimum performance requirements of the contract.
2. Promotes cheapest capital cost options which could have significant life cycle ramifications in terms of cost and durability for the client.

3. The client may need to retain an independent consultant to monitor the design for its compliance with the performance brief. The adjudication of whether the design meets the requirements of the brief can lead to disputes between the client and the contractor.
4. The ability to modify the arrangements or to change contractors in the event of non-performance is extremely problematic due to the issues of design rights and liability. The client must be well advised and the selection of the appropriate building contractor is absolutely critical.

Turnkey

Under a turnkey contract, the builder undertakes all the components of a project. Turnkey places all design, construction and building, and performance responsibilities under a single entity, and the primary advantage to turnkey work is this single point of responsibility. Turnkey projects can be done by a lead company, which subcontracts out the different aspects of the project, or the principal participants join in a consortium or joint venture arrangement. The goal under either framework is to appropriately allocate and assign the risks to the party best able to control that risk. Schedule, performance and price guarantees can thus be provided with a maximum amount of risk sharing and a minimum opportunity for disputes.

With a design and consult strategy the contractor undertakes to design and construct the works to meet the client's time, cost and quality objectives or requirements. In a turnkey contract the contractor may be required to find and purchase a site on behalf of the client and perhaps even provide the required finance. The contractor must then undertake the design and construction phases and commissioning of a building or production line of a factory for a period of time to meet specific performance criteria as required by the client. In contrast, where a design and construct arrangement has been adopted the client will usually only require the contractor to undertake the design and construction aspects of the project.

'Build own operate transfer' (BOOT) strategies

'Build own operate transfer' (BOOT) strategies have become particularly popular with public clients in the provision of infrastructure projects such as roads (tollways) as well as the provision of public services such as hospitals and prisons. These forms of procurement are particularly popular in developing countries where governments use them as a means of securing private sector funding to offset their own capital expenditure requirements. These strategies expand on the design and construct and turnkey concepts and generally include a period of ownership and/ or operation of the facility by the contracting organization. The operation phase is referred to as the Transfer Period.

The Transfer Period commonly varies from between 15 and 25 years depending upon the type of facility and the financial return expected from the completed facility. In some cases the facility may never be transferred to public ownership and the incentive may simply derive from a client's need for a facility which it otherwise would not be in a position to provide. Essentially there are no limitations

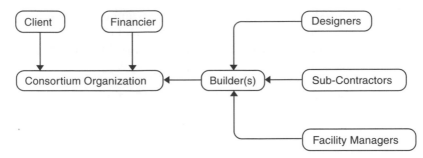

Fig. 4.5 Typical BOOT management structure.

on variations to the standard BOOT arrangements. Figure 4.5 shows a BOOT management structure. Other similar structures are:

BOOT	Build Own Operate Transfer
BOT	Build Operate Transfer
BOO	Build Own Operate
BTO	Build Transfer Operate
BT	Build Transfer
BLT	Build Lease Transfer
ROT	Rehabilitate Operate Transfer

To initiate the project, the client(s) will normally invite consortia to bid for the right to enter into negotiations for the works. The client will normally provide to potential consortia a detailed performance criteria including design, construction and operational requirements.

The consortia bids are normally required to include information that sets out the fundamentals of the project including the design concept, contract form, transfer period, ability of the consortium to provide the required level of service from the facility during the period of operation (including level of fees or charges proposed), financial security to complete and operate the facility and the operational costs of the facility upon transfer. The capital cost of the work is not necessarily part of the proposal at this stage. The client selection of the preferred partner should be based on the conditions which best suit the client's requirements.

The preparation of a feasibility study will be required in the early stages of negotiation following the development of a project financial model and formation of the proponent consortium. BOT arrangements will usually involve intensive negotiations between the client and the consortium members in order to develop an arrangement that provides mutual incentives; this is the key principle of the BOT approach.

In the event of a public project, the land is usually leased from the government on a long-term basis. During the period of operation by the consortia it will be necessary for the client (government) to monitor the performance of the facility with respect to social obligations and the quality of the services being provided. These arrangements provide a single point of responsibility for the client who, through the establishment of clear and concise performance criteria, can procure the design, construction and operation of a facility including all financing simply by creating an investment incentive for interested consortia members. As such, the benefits of the family of BOOT-type arrangements have attracted much interest from the public sector.

4.2.5 Collaborative procurement strategies

Partnering

The concept of partnering has evolved from attempts to overcome the problems of adversarial relationships common on construction projects, particularly as a result of contractual confrontations and disputes. Partnering involves the commitment of two parties to establish a cooperative relationship that promotes a spirit of goodwill and fair dealing with the common view towards the success of a project and is therefore a strategy that is very strongly focused on the win/win principle.

This commitment is formalized by the parties by attending a workshop prior to the contract signing to develop a Partnering Charter or Goal Statement. This document sets out the working relationships between the parties including the organizational structure and moral conduct of the parties. The legal commitment between the parties is established through a traditional works contract that allocates the project risks and sets the legal relationships between the parties.

Partnering is one of the newer more innovative procurement strategies and in recent times has enjoyed increasing attention, particularly in the public sector. In his final report into Productivity in the Building Industry in NSW, Royal Commissioner Gyles recommended that the public sector undertake trials of partnering arrangements as a means of addressing the adversarial nature being experienced within the industry:

> I recommend that the public sector constructing authorities be asked to provide a plan for familiarization with and the trial of partnering. In the course of doing so, the requirements of probity, fairness to contractors and value for money will have to be accommodated. (Gyles, 1992, p. 78)

Subsequently the New South Wales Government embraced the concept of partnering. In its 'Green Paper', *The Construction Industry in New South Wales – Opportunities and Challenges*, listed under the title of 'Industry Development' were cooperative contracting principles and techniques such as partnering and alternative dispute resolution for enhancing business relationships for projects (NSW Government, 1996, p. 4). Early indications following various infrastructure projects, educational institutions, hospitals and civil projects are that the benefits of partnering are real and significant; however, the test of time will be required to assess fully the strategy's performance. According to a senior manager, 'The indications to date are that partnering helps to improve project outcomes, lower costs and increase opportunities for innovations. It increases the prospects for financial success and it certainly reduced exposure to disputation stemming from outdated adversarial attitudes and approaches' (Griffin, 1994, p. 4).

The concept was also supported by the Construction Industry Development Agency (CIDA, 1993), who recognized the potential benefits of partnering arrangements and suggested that such strategies may well be extremely valuable to the industry.

The breaking down of outdated adversarial relationships flowing from improved communication and the development of the win/win philosophy between the parties involved is an attractive prospect for clients looking to procure building or construction projects efficiently. The single point of responsibility that partnering offers is also a significant attraction to clients, particularly those who are looking for a fair deal.

Ronco and Ronco (1996, p. 50) describe the benefits of partnering as reduced costs, usually as a result of compacted or more closely monitored project schedules and time lines, reduced frequency, extent and severity of litigation, increased building quality, and improved safety on the project. The United States Construction Industry Institute Task Force, CII (1991) grouped the benefits of partnering into four areas: an improved ability to respond to changing business conditions; improved quality and safety; reduced cost and project time and improved profit and value; and more effective utilization of resources.

Unfortunately, in practice it is not always possible for a client to find a contractor with the best motivations and, conversely, it is not always easy for a contractor to identify a client who has such pure intentions (Ronco and Ronco, 1996, p. 88). The controlling factor will always be the degree of commitment that both parties apply to maintaining the relationship through honesty and concern for the other's success and, as such, greed is the most significant predator of the partnering agreement.

'Some advocates of partnering . . . have adopted the view that basically all people are trustworthy, want to do things which are valued, and would wish to work in a collaborative "win/win" environment' (Cox and Townsend, 1998, p. 41). Whilst there has been a great deal of support for the use of partnering as a means for clients to achieve better value, it is important to question the potential of such a strategy where the relationship is established on a short-term basis: '. . . nothing other than opportunistic behaviour can result from an adversarial relationship built on the need for short-term gain' (Cox and Townsend, 1998, p. 52).

Joint ventures

A joint venture is the project-specific joining of firms, on a temporary basis, through combined investment of capital and expertise to undertake the works. Gould (1997, p. 59) defines a joint venture as 'the legal binding of two companies for the purposes of providing a competitive advantage that would be difficult to attain alone'. This procurement strategy has been used extensively on large-scale projects around the world since it was first adopted in the United States to construct the Hoover Dam in the 1930s, and is particularly suited to large complex projects where broad ranges of specialized construction services are required over a long period of time.

The formalization of joint venture agreements is particularly important in assigning roles and responsibilities and allocating risks and liabilities between the joint venture parties, including the basis of sharing profits or losses. It is common for joint ventures to be formed between international firms with particular expertise in a specialized field and a local firm with extensive local market knowledge. The ideals behind this arrangement enable larger more complex projects to be taken on using the expertise of the international firm – where in many cases this expertise may not be available locally – and drawing on the local firm's access to locally based suppliers and vast knowledge of local conditions.

With the increasing proportion of the costs of services in many projects it is now common for a builder to join forces with a services contractor in order to benefit from the expertise in both the general and specialized fields. Joint ventures can, therefore, be highly effective for clients with the earlier specialist design input and the single point of responsibility that is achieved.

Strategic alliances

Strategic alliances take the partnering concept one step further, promoting not only cooperative relationships but focusing on the benefits of long-term relationships or alliances between the contracting parties. Effectively, strategic alliances can be described as an extension of a partnering agreement to encompass a number of projects in order to attain the common long-term goals of both parties. As such, strategic alliances have often been referred to as 'Strategic Partnering'. The alliance is characterized by the ongoing involvement of the contractor in the client's business plans and hence strategic alliances have the benefit of giving contractors more certainty about their short- and long-term future in the industry. As a result of this confidence, contractors will naturally be more focused on innovation and investment in new technologies and work practices, the result of which will be commercial advantage over their potential competitors.

For the strategic alliance to be successful it is essential that the parties establish a relationship of trust based on mutually agreed objectives and practices, rather than rely exclusively upon their contractual rights under traditional contractual arrangements. It is therefore essential that the roles and obligations of the parties are clearly defined through the establishment of operating principles, performance objectives and evaluative mechanisms. It should be noted that the alliance will not usually be completely void of formal contractual obligations and the terms and conditions of the alliance may be created through a contract specifically drafted for the particular relationship. The contract will record the intentions and expectations of each of the alliance parties as set out in the mission statement and should be flexible to change within the relationship to reflect best the intentions of the parties at any particular time.

Whilst the benefits of cost plus strategies have previously been discounted, their potential is renewed when considered in light of strategic alliances. Take, for example, the situation where a client has established a long-term relationship through a strategic alliance with a constructor and, as a result, has gained considerable confidence in the ability and motivations of this organization. In the absence of the adversarial relationship and mistrust of the parties the use of a cost-plus arrangement in this circumstance may become highly attractive to both the client and the contractor. This is often called 'open book' contracting.

There are many benefits of open book contracting for the client, the most notable of which will be reduced costs associated with administrating the various contracts. The client will also enjoy the flexibility of design during the project and will be assured of receiving unbiased advice from the contractor concerning design and construction issues.

Similarly, the contractor all but eliminates the usual risks involved in carrying out the work and, as such, is guaranteed to make a profit. Under perfect conditions a strategic alliance with a cost-plus payment arrangement is the most equitable strategy where the client pays no more or no less than is due and the contractor is guaranteed of a reasonable return on investment in the work.

4.3 Tendering strategies

4.3.1 Open tendering

Where an open tendering arrangement is adopted, an invitation to tender is advertised publicly for any interested contractor to submit a tender in accordance with the client's requirements. The number of contractors permitted to tender is unrestricted and, as such, this method provides an excellent basis for ensuring high levels of competition in the tender process. The process has the distinct disadvantage to the client of the necessity to commit considerable resources to evaluating the numerous tenders received. However, this can be managed through the inclusion of rigorous qualification requirements that must be met in order to be able to submit a complying bid.

From the contractor's perspective, open tendering is less desirable as the greater the number of tenders the greater the probability that success will not be forthcoming. It may also have the effect of discouraging contractors who may otherwise be quite suitable for the project.

4.3.2 Selective tendering

Under these arrangements a select group of contractors are invited by the client to tender for the works. Tenderers are pre-selected based on their reputation and ability to carry out the works. A degree of competition is maintained although it is most common for the client to restrict the number of tenders to around three or four. Two distinct alternatives of selective tendering are now briefly discussed.

Single stage selective tendering is where the client invites a number of contractors to submit a tender for the works. The tenders are based on completed designs and specifications or comprehensive performance specifications where the works are being carried out under a single source delivery system such as design and construct. The client will evaluate the tenders and, if accepted, a contract is formed between the client and the contractor.

Two-stage selective tendering is where the client will invite expressions of interest from suitably qualified and experienced contractors. The expression of interest will normally include details of the contractor's relevant experience and capability to handle projects of the size and nature being considered. These are assessed by the client who will then select a number of the tenderers to submit a complete bid.

Following further detailed assessment of the second stage tenders, and possibly some form of negotiation, the client will enter a contract with the preferred contractor.

4.3.3 Negotiated tendering

Where negotiated tendering is adopted the client and a preferred contractor will enter a contract through direct negotiation. This method is ideal where the work is of a unique nature and the client is confident that there is only one contractor suitable to undertake the work or where the client has a strong preference to use a particular contractor.

Negotiated tendering does not facilitate competition in the tender process but the client, to ensure a competitive price is negotiated, can use certain checks. The distinct advantage to both the contractor and the client is that tendering costs are minimized. The method is frequently used amongst private sector clients; however, the accountability constraints of the public sector limit the use of this method to projects of a unique nature.

4.3.4 Value assessment of tendering

The emphasis of the value assessment method of tender selection is to optimize the value for the client. This is achieved through the establishment of weighted criteria that are used by the client to assess the 'value' offered by each of the bids received. Typical criteria used in a value assessment approach are:

- price
- technical, managerial, physical and financial resources of the contractor
- current commitments
- past performance/track record
- claims or disputes record
- industrial relations
- quality management
- environmental protection awareness
- occupational health and safety record.

The criteria adopted for a particular project or by a particular client will vary, as will the importance of each criterion. The importance to the client of each particular criterion is judged by the weighting applied to it. The focus of this approach is not about cost minimizing, but is primarily to compare the 'value added' offered by each tenderer. This approach is one of the key management tools available to clients during the selection of an appropriate procurement strategy, which can be applied to improve value for money and encourage improved performance in their construction projects.

4.3.5 Two-envelope tendering

For many clients, undertaking an objective value assessment of tenders can be made difficult when the tender price is known up front, together with the other value criteria. In many instances, human nature may result in the value assessment being distorted by the client who has knowledge of the tender price. In this regard the benefits of the value assessment may be lost.

To avoid this and to ensure that an objective value assessment is undertaken, clients may adopt a two-envelope tendering process. Tender submissions may be required to be submitted in two sealed envelopes, the first containing only non-price related submissions which may include details of the contractor's experience, expertise, financial capabilities and possibly concept designs under design and construct type arrangements. The second envelope contains only price-related information such as the tender form and priced bill of quantities where appropriate.

Initially the client will open only the envelope containing non-price related information and undertake a detailed value assessment, rejecting any tenders that do not adequately satisfy the client's requirements. Unsuitable contractors will be advised and have their unopened second envelope returned.

From the short-list, the client will then open each tenderer's second envelope containing price-related submissions and complete the value assessment. This process thus ensures that the value assessment is undertaken in an unbiased and purposeful manner, resulting in the client maximizing the chance of selecting the best value tender, through the meaningful comparison of all tenders received.

Another benefit of the process is that contractors are assured that the client is serious about not adopting a low bid selection policy. This in turn may encourage contractors to offer better value services and deter those who rely upon winning work on the low-bid principle.

4.3.6 Two-stage tender

With a two-stage tender, three or four contractors with appropriate experience are separately involved in detailed discussions with the client regarding all aspects of the project, for which only conceptual designs and specifications have been prepared. The contractors are required to price an approximate or notional bill of quantities or schedule of rates prepared on the basis of the notional design. Further selection criteria are then used to determine which contractor will be awarded the contract (Aqua, 1990).

Once selected, the contractor becomes part of the full project team and is involved in the preparation of the detailed designs, from which a complete bill of quantities may be produced. Key subcontractors may also be involved at this stage to provide specialist input into the design process.

4.3.7 Serial tendering

Serial tendering is undertaken where a number of similar projects are awarded to a contractor usually following a competitive tender on a master bill of quantities. This master bill forms a standing offer, open for the client to accept for a number of contracts. Each contract is separate and the price for each is calculated separately (Bentley, 1987). Subject to satisfactory performance the contractor will be awarded further similar projects which will be negotiated on the basis of the first.

Serial contracting is suitable where a number of similar projects are to be completed over a period of time and is commonly employed for projects such as service stations, schools and shop fit-outs. This approach has been trialled by the NSW Department of Public Works and Services, with a number of schools to be built over several years grouped into one contract. The system has many benefits for the contractor, including confidence about future business, and provides a continuous learning curve that should enable the contractor to improve the time and cost performance of subsequent projects (Cook, 1991).

4.4 Contract price strategies

There are three main alternatives for determining the contract price; these are lump sum contracts, measurement contracts, and cost-plus contracts.

4.4.1 Lump sum contracts

With lump sum contracts the price for the work is agreed at the time of signing the contract and is subject to adjustment for: rise and fall (where applicable), variations to the scope of works and prime cost or provisional sum items. It is important that lump sum contracts be used only where the client has a clearly defined vision of exactly what is required and that these requirements are adequately documented, either through a detailed design and specification or through a detailed performance specification.

Lump sum contracts have the distinct advantage of providing the client with an early indication of the costs associated with the project delivery and as such these arrangements are the most popular form of price determination under the traditional procurement strategy. This price certainty can be extremely attractive due to the reduced financial risk; however, clients should be conscious of the need to identify clearly what they require in the project prior to accepting a lump sum price.

4.4.2 Schedule of quantities

Schedule of quantities or schedule of rates contracts are where the method of determining the cost of the works involves the measurement of actual work carried out. The contractor will be required to provide a priced bill of quantities or a schedule of rates with the tender submission which, when accepted by the client, will form the basis for valuing the works. These rates are applied to the agreed quantities measured on the completion of the work to determine the total cost.

Bills of quantities are typically prepared by professional quantity surveyors using standard item descriptions and measurement procedures. The client will provide tenderers with a complete bill of quantities or a schedule of rates to which the contractor must apply unit rates for the various items. In some instances the bill of quantities will be provided for the information of tenderers only, in which case the quantities will not form part of the contract. In these circumstances the contract would be considered to be on a lump sum basis with no adjustment of the contract price as a result of variations to the quantities provided in the bill. Measurement contracts can be of benefit to the client where construction is required to commence prior to the completion of the detailed design; however, clients may also be exposed to considerable risk arising from the uncertainty of not knowing the exact contract sum before work is commenced (Pilcher, 1992).

4.4.3 Cost plus

Under a cost-plus arrangement, or cost reimbursement, the client pays the contractor or contractors the costs incurred by them for the work completed plus a fee. The fee may be a fixed sum or a percentage of the cost of the work completed as agreed prior to the commencement of work. One of the obvious disadvantages of the method is the lack of price competition, particularly where the contractor employs direct labour to undertake the work. This problem can in part be overcome through the subcontracting of trade packages through a system of competitive tendering.

The arrangement is particularly useful on small projects where the client may be required to make frequent design decisions during the course of the works and where changes to the construction must be undertaken (Barrie and Paulson, 1992). The method is also commonly used where it is not possible for the contractor to prepare a realistic price or where the works are of an urgent nature such as in an emergency situation (Pilcher, 1992). The inherent problem with these arrangements however, is that the motivation of the head contractor is not to reduce costs but actually to maximize the cost to the client in order to maximize his reward (profit) for undertaking the works (Cook, 1991). This view is understandable when you consider the adversarial nature adopted by many people within the industry. However, with recent developments in the adoption of long-term 'non-adversarial' relationships, cost-plus strategies can be seen in a different light. The potential of the cost-plus payment arrangement will be driven by the growth of strategic alliances in the industry.

4.5 Forms of private involvement in public projects

There is a widening range of alternative ways in which private enterprise can be involved in public sector building projects, particularly the provision and operation of infrastructure. This discussion focuses on infrastructure, but the forms of private involvement listed below apply equally to building projects. The choice of the appropriate structure is essential for effective collaboration between the public and private sectors, and includes the following four alternatives that have been identified.

1. *Private sector construction of infrastructure owned and operated by government.* Under this arrangement the public sector will design and construct the infrastructure project, and when complete enter into a long-term Operation Agreement (OA) with a private sector company. This OA may or may not include maintenance obligations. Ownership will remain in the public sector.
2. *Private sector Design and Construction (D&C) of infrastructure with ongoing maintenance obligations.* This arrangement comprises a D&C or Turnkey construction agreement coupled with a long-term contract for the operation and maintenance of the infrastructure. The public sector retains ownership of the project, as well as responsibility for design, construction and operation.
3. *Private sector D&C and operation of infrastructure that remains the property of government.* This arrangement comprises a D&C or Turnkey arrangement plus long-term OA. Ownership of the project remains with the public sector.
4. *Equity joint venture between the private sector and government for the delivery of infrastructure.* Under this arrangement, private sector interests and the public sector will enter into a joint venture agreement to produce an infrastructure project required by the public sector but in which it may not hold equity. The public sector often contributes the land as its equity share and the private sector contribution finances the project. The public sector then enters into a long-term lease with the Joint Venture Vehicle at commercially viable rents and the private sector interest then enters an agreement with the Joint Venture Vehicle for the design and construction of the project. Project finance is then arranged, and the work proceeds. On completion of the project, the Joint Venture Vehicle sells the project as a fully tenanted investment. The profit is then split between the public sector and the private sector in proportion to their equities. The client is left with a purpose-built infrastructure project with a long-term lease, with minimal impact on its balance sheet. The principal advantage of this arrangement is that the public sector does not have to fund the project but gets the benefit of a purpose-made, but leased, project.

4.6 Conclusions

From this discussion of procurement strategies it is evident that a number of benefits are available to clients through the appropriate selection of an alternative procurement strategy. Generally, the strategies investigated provide one or both of two key benefits. First, they provide the client with a single point of responsibility for the project outcomes and, secondly, they break down adversarial relationships and facilitate effective communication and coordination between members of the project team.

Moves towards single-source strategies such as design and construct arrangements will enable contractors to look for more innovative ways of construction, because they have a greater input during the design process and hence much greater scope to apply innovative construction solutions to the process. Collaborative strategies such as partnering may provide significant benefits to clients in terms of improvements to relationships and minimization of adversarial relationships that have dominated in the past.

In response to the poor performance of the traditional procurement strategy clients have looked towards new and innovative strategies in order to overcome these deficiencies. As such, these alternative procurement strategies have been developed in response to the limitations of strategies previously adopted.

It has also been demonstrated that each type of strategy, whether traditional, single source or collaborative, has inherent strengths and weaknesses and, as such, each strategy may be more suitable to any particular application than another strategy.

This in turn suggests that there is no such thing as a best or a better strategy for blanket application across all projects. The selection of an 'appropriate' procurement strategy is a real challenge for clients.

Comparisons of the various project delivery methods show the traditional method is generally considered the safest but is also the slowest, as the design must be completed before the project is bid, design-build is faster and more convenient and design-build finance provides expertise, value, and money for undercapitalized owners. The quality of the design brief and project specifications appears to be fundamental to the success of alternative delivery systems.

References and bibliography

Aqua (1990) *Tenders and Contracts for Building*. Second edition. The Aqua Group (London: BSP Professional Books).

Barrie, D.S. and Paulson, B.C. (1992) *Professional Construction Management* (McGraw-Hill).

Bentley, J.I.W. (1987) *Construction Tendering and Estimating* (London: E & FN Spon).

CIB (1992) *Code of Practice for Project Management for Construction and Development* (London: The Chartered Institute of Building).

CIDA (1993) *Partnering: A Strategy for Excellence*. Construction Industry Development Agency (Canberra: Master Builders Association).

CII (1991) *In Search of Partnering Excellence* (Construction Industry Institute, University of Texas at Austin).

Cook, A.E. (1991) *Construction Tendering – Theory and Practice*. Construction Technology and Management (London: B.T. Batsford Ltd).

Collins, R.G. (1991) Strategies for Efficient Project Delivery. In: *BSFA Seminar, Contracts – Obligations of Clients, Contractors & Consultants*. Sydney.

Cox, A. and Townsend, M. (1998) *Strategic Procurement in Construction* (London: Thomas Telford Publishing).

Franks, J. (1984) *Building Procurement Systems – a Guide to Building Project Management* (London: The Chartered Institute of Building).

Gould, F.E. (1997) *Managing the Construction Process – Estimating, Scheduling and Project Control* (New Jersey: Prentice Hall).

Griffin, A. (1994) *Partnering: the New South Wales Government's Perspective*. Deputy Director – Industry Policy Division, NSW Public Works, CPLI Seminar.

Gyles, R. (1992) *Final Report of the Royal Commission into Productivity into the Building Industry in New South Wales* (NSW Government).

Hawk, D.L. (1992) *Forming a New Industry – International Building Production* (Institute of International Business, Stockholm School of Economics).

Latham, M. (1994) *Constructing the Team*. Final Report of the Government/Industry Review of Procurement and Contractual Arrangements in the UK Construction Industry (London: HMSO).

Moxley, R. (1993) *Building Management by Professionals*. Butterworth Architecture Management Guides (London: Butterworth-Heinemann).

NSW Government (1996) *The Construction Industry in New South Wales – Opportunities and Challenges*, NSW Government Green Paper (Sydney: Department of Public Works and Services, Policy Division).

Pilcher, R. (1992) *Principles of Construction Management*, Third edition (New York: McGraw-Hill).

Ronco, W.C. and Ronco, J.S. (1996) *Partnering Manual for Design and Construction* (McGraw-Hill).

Rougvie, A. (1987) *Project Evaluation and Development*. Construction Technology and Management (London: B.T. Batsford Ltd).

Uher, T.E., Levido, G.E. and Davenport, P. (1993) *Building Contract Administration*. Construction Management and Economics Unit, School of Building, The University of New South Wales.

Venmore-Rowland, P., Brandon, P. and Mole, T. (1991) Investment Procurement and Performance in Construction. In: *Proceedings of the First National RICS Research Conference*. The Royal Institute of Chartered Surveyors.

5

Planning for conflict

Martin Loosemore†

Editorial comment

The ultimate success or failure of a construction project often depends on the manner in which disputes and conflicts between participants are managed and resolved. In Chapter 22, John Twyford examines in some detail a number of practical methods for minimizing disputes and for resolving those that do arise, but in this chapter a more philosophical approach is adopted as the author looks at various aspects of the mechanisms which lead to conflict in construction projects.

Understanding why conflict arises is fundamental if any successful attempt is to be made to plan for the possibility of conflict during the life of a project. Conflict, whether between client and contractor, contractor and workforce, contractor and consultant, or in any situation, is costly. Delays, rework, programming difficulties, quality control problems – all these factors can result from conflict, and will inevitably result in financial loss, for some or all of those involved. It is for this reason that planning for conflict is an important concern for clients who wish to achieve the best result when they invest in construction works.

The question is not so much how to avoid conflict, as it is recognizing that some degree of conflict will occur during the design and construction of a building. The aim of clients and their advisers should be to have suitable methods for identifying and resolving conflict as early as possible in place in their management framework from the outset.

Managing conflict before it materializes allows for the minimization of financial impact on participants, as disputes can be identified and settled before costs escalate. Obviously, minimizing costs associated with disputes will contribute to cost minimization for the entire project and therefore increase the value which the client gains. There is, however, more to be considered: conflict which causes delay can also lead to acceleration of the construction programme with attendant problems of quality control, disruption of supply chains, unavailability of sub-contractors and so on, all of which can increase the economic loss attributable to the conflict.

In the following chapter, Martin Loosemore takes a broad view of conflict in the construction industry, and provides a valuable insight into the mechanics of conflict and rational approaches to its management, all of which can assist clients in minimizing both the disruption and the financial damage which conflict can produce.

† University of New South Wales, Australia

5.1 Introduction

Fenn and Speck (1995) quoted a fourfold increase in the level of litigation over the previous 20 years, suggesting that the construction industry was developing dysfunctional levels of conflict. Care is needed in the interpretation of such evidence because of measurement problems that are posed by the diversity of the construction industry, the informal manner in which many disputes are resolved and the difficulty in defining conflict. Furthermore, market forces, increasing educational standards within the industry, the emergence of specialist construction lawyers and changes in the law which make it easier to litigate, may have merely resulted in an increase in formally recorded conflicts that would have previously gone unpublicized. Despite these limitations, the issue of conflict has been elevated to one of the most important contemporary challenges facing the construction industry. Considerable momentum has been added by reports such as AFCC (1988), NPWC (1990), CSSC (1988) and Latham (1994) which have portrayed conflict as a cancerous force that needs to be reduced and ideally eliminated from the construction process. The result has been the diversion of considerable resources towards its reduction.

Gardiner and Simmons (1995) demonstrated that conflict occurs in all phases of the construction process and is often most prevalent during design. They argued that managers should plan for conflict as early as possible in the project life cycle because the costs of handling it escalate as time goes by. The aim of this chapter is to demonstrate the benefits of developing a conflict management plan during the pre-design stage of a construction project and to discuss the methods which should be employed in its formulation.

5.2 Conflict management methods

Methods of dealing with organizational conflict have changed over time.

5.2.1 Early methods

Arguably, the earliest records of managerial concern with organizational conflict are found in the work of Adam Smith, Robert Owen and Charles Babbage (Sheldrake, 1996). They recognized the productivity benefits that could be achieved from specialization and mechanization but they also experienced the industrial conflict which resulted. Adam Smith chose to live with the problem, considering that the advantages to society outweighed the disadvantages to individual workers. However, during the industrial revolution, the Luddist campaigns of intimidation and machine-breaking produced a more conscious effort to deal with the problem of industrial conflict. In particular, Robert Owen experimented with a paternalistic style of management and demonstrated that a healthy workforce was good for business. Similarly, Charles Babbage used profit-sharing schemes to create a mutuality of interests between managers and employees which could transcend their differences.

5.2.2 Traditional methods

During the early 20th century, the rise of American industry and the increased scale of operations which accompanied it, forced managers to explore new methods of controlling labour. Initially, Frederick Taylor, the Gilbreths and Henry Gantt sought to apply the principles of engineering in a human setting and replaced a traditional craft and judgement based system with the certainty and clinical nature of what became known as scientific management. Conflict was seen as wholly undesirable and arising from indiscipline and poor supervision. Consequently, every effort was directed at its elimination, primarily through the provision of monetary rewards and penalties, the scientific selection and development of workers, greater standardization, specialization, mechanization and close control. Although this was an attempt to produce harmonious organizations of cooperative structures the result was reduced worker autonomy, de-skilling, the alienation of supervisors from workers, and increased levels of antagonism between unions and employers.

5.2.3 Contemporary methods

The human costs of scientific management prompted a change in management thinking to that founded upon a greater sensitivity to human needs. Scholars like Henri Fayol, Elton Mayo and Abraham Maslow sought to redefine management by providing it with a psychological and moral base. This was an attempt to reunite workers and managers in a more open, interactive and considerate environment. There emerged the view that conflict was an inevitable aspect of organizational life to be accommodated rather than suppressed. Conflict had psychological as well as economic origins and organizations were seen as arenas where competing interest groups battled through formal and informal systems, for limited resources. This laid the foundations for the political science tradition in organizational theory which first challenged the view that conflict was a wholly disruptive force in organizations (Simon, 1948; Selznick, 1957; Crozier, 1964). The contemporary view of organizational conflict is a direct descendant of this position and it is that conflict is the norm rather than the exception in organizations and that absolute harmony is impossible. Conflict develops through a number of phases and is an essential and healthy part of organizational life in that it performs many important and positive functions for its participants (Robins, 1974; Likert and Likert, 1976; Argyris, 1990; Handy, 1993). The managerial challenge is to harness these potential positives rather than to minimize conflict.

5.2.4 Radical methods

The most radical method of dealing with conflict is to positively provoke it (Frazer and Hippel, 1984; Pascale, 1991; Furze and Gale, 1996). Those who advocate this interactionist approach see conflict as a doorway of opportunity to organizational learning, vibrancy, vitality and to the fulfilment of organizational and individual potential.

5.3 Construction conflict management methods

The historical review of conflict management provided above illustrates that a manager's beliefs will determine the mix of prevention, accommodation and provocation in their conflict management plans. In the construction industry, traditional methods of prevention are likely to dominate because of a scientific value system that is reflected in the popularity of management trends such as lean construction, business process re-engineering, benchmarking and value-engineering (McGeorge and Palmer, 1997). Each of these trends is based upon the scientific principles of measurement and control. The emphasis upon prevention is also reflected in the preponderance of literature seeking to reduce construction conflict through the development of predictive models based upon an understanding of its incidence and causes, by the relative lack of behavioural research and by the general perception that conflict is, most fundamentally, an economic rather than psychological phenomenon (Conlin et al., 1996; Fellows et al., 1994; Barnes, 1991; Clegg, 1992; Gardiner and Simmons, 1992; Yogeswaran and Kumeraswamy, 1997).

Despite the apparent seductiveness of science to construction managers, the contemporary view that conflict is inevitable has begun to emerge. Not only has there been significant interest in the relative attributes of litigation, arbitration and alternative dispute resolution techniques (Brooker and Lavers, 1995; Fenn and Speck, 1995), but some have begun to argue that conflict is not necessarily dysfunctional for construction projects (Loosemore and Djebarni, 1994; Rhys Jones, 1994; Chew and Lim, 1995; Osborne et al., 1995). The emerging view is that the construction industry's problem is not in the incidence of conflict but in the way it is managed.

The most radical view, that conflict should be positively provoked, has little support within construction, although Hughes (1994) has argued that in complex, creative and high-risk construction projects, conflict is a necessity. While this may be true, it is unclear whether construction managers have the desire, attitudes or skills to manage conflict constructively, and calls for the encouragement of conflict could be premature and potentially damaging.

5.4 Conflict management plans in construction projects

While the current emphasis upon conflict reduction may be justified for the time being, it would be dangerous to neglect the need for organizational resilience in dealing with the unexpected. Wildavsky (1988) and Rosenthal and Kouzmin (1993) warn that managers who focus upon prevention tend to create a culture of invincibility which reduces their ability to mitigate the effects of conflicts when they do arise. Therefore, although prevention is better than cure, it is important that conflict management plans have both preventative and reactive elements. Reactive elements do not refer to increased ingenuity in the development of alternative dispute resolution techniques, because they deal with the consequences of failed managerial efforts. Rather, they refer to the ability to manage conflict to positive effect, preventing it escalating to the point where expensive and time-consuming third party intervention is required.

5.5 The preventative element of conflict management plans

The key to prevention is the discovery and elimination of causes and there is no shortage of literature identifying the potential causes of construction conflict. What follows is a brief review of the main threads which hold this voluminous and repetitive literature together. In essence, the causes of conflict can be grouped under three main headings; namely, manufactured, institutional and psychological.

5.5.1 Manufactured causes

Manufactured causes relate to those created by managerial practices.

Mistrust

Latham (1994) argued that a culture of mutual trust and collective responsibility is needed to reduce conflict in the construction industry. Such a culture would improve communication and could be achieved by greater attention to the language and structure of contracts and to the equitable and clear distribution of risks within them (Barnes, 1989, 1991; Uff, 1995; Hancock and Root, 1996). Further contributions could be made by reducing abuses of competitive tendering which force down margins to restrictively low levels and by placing less emphasis upon price as a team selection criterion. This latter argument has led to calls for more intelligent team building practices and for a movement to negotiated contracts, often within continuous partnering arrangements (Luck and Newcombe, 1996; Hatush and Skitmore, 1997).

The client's role

There is much evidence to support the contention that the client's attitude towards paying for risks and their integration into the project team is essential in reducing conflict (Morris, 1972; NEDC, 1983; Kometa et al., 1994). In particular, the importance of the briefing process in defining the project and its boundaries has received considerable attention for some time and it is accepted that rushing this pre-design phase can sow the seeds of problems which may manifest themselves later in a project (Kelly et al., 1992).

A lack of participation

Newcombe (1994) has criticized the hierarchical, class-based structure of traditional construction projects and argued for flatter organizations that are receptive to more participatory management styles. To facilitate greater participation, some have questioned the role of standard contracts, arguing that they discourage people talking about their roles, relationships and responsibilities. By doing this they build misunderstanding into construction projects from a very early point (Murdoch and Hughes, 1992). Indeed, Murdoch and Hughes attributed the early success of construction management to the very fact that there was no standard form of contract. This meant that people were forced to talk through their contractual responsibilities at an early point in their relationship.

Design

Gardiner and Simmons (1995) dispelled the traditional perception that conflict is primarily a construction stage phenomenon. Indeed, they showed that conflict is more prevalent in design, this being correlated to the large amount of change which characterizes this process. This relatively high level of conflict may also be related to the emotive and creative nature of design, to the high degree of reciprocal interdependency between those involved and to the volume of decisions that have to be made from first principles (Gray *et al.*, 1994; Austin *et al.*, 1994). Conflict within the design phase can also affect later project stages since design-based conflicts that remain unresolved manifest themselves in poor quality contract documentation, which is a common cause of conflict during the construction stage (Crawshaw, 1976).

Sub-contracting

The growth of sub-contracting has also been identified as a source of construction conflict in the complexity of the contractual arrangements and the managerial challenges it produces (NEDC, 1983, 1988). With the growth of sub-contracting has come fragmentation, instability, short-termism, reduced customer orientation and problems of communication, coordination, motivation and quality control.

Procurement systems

There has been some inconclusive debate about the relationship between procurement systems and conflict (Naoum, 1991; Conlin *et al.*, 1996). While the possibility of a simple explanation for construction conflict is attractive, such research could grossly over-simplify the complexity and variability of construction project organizations and produce overly simplistic answers to complex questions. The concept of a procurement system is one that has been much maligned and, although it performs a convenient categorization function, an over-reliance upon it could mask the true variability of construction project organizations and thereby induce inflexibility and complacency in their management. In this sense, the concept of procurement may be more damaging than helpful in the conflict debate.

Market size

One of the simplest reasons offered for the relatively high levels of conflict in the construction industry is that the industry is too big for its market. The temptation for clients who are confronted with an over-supply of competition is to secure the lowest possible price from tenderers. The consequence of this behaviour is lean, under-resourced projects that are susceptible to conflict because of their inflexibility in coping with unexpected problems which impose extra resourcing demands (Richardson, 1996). Within such organizations survival is the priority and goodwill is an unaffordable luxury. This is particularly so in high-risk projects where the increased occurrence of problems erodes contingency allowances more rapidly and reduces the ability and willingness of companies to accept their resourcing responsibilities. Unfortunately, the companies that thrive in such an environment are the unscrupulous ones who 'go in' low with the aim of 'coming out' high. In this sense, it is an environment that encourages mediocrity and, at worst, corruption, and which reduces the behaviour of the industry to the lowest common denominator. In many respects this is where the role of clients is critical because they determine the way in which the market operates.

5.5.2 Institutional causes

Institutional causes relate to those causes that are inherent within the industry's culture as a result of its historical development.

The Australian construction industry has the same origins as the UK construction industry and therefore reflects its hierarchical, class-based structure. This has resulted in a construction process that is divided along functional lines, particularly between design and construction. Although these divisions have been eroded in recent times with the development of more integrated procurement systems such as design and build and construction management, they still remain strong. This is because they have been cemented into place by professional institutions who have vested interests in maintaining their distinct identities and divisions (Emmerson, 1962; Banwell, 1964; Bowley, 1966). There have been two mechanisms for achieving this; namely, the educational system, over which the professional institutions have a hold through accreditation systems, and the standard contracts, which are widely used and which reflect the divisions within the representative body of professions that produced them. The self-imposed isolation of the construction professions has produced an industry that is characterized by a diverse range of languages, cultures, aspirations, needs and interests. This is unconducive to the effective communication, teamwork, and sense of trust and collective responsibility that is necessary for a smooth-running project. While each of the construction professions has developed its own unique language, practices and culture, they all have their engineering origins in common (Seymour and Rooke, 1995). The culture of faith in scientific methods that this has produced also contributes to the industry's incidence of conflict by subjecting its personnel to rigid systems that fail to satisfy their need for autonomy in performance of non-routine, non-mechanized tasks in an uncertain environment. It has been known for some time that the consequence of doing this is likely to be dysfunctional behaviour (Burns and Stalker, 1961).

5.5.3 Psychological causes

Both De Bono (1991) and Covey (1994) have argued that the most fundamental cause of organizational conflict operates at a psychological level. In particular, it is to be found in the different ways people see the world, because most relationship difficulties stem from people's differing expectations about roles and goals, which in turn causes misunderstanding, disappointment and a withdrawal of trust. To resolve conflict successfully, a manager must seek to understand the different paradigms through which people see a problem, since this determines their own behaviour and the way they interpret that of others. The forces which shape these paradigms are discussed below.

Personality, education and past experience

Past experiences, personal, practical and educational, represent a powerful shaper of opinions and perceptions about situations and other people. This is particularly so in the construction industry where professional institutions have successfully nurtured strong stereotype images of each other in order to defend their traditional roles within the procurement process. Seymour and Rooke (1995) argue that such images play a particularly important role in construction projects because, at the start of a project, the newness of the team means that there is little else to base relationships upon. All construction projects are therefore characterized, in their early stages, by an uncertain and cautious period of relationship building, which is important in reinforcing or dispelling these stereotype images.

Information, truth, falsity and corruption

De Bono recognizes that everyone enters a dispute with different information, which shapes the way they perceive and define the dispute. For example, Touskas (1995) points to the complex and unpredictable way in which information is dispersed within organizations and how this makes it impossible for anyone, including the manager, to understand the different perceptions of a conflict that those involved hold at any point in time. The challenge for conflict managers is to equalize information differences between people and thereby allow a common definition of a problem to emerge. In the highly dynamic environment of a construction project, the difficulty of this task is likely to be heightened. Further information discrepancies may be caused by the habit of separately distributing construction risks. This makes people more likely to hold onto information as a potentially important source of power in negotiations. Indeed, De Bono argues that in extreme circumstances, desperation may induce people to fabricate or falsify evidence in order positively to deceive and distort the truth in their favour.

Human desire for contradiction and consistency

De Bono argues that people have a natural desire for consistency and certainty and therefore tend to think in 'black and white' terms by deliberately searching out contradictions, opposites and categorizations. Unfortunately, during a conflict, this results in aggressive, non-compromising signals, which during negotiations can increase the chances of polarization, stalemate and escalation. In this sense, it is the responsibility of managers to encourage people to think in 'tones of grey'. However, this is made difficult by the separation of project risks by construction contracts, the time and cost pressures that increasingly characterize construction projects and the inherently 'macho' and confrontational culture of the industry (Gale, 1992).

5.6 The reactive element of conflict management plans

The need for reactive capabilities in organizations is particularly acute on high-risk projects where there is an increased chance of problems slipping through the preventative net. In dealing with conflict the aim should be to minimize potential costs and maximize potential benefits and, in doing so, efficient detection, diagnosis and intervention are essential.

5.6.1 Detection

Philips (1988) has shown the importance of early conflict detection by illustrating an exponential increase over time of the potential costs of conflict and the potential benefits that can be gained from its management. However, there are many forces within a construction project that reduce the likelihood of conflict detection and these are discussed below.

Natural resistance to change

The end result of conflict may involve change to the reward structure of an organization, and the separation of risks by most standard construction contracts ensures that there are distinct winners and losers in the new order that emerges. While potential winners are likely to be effective detectors of conflicts, Machiavelli suggests that potential losers are likely to be a much stronger force against detection (Plamantez, 1972). Thus, to ensure early intervention, managers must seek to identify, protect and manage potential losers. However, this is difficult when conflict remains undetected and a more beneficial strategy may be to eliminate imbalances in resource redistributions when a conflict arises, by sharing project risks.

The progressive and ill-defined nature of conflict

One of the main difficulties in detecting conflict is knowing when it has occurred. It is well known that conflict progresses through a number of stages from *just begun* (where parties are simply discovering and discussing their differences) to *contention*, to *dispute*, to *limited warfare* and ultimately to *all-out warfare* (where parties are trying to destroy each other at any cost). The vast majority of conflicts do not escalate to the all-out warfare stage and are resolved informally at an earlier stage. While a project can potentially benefit from managerial intervention in every conflict, time pressures ensure that project managers should only be informed of those that reach the dispute stage, since it is here that the potential for damage starts to accumulate, and the potential for benefits begins to depreciate. The problem in judging this critical point of managerial intervention rests in the unmeasurable nature of these stages and the gradual way in which they develop.

Feedback

Information about potential problems is fed through dedicated management information systems which provide channels for specialist information relating to specific project goals. Most construction problems affect a range of project goals and the lack of interaction between distinct management information systems in construction projects can result in potential problems going undetected or important aspects of problems being ignored (Baxendale, 1991). The skill of a project manager in overcoming this problem depends upon an ability to synthesize possibly conflicting information from a variety of management information systems and which is likely to arrive at different times.

Dynamic team membership

The dynamic nature of project team membership can have an impact upon receptivity to warning in three ways. First, as people leave a project they take important knowledge with them which can provide clues to potential problems. Secondly, new project members move along a learning curve during which their energies are likely to be focused upon issues other than the detection of potential problems. Finally, when team membership is not consistent throughout a project, people are more likely to suppress problems in the hope that they can be dealt with by others, later in the project (Loosemore, 1996). The result of this 'buck-passing' behaviour can be disproportionately damaging, particularly when it occurs early in the project life cycle and the problem has a high interdependence with other activities.

Unstable sensitivity

Loosemore (1996) found that receptivity to warning may vary throughout the life of a construction project as it moves from one stage to the next. In traditionally procured projects, low sensitivity to potential problems was found to be a particular problem during the dormant period between design and construction. During this period of relative inactivity, people's attentions were temporarily focused elsewhere and the team structure was in a state of transition. Project managers must seek to identify these dangerous transitionary periods and maintain the project team's diligence.

Defensiveness

Argyris (1990) has shown that people implicated in blame for a problem will ignore it, play it down or deliberately craft messages to distort it. In many instances, this behaviour is encouraged by managers' conflict handling styles and by organizational policies. The result is an illusion of control which eventually results in simple problems becoming crises that can threaten an organization's existence.

5.6.2 Diagnosis

At this stage, detected conflicts should be assessed in terms of their potential costs and benefits, and intervention plans should be developed to minimize the former and maximize the latter. The organizational forces that act to reduce the efficiency of this process are similar in nature to those affecting the detection phase but they operate in a subtly different manner.

Defensiveness

Project managers who have an interest in a dispute are unlikely to see it in the independent fashion that is needed for reliable diagnosis. Rather, they are likely to employ the defensive routines which Argyris has identified in an attempt to defend their own interests. While an independent project manager is essential to the effective management of conflict, such independence is difficult to achieve in practice since a project manager's involvement throughout a project's life cycle ensures that they are more likely than most to have a vested interest in the decisions that shape it. One solution would be to have a number of project managers with overlapping responsibilities, but in the under-resourced environments of most construction projects this is an unrealistic expectation.

Diverse information

A project manager investigating a conflict is likely to be presented with a wide range of conflicting information from diverse sources (Walker, 1989). Furthermore, the separation of project risks in many construction contracts, coupled with the matrix structure of project organizations, ensures that some project participants with important diagnostic information will have no direct interest in communicating it. In overcoming information shortages and in reconciling different perspectives a project manager will have to embark upon a reiterative and time-consuming process of investigation. In the case of crises, there is little time for such a process and the experience and judgement of a project manager become critical.

Centralized decision-making

In the case of serious problems, where reference to high authority is justified, or in highly centralized organizations where even the smallest decisions require reference upwards through the organizational hierarchy, diagnostic problems can arise. In particular, centralized decision-making slows down the speed of response and provides more opportunities for information filtering, which can distort the appearance of a problem. Rapid and accurate responses are vital in situations where a conflict is escalating rapidly out of control and managers should design their organizational structures in a decentralized fashion.

5.6.3 Intervention

Intervention is the stage at which a project manager mediates within a conflict to bring it to a successful conclusion. The ability to do this is largely dependent upon an appropriate conflict-handling style.

Conflict-handling styles

Rahim (1983) has argued that a manager's conflict-handling style varies along two dimensions; namely, assertiveness (the degree to which a manager pursues his or her own interests) and cooperativeness (the degree to which a manager attempts to satisfy his or her opponent's interests). His typology of five conflict-handling styles, which vary along these two dimensions, is illustrated in Fig. 5.1.

Competitive managers seek to further their own interests regardless of the impact upon their opponents. Such managers create a highly centralized, authoritarian, individualistic, confrontational and formal environment. Solutions to conflicts are always win–lose (I win, you lose) in nature and the winner is normally the one with the greatest power to force through their individual agenda. Valuable ideas and legitimate entitlements have little influence over conflict outcomes. In the long term, tensions are forced under the surface only to re-emerge later in a seemingly unrelated dispute. In this sense, the solution which results is lose–lose.

The avoiding style of conflict management is characterized by withdrawal, ignorance or suppression. This type of style creates a non-communicative, overly optimistic environment, which provides the illusion of control but results in no conflict resolution. In the long term, this strategy is dangerous since it permits

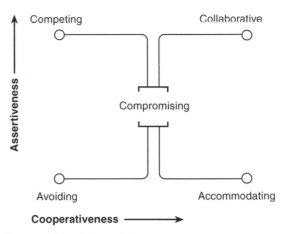

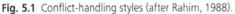

Fig. 5.1 Conflict-handling styles (after Rahim, 1988).

simple problems to escalate into serious crises that can threaten the viability of an organization. The compromising style is characterized by an emphasis upon sharing and it results in a lose–lose solution to conflict where both parties give up something. The accommodating style is characterized by one party giving something up for nothing in return and it results in a lose–win solution ('I lose, you win'). The environment that is created by this type of manager is similar to that created by the avoiding style because accommodating an opponent is an effective way of avoiding conflict. Finally, the collaborative style is characterized by a search for mutually beneficial solutions. The emphasis is upon investigating new solutions that are outside those which are preconceived by both parties at the commencement of a conflict. This style of management creates an explorative, participative, cooperative, group-based environment with a wide sense of collective responsibility and healthy levels of communication. While the collaborative approach demands a great deal of courage, imagination and energy, the solutions produced are win–win in nature and result in a real and long-term diffusion of organizational tension. While it is not always appropriate to pursue a collaborative style of conflict management, project managers should do so whenever possible and should structure their project teams to ensure they are made up of organizations and individuals who are receptive to such an approach.

5.7 Conclusions

This chapter has argued that it is essential to plan for conflict in today's increasingly competitive and technologically complex world. It is most beneficial to construct conflict management plans during the pre-design stages of a project and such plans should incorporate preventative and reactive elements.

Preventative elements should aim to:

- nurture a sense of trust and collective responsibility through more equitable employment and contractual practices

- involve clients and educate them about the importance of paying for risk transfer
- encourage decentralized organizational structures and participative management styles
- discourage rushing the early stages of projects
- provide opportunities for project participants to talk about their interdependencies and responsibilities
- minimize the potential for change
- minimize the number of organizational interfaces
- encourage intelligent team selection, paying attention to the compatibility of personalities, organizational cultures, experiences, educations, attitudes and values
- provide people with flexibility and autonomy, particularly in high-risk and innovative projects carried out in uncertain environments
- ensure effective communication by reducing organizational complexity and creating a trusting, decentralized environment characterized by shared project risks.

Reactive elements should aim to:

- nip a conflict in the bud
- avoid penal environments of winners and losers by sharing risks
- avoid recriminatory environments by sharing project risks
- monitor all potential conflicts to avoid uncontrolled escalations
- clarify reporting responsibilities for potential problems
- create clear reporting systems with effective lines of communication from monitors to decision-makers
- create a culture of collective responsibility for problem detection
- avoid matrix structures
- avoid changes in team membership
- encourage vigilance in dormant project periods
- maintain independent project management
- encourage thorough problem investigation before decision-making
- avoid centralized decision-making structures
- encourage collaborative conflict management styles whenever possible
- create teams of people and organizations who are receptive to collaborative management styles.

References and bibliography

AFCC (1998) *Strategies for the Reduction of Claims and Disputes in the Construction Industry: A research report* (Sydney: Australian Federation of Construction Contractors).

Austin, S., Baldwin, A. and Newton, A. (1994) Manipulating the flow of design information to improve the programming of building design. *Construction Management and Economics*, **12** (5), 457–65.

Argyris, C. (1990) *Overcoming Organizational Defences* (London: Allyn and Bacon).

Banwell, G.H. (1964) *The Placing and Management of Building Contracts for Building and Civil Engineering Works* (London: HMSO).

Barnes, M. (1989) The role of contracts in management. In: *Construction Contract Policy – Improved Procedures and Practice* (Kings College, London: Centre for Construction Law and Management), 119–38.

Barnes, M. (1991) Risk sharing in contracts. In: *Civil Engineering Project Procedure in the EC.* Proceedings of the conference organized by the Institution of Civil Engineers, Heathrow, London, 24–25 January, pp. 7–15.

Baxendale, A.T. (1991) Management information systems: the use of work breakdown structure in the integration of time and cost. In: Bezelga, A. and Brandon, P. (eds) *Management, Quality and Economics in Building.* Transactions of CIB Symposium on management, quality and economics in housing and other building sectors, Lisbon, 30 September to 4 October, pp. 14–23.

Bowley, M. (1966) *The British Building Industry: Four Studies in Response and Resistance to Change* (Cambridge University Press).

Bresnen, M. (1990) *Organising Construction – Project Organisation and Matrix Management* (Routledge).

Brooker, P. and Lavers, A. (1995) Perceptions of the role of alternative dispute resolution in the settlement of construction disputes: lessons from the UK and US experience. In *Proceedings of TG15 Conference on Construction Conflict, Management and Resolution*, CIB Publication 171, October, Lexington, Kentucky, USA, pp. 49–69.

Burns, T. and Stalker, G. (1961) *The Management of Innovation* (Tavistock).

Centre for Strategic Studies in Construction (1988) *Building Britain 2001* (Reading: University of Reading).

Chew, I.K.H. and Lim, C. (1995) A confucian perspective on conflict resolution. *International Journal of Human Resource Management*, **6** (1), 143 57.

Clegg, S.R. (1992) Contracts cause conflicts. In: Fenn, P. and Gameson, R. (eds) *Construction Conflict and Resolution.* Proceedings of the first international conference on construction conflict management and resolution, UMIST, UK, September (E and F N Spon), pp. 128–44.

Conlin, J.T., Langford, D.A. and Kennedy, P. (1996) The relationship between construction procurement strategies and construction contract disputes. In: Taylor, R.G. (ed.) *Proceedings of the CIB W92 North Meets South.* Procurement systems symposium, Durban, South Africa, January, pp. 66–82.

Covey, S.R. (1994) *Daily Reflections for Highly Effective People* (Fireside, Simon and Schuster).

Crawshaw, D.T. (1976) *Coordinating Working Drawings.* BRE Current Paper, 60/76, DOE.

Crozier, M. (1964) *The Bureaucratic Phenomenon* (University of Chicago Press).

De Bono, E. (1991) *Conflicts: A Better Way to Resolve Them* (Penguin Books).

Emmerson, H. (1962) *Survey of the Problems Before the Construction Industry* (London: HMSO).

Fellows, R.F., Hancock, M.R. and Seymour, D. (1994) Conflict resulting from cultural differentiation. In: *Proceedings of TG15 Conference on Construction Conflict, Management and Resolution.* CIB Publication 171, October, Lexington, Kentucky, USA, pp. 259–67.

Fenn, P. and Speck, C. (1995) The occurrence of disputes on United Kingdom construction projects. In: *Proceedings of the ARCOM 11th Annual Conference*, University of York, September, pp. 581–91.

Frazer, N.M. and Hippel, K.W. (1996) *Conflict Analysis: Models and Resolutions* (New York: North Holland).

Furze, D. and Gale, C. (1996) *Interpreting Management: Exploring Change and Complexity* (London: International Thomson Business Press).

Gale, A.W. (1992) The construction industry's male culture must feminise if conflict is to be reduced: the role of education as a gate-keeper to a male construction industry. In: Fenn, P. and Gameson, R. (eds) *Construction Conflict and Resolution.* Proceedings of the first international conference on construction conflict management and resolution, UMIST, UK, September (E and F N Spon), pp. 416–27.

Gardiner, P.D. and Simmons, J.E.L. (1992) Analysis of conflict and change in construction projects. *Construction Management and Economics,* **10**, 459–78.

Gardiner, P.D. and Simmons, J.E.L. (1995) Case explorations in construction conflict management. *Construction Management and Economics*, **13**, 219–34.

Gray, C., Hughes, W.P. and Bennett, J. (1994) *The Successful Management of Design: A Handbook of Building Design Management* (Reading, UK: University of Reading, Centre for Strategic Studies).

Handy, C. (1993) *The Future of Work. A Guide to a Changing Society* (Oxford: Basil Blackwell).

Hancock, M.R. and Root, D. (1996) Standard forms and conditions of contract: the imposition of roles, rules and rationality. In: *Proceedings of the Twelfth Annual ARCOM Conference*, September, Sheffield Hallam University, UK, pp. 160–9.

Hatush, Z. and Skitmore, M. (1997) Evaluating contractor prequalification data: selection criteria and project success factors. *Construction Management and Economics*, **15** (3), 129–47.

Hughes, W.P. (1994) Improving the relationship between construction law and construction management. In: *Proceedings of TG15 Construction Conflict: Management and Resolution,* Lexington, Kentucky, USA, October, pp. 278–94.

Kelly, J., MacPherson, S. and Male, S. (1992) *The Briefing Process; A Review and Critique.* Paper Number 12, The Royal Institution of Chartered Surveyors.

Kometa, S.T., Olomolaiye, P.O. and Harris, F.C. (1994) Attributes of UK clients influencing project consultant's performance. *Construction Management and Economics*, **12** (5), 433–43.

Latham, M. (1994) *Constructing the Team.* Final Report of the Government/Industry Review of Procurement and Contractual Arrangements in the UK Construction Industry (London: HMSO).

Likert, R. and Likert, R.G. (1976) *New Ways of Managing Conflicts* (New York: McGraw-Hill).

Loosemore, M. (1996) Crisis management in building projects. A longitudinal investigation of communication and behaviour patterns within a grounded theory framework. University of Reading, UK: unpublished PhD thesis.

Loosemore, M. and Djebarni, R. (1994) Tension, problems and conflict behaviour. In: Rowlinson, S. (ed.) *East Meets West.* Proceedings of CIB W92 Procurement Systems Symposium, University of Hong Kong, pp. 187–95.

Luck, R.A.C. and Newcombe, R. (1996) Integration of the project participant's activities within a construction project environment. In: Langford, D.A. and Retik, A. (eds) *The Organisation and Management of Construction: Shaping Theory and Practice* (London: E and F N Spon).

McGeorge, D. and Palmer, A. (1997) *Construction Management: New Directions* (London: Blackwell Science).

Morris, P.W.G. (1972) A study of selected building projects in the context of the theories of organisations. Department of Building, UMIST, PhD thesis.

Murdoch, J. and Hughes, W. (1992) *Construction Contracts – Law and Management* (London: E and F N Spon).

Naoum, S.G. (1991) *Procurement and Project Performance*, Occasional Paper, No. 45 (Ascot, UK: CIOB Publications).

National Economic Development Council (1983) *Faster Building for Industry* (London: HMSO).

National Economic Development Council (1988) *Faster Building for Commerce* (London: HMSO).

National Public Works Council (1990) *No Dispute: Strategies for Improvement in the Australian Building and Construction Industry* (Sydney: NPWC).

Newcombe, B. (1994) Procurement paths: a power paradigm. In: Rowlinson, S. (ed.), *East Meets West.* Proceedings of the CIB W92 procurement systems symposium, CIB Publication 175, University of Hong Kong, pp. 243–51.

Osbourne, A.N., Greenwood, D.J. and Maguire, S. (1995) A comparative analysis of inter-organisational conflict in the United Kingdom construction industry. In: *Proceedings of the ARCOM 11th Annual Conference*, University of York, September, pp. 591–603.

Pascale, R.T. (1991) *Managing on the Edge* (Harmondsworth, UK: Penguin Books).

Philips, R.C. (1988) Managing changes before they destroy your business: a non-traditional approach. *Training and Development Journal*, **42** (9), 66–71.

Plamenatz, J. (ed.) (1972) *Machiavelli* (London: Fontana/Collins).

Rahim, M.A. (1983) A measure of styles of handling interpersonal conflict. *Academy of Management Journal*, **26** (2), June, 368–76.

Rhys Jones, S. (1994) How constructive is construction law? *Construction Law Journal*, **10**, 28–38.

Richardson, B. (1996) Modern management's role in the demise of a sustainable society. *Journal of Contingencies and Crisis Management*, **4** (1), 20–31.

Robins, S.R. (1974) *Managing Organizational Conflict: A Non-traditional Approach* (Englewood Cliffs, New Jersey: Prentice Hall).

Rosenthal, U. and Kouzmin, A. (1993) Globalization an addenda for contingencies and crisis management – an editorial statement. *Journal of Contingencies and Crisis Management*, **1** (1), 1–11.

Selznick, P. (1957) *Leadership in Administration: a Sociological Interpretation* (Evanston, Illinois: Harper & Row).

Seymour, D. and Rooke, J. (1995) The culture of the construction industry and the culture of research. *Construction Management and Economics*, **13**, 511–23.

Sheldrake, J. (1996) *Management Theory: from Taylorism to Japanization* (London: Thomson Business Press).

Simon, M.A. (1948) *Administrative Behaviour* (New York: Macmillan).

Touskas, H. (1995) *New Thinking in Organisational Behaviour* (Oxford, UK: Butterworth-Heinemann).

Uff, J. (1995) Contract documents and the division of risk. In: Uff, J. and Odams A.M. (eds) *Risk Management and Procurement in Construction* (Kings College, London: Centre for Construction Law and Management), 49–69.

Walker, A. (1989) *Project Management in Construction*. Second edition (Oxford, UK: BSP Professional Books).

Wildavsky, A. (1988) *Searching for Safety* (New Brunswick: Transaction).

Yogeswaran, K. and Kumaraswamy, M.M. (1997) Perceived sources and causes of construction claims. *Journal of Construction Procurement*, **3** (3), 3–26.

6

Environmental impact assessment

Alan Gilpin†

Editorial comment

Concern about the effects that human developments have on their environments, whether local, regional or global, has increased dramatically in recent years. Part of that concern has been manifested in the worldwide adoption of assessment procedures that are intended to predict the environmental impacts which proposed projects would cause.

A common outcome of such assessments is the granting of permission for projects to proceed subject to a range of conditions being met. Typically, these conditions include things such as control of stormwater run-off, protection of natural habitats, or requirements for site rehabilitation in the future.

The great majority of such conditions will impose additional costs on those initiating a project, and therefore these conditions need to be identified and their likely costs factored into the economic evaluation of the proposed project.

Conditions may be imposed in response to a variety of statutory controls, local by-laws, or conservation and management plans, or they may occur as a result of public opposition or other project-specific factors.

Regardless of the process which produces the conditions, any appraisal of a proposed project must take clear account of the cost implications, as increased initial cost must affect the final 'value for money' result which those proposing a development hope to achieve.

Clients should also give some thought to the possible effects of future changes in environmental regulation; these effects may be difficult to predict, however, and will be of little interest to speculative developers who aim to build and sell in quick succession. These considerations will be of much greater concern to owner/occupiers or investors who intend to maintain an interest in a project for a considerable time after construction is complete. Future changes could include reduced building energy consumption targets (building energy codes have been introduced in many countries and such codes may be expected to become more stringent in the years ahead), phasing out of refrigerants (as has happened with CFCs and is happening with

† University of New South Wales, Australia

HCFCs), or the imposition of carbon or emissions taxes (as has happened in some European countries). Any changes of this nature can be expected to increase costs for building owners.

Thus, while the connection between environmental impact studies and assessments and value for money in building may not be immediately obvious, there are sound financial reasons, quite apart from any legal requirements, for a thorough appraisal and evaluation of these factors during the earliest stages of project procurement.

The following chapter provides a general outline of environmental assessment procedures; those requiring detailed information regarding Environmental Impact Assessment (EIA) requirements in specific places should seek pertinent information from the region or country concerned.

6.1 Evolution of environmental law

In early times, the world appeared large and people few. Today, the world is small and people many. Initially, the prevalent environmental influences and impacts were essentially natural: forest fires, volcanic eruptions, floods, droughts, earthquakes, with dramatic variations in climate, fauna and flora. Human influences were relatively minor.

With the growth of towns and the increasing urbanization of an expanding population, the afflictions of humanity intensified, turning on insanitary dwellings, water pollution, vector breeding, congestion, overcrowding, contaminated and adulterated food, poor hygiene and inadequate waste disposal procedures. Humanity was racked with smallpox, cholera, typhoid, typhus, dysentery, malaria, yellow fever, tuberculosis, diphtheria, schistosomiasis, and a raft of other diseases. The Black Death of the 14th century, transmitted by the rat flea, killed between one-third and one-half of the European population, and over 1000 towns and villages in England were depopulated. Countermeasures, without germ theory or vaccines, were essentially crude; and public health measures were slow to evolve. Isolated steps included closing suspect wells, and separating animals from dwellings.

In 1273, the use of coal was prohibited in London as being 'prejudicial to health'. In 1306, a Royal Proclamation prohibited artificers from using sea-coal (coal from Newcastle) in their furnaces; the execution of one offender is recorded. A further proclamation during the reign of Queen Elizabeth I made illegal the burning of coal while parliament was sitting. However, the use of raw coal continued to expand for the great forests were dwindling rapidly and wood was becoming both scarce and expensive. In 1648, Londoners presented a petition to Parliament in an attempt to prohibit the importation of coal from Newcastle because of its ill-effects. In 1661, John Evelyn submitted his famous 'Fumifugium, or the smoke of London dissipated' to King Charles II. He wrote: 'London's inhabitants breathe nothing but an impure and thick mist accompanied with a fuliginous and filthy vapour which renders them obnoxious to a thousand inconveniences, corrupting the lungs so that catharrs, phthisicks, coughs and consumption are more in this city than the whole earth besides.'

Eventually, a smoke abatement section was embodied in the Public Health Act 1875, known as the Charter of Public Health, and the first industrial inspectors were appointed. However, not until the 1952 London smog disaster, which claimed up to 4000 lives, were comprehensive national measures adopted.

In the meantime, vaccinations, anti-vector measures, better public health measures, the elimination of noxious 'accumulations and deposits', the appointment of environmental health officers, improved food hygiene, and a host of other measures gradually reduced the mortality and morbidity rates in many industrially advanced countries. However, many great cities in the developing world today remain largely unsewered.

Despite many successes and an increasing expectation of life in all countries, environmental problems have tended to intensify. This has led in more recent years to a whole series of United Nations Conferences relating specifically to the condition of the human environment and the quality of life. The explosion in world population; the massive movement of people from country to town; increasing industrialization; the burgeoning of the automobile; the increasingly widespread use of chemicals and fertilizers; noise problems; hazardous wastes, including radioactive wastes; the contamination of the air, water, land and food; major industrial and marine mishaps; soil erosion; the destruction of forests; over-fishing; a general over-exploitation of natural resources; and evidence of global climate change with progressive loss of biodiversity, has led to a worldwide recognition that development cannot be sustained in the longer term, in its present form, without a significant depletion of the resource base upon which all development ultimately depends. Industry itself, as well as government and its agencies, have become aware of new responsibilities that go beyond traditional concerns: a duty of care to the planet itself.

Pollution control measures have been adopted by all countries over the years, with varying degrees of effectiveness. Since 1972, many countries have created environmental departments and agencies to manage the whole spectrum of air and water pollution, noise pollution, land contamination, solid and liquid waste disposal, marine pollution, and protection of the environment generally. Environmental planning has become a strong feature in many advanced countries; while much more attention is given to the economic base of natural resources. The concept of *sustainable development* is now referred to in much legislation, while traditional regulation by laws and regulations is now being increasingly supplemented by what are called *economic instruments*. Above all, the principle of *environmental impact assessment* (EIA) has been accepted in many countries, as Table 6.1 indicates. EIA allows a comprehensive view to be taken of all major developments, so that the full implications may be taken into account by the decision makers, usually local, state or national governments. This is now widely practised in Europe, North America, and all Asian countries.

6.2 The environmental impact statement (EIS)

The EIS, sometimes known as the environmental effects statement (EES), the environmental statement (ES), or as the environmental impact assessment (EIA),

Table 6.1 Introduction of statutory EIA requirements

Country	Year
Australia	1974
Belgium	1985
Britain	1988
Canada	1973
China	1979
Czechoslovakia (the Czeck and Slovak Republics, 1992)	1991
Denmark	1989
France	1976
Germany	1975
Greece	1986
Hong Kong	1972
Indonesia	1982
Ireland	1988
Italy	1988
Japan	1972
Korea, South	1981
Luxembourg	1990
Malaysia	1985
Netherlands	1986
New Zealand	1991
Norway	1989
Philippines	1977
Poland	1989
Portugal	1987
Singapore	1972
Spain	1986
Sweden	1987
Switzerland	1983
Taiwan	1979
Thailand	1984
United States of America	1969

depending upon country, is a document, prepared by a proponent, describing a proposed development, or activity (or a plan, or programme) and disclosing the possible, probable, or certain effects of that proposal on the environment. An EIS should be comprehensive in its treatment of the subject matter, objective in its approach, and should be sufficiently specific for a reasonably intelligent mind to examine the potential environmental consequences, good and bad, of carrying out, or not carrying out, that proposal.

An EIS should meet the requirement that it alerts the decision maker, the proponent, members of the public, and the government, to the consequences for the community; it should also explore possible alternatives to the project that might maximize the benefits while minimizing the disbenefits (disadvantages). The primary purpose of an EIS, however, is to assist the decision maker (usually government at some level, or a government agency) to arrive at a better informed decision than would otherwise have been the case. A decision might involve the outright rejection of the proposal or its deferment for further studies or revision, although more usually the project is approved, subject to a range of legal conditions and requirements that are attached to the development consent, approval, or permit.

In addition to an EIS and its assessment or review by a responsible agency or branch of government, a controversial and highly sensitive issue might involve an independent public inquiry by commissioners specially appointed for that purpose. The principle of the EIS was introduced to the USA through the National Environmental Policy Act 1969, and has since been widely adopted in many countries including Australia, Canada, the members of the European Union, the Nordic countries, and many Asian countries.

An EIS usually includes the following: a full description of the proposed project or activity; a statement of the objectives of the proposal; an adequate description of the existing environment likely to be affected by the proposal; the identification and analysis of the likely environmental interactions between the proposal and the environment; the justification of the proposal; economic, social, and environmental considerations; the measures to be taken with the proposal for the protection of the environment, and an assessment of the likely effects of those measures; any feasible alternatives to the proposal; and the consequences of not carrying out the proposal for the proponent, community, region and state. The following points summarize the characteristics of a good EIS.

1. A summary of the EIS, intelligible to non-specialists and the public, should precede the main text.
2. Acronyms and initials should be defined; a glossary of technical terms can be relegated to an appendix.
3. The list of contents should permit quick identification of the main issues.
4. The authors of the EIS should be clearly identified.
5. A brief outline of the history of the proposed development should be given, including details of early consultations.
6. A full description of the proposed project or activity, its objectives and geographical boundaries; its inputs and outputs and the movement of these; also the inputs and outputs specifically during the construction phase. Diagrams, plans, and maps will be necessary to illustrate these features, with a clear presentation of the likely appearance of the finished project.
7. A full description of the existing environment likely to be affected by the proposal; the baseline conditions; deficiencies in information; data sources; the proximity of people, other enterprises, and characteristics of the area of ecological or cultural importance.
8. The alternative locations considered, or alternative processes, resulting in the preferred choice of site; evidence of credible studies will be needed here.
9. The justification of the proposal in terms of economic, social, and environmental considerations; the consequences of not carrying out the proposal for the proponent, the locality, the region, and the nation.
10. The planning framework, relevant statutory planning instruments, zoning, planning, and environmental objectives.
11. The identification and analysis of the likely environmental interactions between the proposed activity and the environment.
12. The measures to be taken with the proposal for the protection of the environment, and an assessment of their likely effectiveness, particularly about pollution control, land management, erosion, aesthetics, rehabilitation, eco-

logical protection measures, and decommissioning. Measures to achieve clean production and recycling; the management of residuals.

13. The effect on the transport system of carrying people, goods, services, and raw materials, to and from the project.
14. The duration of the construction phase, operational phase, and decommissioning phase; housing the workforce, both construction and permanent.
15. The implications for public infrastructure such as housing, schools, hospitals, water supply, garbage removal, sewerage, electricity, roads, recreational facilities, fire, police, emergency services, parks, gardens, and nature reserves; the implications for endangered species and threatened ecological features and ecosystems; the prospective financial contributions of the proponent.
16. Any transboundary or transborder implications of the proposal.
17. Any cumulative effects from similar enterprises should be considered, being either short term or long term, permanent or temporary, direct or indirect.
18. Proposals for annual reporting to the decision-making body on the implementation of the conditions of consent; post-project analysis and environmental auditing.
19. Arrangements for consultation with the relevant government agencies, planners, the public and interested bodies during the concept, preliminary, screening, scoping phases, the preparation of the EIS, the EIA stage, the construction, operational, and decommissioning stages; the communication of results.
20. Any unique features of the proposal of national or community importance, such as technology, employment characteristics, training, contributions to exports or import replacement, defence, landscaping, recreational facilities; foreign investment, or multiplier effects.
21. The contribution to sustainable development, and the containment of global environmental problems.

The EIS is an essential input to the EIA, which may be conducted by a government department, or by a Commissioner of Inquiry. It should be stressed that the EIS is but one input to the EIA; it is in the end, the proponent's view of the world, and often many will disagree. An EIA involves submissions from government departments, non-government organizations and, above all, by members of the general public, usually those most likely to be affected. An EIA aims to take account of all viewpoints, and to arrive at a balanced conclusion.

6.3 The environmental impact assessment (EIA)

The EIA is the critical appraisal of the likely effects of a policy, plan, programme, project or activity, on the environment. To assist the decision-making authority, assessments are carried out independently of the proponent, who may have prepared an environmental impact statement. The decision-making authority may be a level of government (local, state or national) or a government agency (at local, state or national level). Assessments take account of any adverse environmental effects on the community; any environmental impact on the ecosystems of the locality; any diminution of the aesthetic, recreational, scientific or other environmental values of a locality; the endangering of any species of fauna or flora; any adverse effects on any place or building having aesthetic, anthropological, archaeological, cultural, historical, scientific or social significance; any long-term or cumulative effects on the environment; any curtailing of the range of beneficial uses; any environmental problems associated with the disposal of wastes; any implications for natural resources; and the implications for the concept of sustainable development. The EIA extends to the entire process from the inception of a proposal to environmental auditing and post-project analysis.

In considering and determining a development application, whether from the public or private sector, a decision-making or approving authority should take into consideration such of the following matters as are of relevance to the proposed development or activity (Gilpin, 1995).

1. The provisions of:
   a. any environmental planning instrument, national, state, regional, provincial, or local plan relevant to any proposed preferred and alternative sites, together with any draft instruments or policies, conservation or management plans
   b. any conservation plan creating national parks or reserves, protecting rare or endangered species of flora or fauna, safeguarding elements of biodiversity, preserving heritage assets or natural resources, including mangroves, wetlands, bushland, waterways, coral and marine life
   c. the provisions of any government strategies for the development of roads and transport systems, energy facilities, transmission systems, mining or extracting natural resources and minerals, forestry activities and logging, agriculture and infrastructure.
2. The results of any specific studies such as ecological, social, hazard, health, urban, resource, or economic impact assessments.
3. The nature and character of the existing environment and something of its natural or industrial history, the nature and character of alternative sites or techniques which might be available.
4. The nature and character of the proposed development or activity.
5. The certain, likely or possible impact of the proposed development or activity on the environment in the short, medium or long term; the means proposed to control pollution and protect the environment; and the means of monitoring and auditing such measures within an environmental management plan.
6. The environmental consequences of the construction or development phase; and the decommissioning phase.
7. The effect of the proposed development or activity on the landscape or scenic quality of the locality; on any wilderness area or other natural asset.
8. The social effect and the economic effect of the proposal.
9. The physical character, location, siting, bulk, scale, shape, size, height, density, design and external appearance of the proposal.
10. The size and shape of the land to which the development application relates, its zoning, the siting of any building or facility thereon and the area to be actually used.

11. Whether the land involved is subject to flooding, tidal inundation, subsidence, slip, soil erosion, or has been contaminated by previous activity.
12. The relationship of the proposed site to developments on adjoining land, existing or already approved.
13. Whether the proposed means of entrance and exit from the subject land are adequate, and whether ample provision has been made for the loading, unloading, manoeuvring and parking of vehicles within that development.
14. The amount of traffic likely to be generated by the development or activity, particularly in relation to the capacity of the road system in the locality, and the effect on traffic generally in the regional road system; the alternative of rail transport.
15. Whether additional public transport services are needed.
16. Whether other infrastructure, such as water and sewerage facilities, housing and workforce accommodation, are also needed.
17. Whether the developer or proponent has agreed to make financial contributions towards the improvement of public infrastructure requirements.
18. Whether adequate provision has been made for the landscaping of the subject land and whether any trees or other vegetation already on the site should be preserved.
19. Whether the proposal is likely to cause soil erosion, or chemical contamination.
20. The contribution of the proposal to clean technology, waste minimization, recycling and reuse of materials, and acceptable disposal of unavoidable residues.
21. Any regional cumulative effects that may occur.
22. The likely contribution to the greenhouse effect.
23. In respect of mining, the adoption of progressive rehabilitation.
24. The environmental implications of the importation of natural resources to serve the proposed development or activity; and the environmental implications of the use of the final product.
25. Any representations by public authorities, corporations, private bodies, commercial interests, local councils, non-government organizations, and individuals or groups of residents, in relation to the proposal, either favouring or opposing.
26. Any special circumstances of the case, particularly in the context of the public interest.
27. The proposal in relation to the precautionary principle, polluter-pays principle; intragenerational and intergenerational equity.
28. The proposal within the context of sustainable development.

The outcome of an EIA is a recommendation to the decision maker either to approve the project in its entirety, subject to a range of recommended statutory conditions or requirements; or to reject the proposal, outright and completely. Outright rejection is rare, but it does occur. For example, in the case of the proposed Holmepierrepont 2000 MW power station in Britain in 1960, and the proposed Bayer plant for Kurnell in Sydney, Australia, in 1987.

6.3.1 Holmepierrepont, Nottingham

In respect of the former, the Central Electricity Generating Board (CEGB) for England and Wales had selected a site at Holmepierrepont on the east side of the city of Nottingham, in the East Midlands, for a major power station, the first of ten such installations. The power station was to be the latest design, comprising four 500 MW units. The coal-fired plant would be connected to a single multi-flue stack some 200 m in height, ensuring adequate dispersal of the flue gases during most meteorological conditions. Some 600 tonnes of sulphur dioxide would be dispersed from the daily consumption of 20 000 tonnes of coal. Dust-arresting equipment would remove over 99.3% of the fly ash carried forward to the stack. High efflux velocities would be maintained. The development application was placed before the Minister for Power who ordered a public inquiry. The public inquiry was held in Nottingham, attended by many objectors and representatives of the CEGB. The procedures allowed full legal representation by all parties, with extensive cross-examination of witnesses, a normal British procedure.

Witnesses for the CEGB argued that this excellent plant would have minimal adverse environmental impacts. However, searching questions remained unanswered. What was the ultimate fate of the 600 tonnes of sulphur dioxide released each day? Of the 40 tonnes of dust emitted per day, how much was in the respirable range, reaching the lungs of residents? What was the effect of 20 trains a day bringing coal to the power station on the level crossings of the city? How would 4000 tonnes of ash produced each day be adequately handled? How could the CEGB disguise the presence of huge steam generators and eight immense cooling towers on a flat site open to the city? Witnesses stumbled on some of these questions.

The decision of the Minister of Power stunned the CEGB. The development application had been rejected in its entirety, leaving a 2000 MW hole in the programme of an industry beset with electricity shortages, black-outs and brownouts. The Board was stunned as the reasons for the rejection were entirely environmental.

The report of the commission of inquiry stressed the openness of the site in relation to the city, the wind directions for most of the year, the uncertainties regarding plume dispersal, the untested claims in respect of dust-arresting equipment, the absence of knowledge regarding the fate of large amounts of sulphur dioxide, the possible effects of warmed water on the river, and the poor record of the CEGB in respect of pollution control generally. All these factors militated against this development application, notwithstanding the technical merits of the station, being the first power station in Britain to have a high single stack with high-efficiency electrostatic precipitators for dust removal. A dramatic revision of site selection procedures was undertaken. It became essential to select sites more carefully with regard to possible resident reactions, and to research some of the more pressing questions not yet fully answered. In the end it was concluded that much of Britain's sulphur pollution was reaching Scandinavia.

The CEGB was back to Nottingham quite quickly. An alternative site at

Ratcliffe-on-Soar was to the south of the city, further away, screened by hills. Being less intrusive, it attracted less opposition. The sulphur in the fuel was restricted to 1%, and the coal trains would no longer run through the city of Nottingham. Notwithstanding a long and exhaustive public inquiry, the development application was successful without modification. Following this success, subsequent proposals were allowed including Fiddlers Ferry, Didcot, and a chain of others.

6.3.2 Bayer Australia, Kurnell

The second case involved a development application by Bayer Australia for permission to construct facilities on the Kurnell peninsula at Botany Bay in Sydney, Australia, for the formulation and storage of agricultural and veterinary products. In 1987 the Minister for Planning appointed a commission of inquiry to investigate the proposal and examine objections. The products of the plant would include insecticides, fungicides, herbicides, cattle dips and sprays, and sheep drench, many of which would contain toxic chemicals.

The outcome of the review of evidence by the commissioners was to recommend to the Minister that development consent be refused. In essence, the commissioners concluded that they were not satisfied on the evidence put to the public inquiry that the Bayer project would operate without harmful effects on the local environment of the Kurnell peninsula; in particular, nature reserves, oyster leases, fishing and prawn feeding grounds, and the wetlands of Botany Bay generally. It could not be guaranteed that contaminated waters would not enter the aquatic environment. Spillages were likely to occur on the street approaching the plant. Bayer also indicated that it would not accept certain requirements put forward by the Department of Planning and the State Pollution Control Commission. Difficulties also surrounded the proposed construction of a high-temperature incinerator. Furthermore, alternative sites had not been adequately explored. The commissioners were not convinced that the plant could operate without adverse effects, particularly as the company was objecting strongly to a number of controls and contributions to improved infrastructure. In the light of the report, the Minister for Planning refused the application.

6.3.3 Procedures

Usually, a project will proceed. Much of an EIA is then devoted to the question of the kind of conditions that should be imposed. Table 6.2 reviews the various sorts of conditions. These conditions become part of the statutory consent issued, and are enforceable. Their range is wide, and will often embrace requirements not anticipated by the proponent or developer. Many of the conditions will be based on suggestions by government departments and agencies, by local councils, and by members of the public. The more controversial issues are often examined in a public inquiry, independently chaired.

Table 6.2 EIA – typical matters dealt with in conditions of consent, binding on the developer

Approvals, licences and permits to be obtained from statutory authorities, boards, departments and local councils
Conformity with certain specifications contained in the environmental impact statement (EIS)
Compatibility with all applicable planning instruments
Control of air, water and noise pollution including discharges to catchments, protection of aquifers, control of leachates, blasting controls, incineration, waste disposal, oil contamination, sewage treatment, drainage, stormwater and run-off management, dust suppression, and noise insulation
Life of project
Location of buildings and individual items of equipment
Sequence of mining, quarrying and extractive operations
Working hours
Buffer zones
Access roads, junctions, traffic, rail pipeline and transmission routes
Risks and hazards
Emergency procedures: fire fighting, evacuation
Water supplies and storage
Heritage items
Visual amenity, trees, vegetation, screening
Rehabilitation
Social and economic effects of proposal
Housing of workforce
Monitoring and recording of results
Environmentally hazardous chemicals
Heights of buildings and stacks
Acquisition of properties
Protection of wetlands, parks and reserves, oyster leases, mangroves, rainforest, and other natural resources
Subsidence
Closure of existing plant and replacement of less efficient plant
Effects on residents, schools and hospitals, industries
Lodging of guarantee funds in respect of performance, payment of levies towards future management of the site, contributions towards infrastructure costs and road improvements
Appointment of environmental management officers by proponent
Independent auditing of risks and hazards
Annual reports to the Department of Planning
Arrangements for continuous liaison with the public, local councils, conservation bodies and resident action groups
Matters relating to sustainable development, sustainable yield, or cross-frontier or global environmental protection
Compliance with international conventions

Source: Gilpin (1995).

6.3.4 Example: Australia

Procedures vary from country to country. Table 6.3 lists the procedures adopted at most such inquiries in Australia.

Following the conclusion of the inquiry, the Commission prepares its report to the Minister, setting out and discussing the issues, summarizing the views of the parties, concluding with findings and a specific recommendation. Options may be discussed but only one is recommended as a course of action for government. The report must take account of criteria laid down in the legislation, State environmental policies, the stipulations of planning instruments, and the public interest as perceived by the Commission.

Table 6.3 Public inquiry procedures in Australia

The Minister for Planning or Environment at state or national level directs that a Commission of Inquiry be held into a proposed development and appoints a Commissioner of Inquiry for this purpose.

Notice of the Commission of Inquiry appears in newspapers indicating where and when the inquiry will commence; known interested parties are advised directly. Procedures to be followed are included in the notice.

Persons seeking to make submissions to the inquiry are required to register by lodging a primary submission with the Registrar of the Office of Commissioners of Inquiry.

Before the inquiry and the expiry date for primary submissions any person may examine the development application, the EIS, the initial EIA, and other related documents at specified venues.

Questions in written form should be available at the commencement of the inquiry. Further questions during the inquiry are dealt with at the Commission's direction.

Primary and subsequent submissions are to be made available to all parties.

Proceedings are generally as follows:
- Opening statement by Commissioner
- Preliminary matters such as procedures and personal difficulties
- Primary submissions (in stated order)
- Questions and replies to questions, in writing
- Submissions in reply (in reverse order)

Inquiries are conducted in accordance with the rules of natural justice; each person is treated on an equal footing whether legally represented or not; evidence is not generally on oath; cross-examination is rarely allowed.

Adjournments, usually of short duration, may be granted following an application from a party and argument by other parties.

All communication with the Commission is public and queries are through the Registrar. No private communication with a Commissioner may occur.

The Commission presents its report to the Minister, usually personally and with a verbal briefing. However, the report is final and becomes a public document the moment it passes over the Minister's desk.

The Minister is not in any way bound by the recommendation of the report and may consult others; however, departure from such recommendations is rare, including the often quite stringent conditions laid down in respect of projects thought fit to proceed.

6.3.5 International examples

The general effect of EIA procedures has been to improve the quality of proposals, and also the quality of decision making. The final test is, of course, what it does for the quality of the environment, and how local residents feel about it. Figure 6.1 offers a simplified flow chart for environmental impact assessment (EIA)

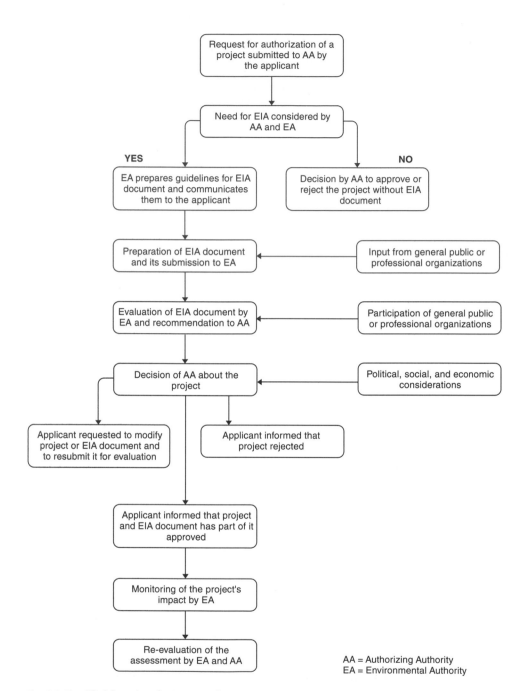

Fig. 6.1 Simplified flow chart for EIA procedure.

Within the figure:

Request for authorization of a project submitted to AA by the applicant

Need for EIA considered by AA and EA

YES

NO

EA prepares guidelines for EIA document and communicates them to the applicant

Decision by AA to approve or reject the project without EIA document

Preparation of EIA document and its submission to EA

Input from general public or professional organizations

Evaluation of EIA document by EA and recommendation to AA

Participation of general public or professional organizations

Decision of AA about the project

Political, social, and economic considerations

Applicant requested to modify project or EIA document and to resubmit it for evaluation

Applicant informed that project rejected

Applicant informed that project and EIA document has part of it approved

Monitoring of the project's impact by EA

Re-evaluation of the assessment by EA and AA

AA = Authorizing Authority
EA = Environmental Authority

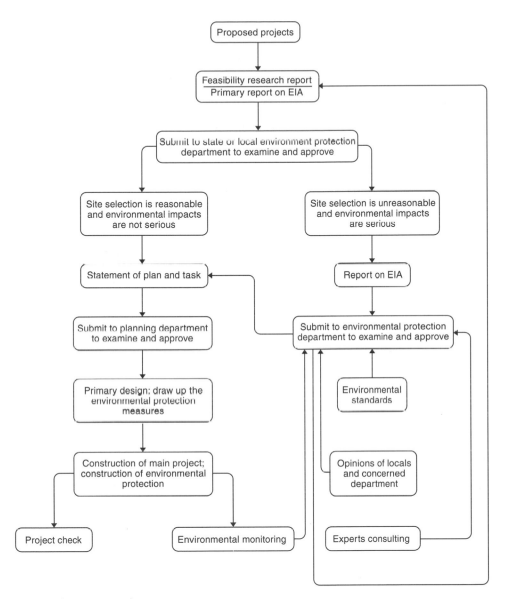

Fig. 6.2 China: EIA procedures.

procedures as drawn up by the United Nations Environment Program (UNEP) based on worldwide practice. Figure 6.2 outlines the EIA procedures now common throughout the People's Republic of China while Figs 6.3 and 6.4 outline EIA procedures in Japan and Indonesia. The diagrams illustrate the variations that occur at national level.

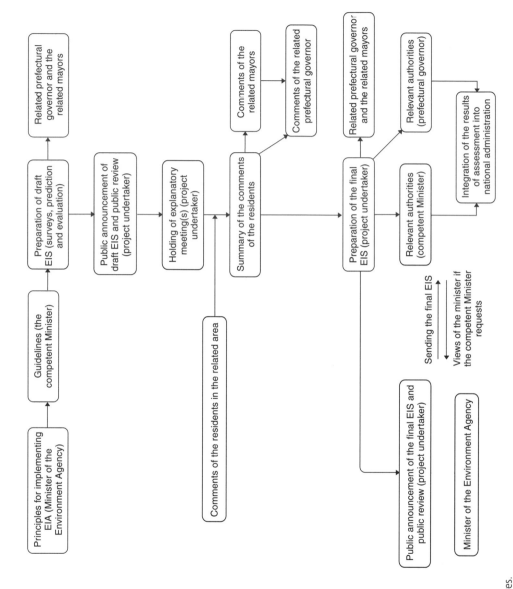

Fig. 6.3 Japan: EIA procedures.

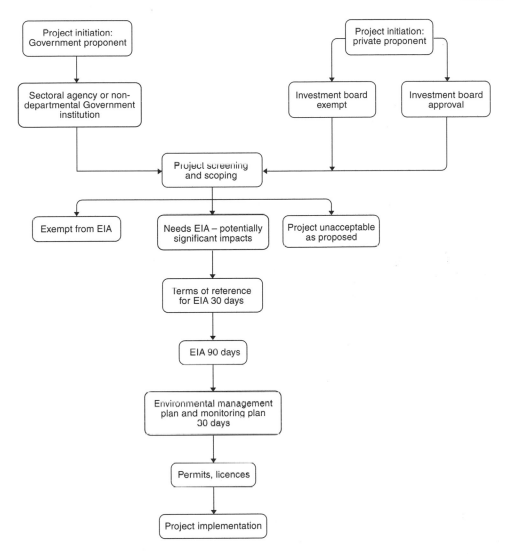

Fig. 6.4 Indonesia: EIA procedures.

6.4 Conclusions

Over the last 30 years, the principle of environmental impact assessment (EIA) has emerged in many countries. Signal events have been the passing of the National Environmental Policy Act 1969 in the USA and the directive of the European Union on environmental impact assessment in 1985. The principle has now been widely adopted in Asia.

The reason for the success of the principle, and the associated principle of the public inquiry, has been the increasing complexity of public decision-making. Only a systematic approach enables decision-makers to receive balanced advice on whether a project, programme or policy should proceed or not and, in particular, what conditions should be imposed on the construction, operation, and decommissioning of the plant, with provisions for site rehabilitation. Quite often, such decisions are beyond local capability and require attention at state, provincial, or national level. Occasionally, development applications have been refused outright, while in other cases applicants have withdrawn their applications in the presence of strong public opposition. The central problem in most cases is to establish a range of conditions with which the successful applicant must comply. Such conditions relate to timing, air and water pollution control, noise management, transportation issues, working hours, workforce facilities, monitoring, risks and hazards, buffer zones, emergency procedures, visual amenity, infrastructure, auditing arrangements, compliance with laws, regulations and conventions, and relationships with the local community.

Public inquiries take time, although often no more than three months. While the costs of delays are expensive, an EIA procedure carefully implemented tends to minimize long-term problems and yields a better outcome for the general public than might otherwise have been the case. Industry has progressively improved its EIA skills producing environmental impact statements of quality, and addressing fundamental issues head-on. Proponents often contribute at public inquiries to the framing of statutory conditions with which they will have to comply. Annual reporting to government agencies is often required in more controversial cases.

EIA has emerged as an important environmental planning tool for the 21st century.

References and bibliography

BAPEDAL (1991) *EIA: A Guide to Environmental Assessment in Indonesia.* Jakarta: Environmental Impact Management Agency, Republic of Indonesia.

Gilpin, A. (1995) *Environmental Impact Assessment: Cutting Edge for the Twenty-first Century.* Cambridge University Press.

Japan National Report (1992) *UN Conference on Environment and Development.* Rio de Janeiro.

Porter, A.L. and Fittipaldi, J.J. (1998) *Environmental Methods Review: Retooling Impact Assessment for the New Century.* US Army Environmental Policy Institute, Georgia Institute of Technology.

UNEP (1990) *Simplified Flow Chart for EIA Procedure.* Nairobi, Kenya: United Nations Environment Program.

Wang, H. and Ma, X. (1987) Progress of environmental impact assessment in China. In: *Proceedings of the International Symposium on Environmental Impact Assessment.* Beijing Normal University, p. 22.

7

Environmental planning

Margaret Durham†

Editorial comment

There has been a radical change in thinking about the environment by the community and governments over the last decade. International agreements, new legislation, policy development and environmental liability require the building and construction industry to rethink attitudes toward the environment and the impact of the industry's activities on it. Section 57 of the Environment Act 1995 in the UK and the NSW Protection of the Environment Operations Act 1997 in Australia are good examples of the increasingly rigorous standards the industry is expected to meet.

There is also a growing role for specialists who understand environmental risk and can bring specific skills to the analysis of building projects. Managing environmental risk and understanding the implications of it for buildings needs a combination of technical knowledge, design awareness and project management skills. People who have this expertise and can do building performance and environmental impact analysis will be in demand, by both the industry and its clients.

The importance of environmental requirements to the building industry cannot be understated. In the 1990s the fastest growing area of legislation worldwide is in environmental protection and performance, and a significant proportion of that legislation concerns the production and disposal of building materials or the performance of buildings. This chapter provides background to the current situation by showing the origins of environmental regulation and discussing some of the current international agreements that are guiding individual countries in the development of environmental policies.

As environmental planning regimes become more rigorous one of the tasks a project team has to deal with early on is compliance with environmental regulation. If a design is submitted for approval without having the requisite environmental standards, and is then rejected, the cost of redrawing and resubmitting the plans can be significant. On top of these direct costs there will be a loss of time and momentum on the project.

One way of adding value to projects is to avoid losses such as these, by ensuring compliance with the increasingly strict requirements of environmental regulation and performance standards.

† University of Technology, Sydney, Australia

7.1 Introduction

One of the greatest problems of western culture is that many thoughts, ethics and beliefs are based on the centuries-old perceived right of people to assert dominion over the environment. We are therefore shocked when people, animals and property are destroyed by natural disasters, because the event affronts the community's perception of technological control over the environment.

With the rapid deterioration of the global environment people have become aware of the need to protect the environment on both a national and international level. To develop policies to protect the environment it is necessary for governments to put in place specific legislation to deal with impacts on land use and the range of externalities. As more information becomes available of the sources of environmental degradation, governments must meet the challenge of establishing appropriate environmental legislation. This means that the body of environmental law is relatively new and will continue to develop.

Environmental law differs from other areas of law as it often touches on people's feelings and values about what aspects of the environment should be protected. As such, it is designed to have an impact on people's attitudes and behaviour. This means that environmental law is often grouped with social welfare legislation such as anti-discrimination and equal opportunity legislation. Whereas a breach of contract can be easily identified, an environmental wrong becomes more difficult to classify. To understand the development of environmental law it is necessary to appreciate the arguments that developed for the protection of the environment and the philosophical and economic justifications for environmental legislation.

7.2 The development of an environmental ethic

The meaning of 'environment' has evolved with an appreciation of the value of conservation, the understanding of the impact of humans on the environment and with concerns for the future of this planet. This implies that a definition must be dynamic. Whilst there may be some frustration and considerable uncertainty about finding a precise definition for the environment it is felt that there are many benefits in a flexible definition. This is because matters that affect the environment touch on human feelings.

Philosophers argue that if humans refuse to accept their place in the fragile structure we will do irreparable damage to the earth and ultimately to our species. O'Riordan and Turner (1983) identified two philosophical groupings:

1. Technocentrics or anthropocentrics: who perceive the relationship between the human and non-human world as one of human dominance. The 'enlightened' technocentric adopts a compromise approach between creating wealth and safeguarding against environmental damage and loss of heritage.
2. Ecocentrics: who embrace a philosophy of interdependence, with nature having an intrinsic value in itself. Many of their philosophies have been adopted in terms of environmental impact assessment.

Elements of these approaches can be found in all aspects of environmental law. In 1974, Tribe suggested a synthesis:

> . . . to recognize that humanity is a part of nature and the natural order a constituent part of humanity is to acknowledge that something deeper and more complex than the customary polarities must be articulated and experienced if the immanent and transcendent are somehow to be united.

Tribe's approach was tremendously important because he presented a new rationale and framework to consider our obligations to natural objects and how they should be protected. He argued that it was necessary to expand the value system so that informed decisions could be made.

Ten years later Boer (1984) used an ecocentric definition of 'environment' which at the time was regarded as radical, but is now thought to be flawed because it lacked reference to the biosphere. He identified a paradigm shift with a transformation from an 'economic human' of the post-industrial revolution era to the 'ecological human'. He felt that ecological values were often excluded from decision-making because of difficulties evaluating what they include. This allowed governments to use narrowly based assessment schemes particularly in hard economic times to justify more rapid development. Boer argued that there must be more opportunity for communities to put forward ecological values and develop an environmental ethic.

More recently, Dobson (1990) adopted an anarchist approach and proposed the best way to protect the planet was by developing a sustainable society with reduced consumption. He questioned whether a commitment to economic growth was appropriate given the finite nature of the earth's resources.

Another approach suggested by Hardin (in O'Riordan and Turner, 1983) is the concept of 'commons'. This is based on the village use of common land, largely for the use of grazing. Those who enjoyed the benefit of the land also had an ongoing responsibility to maintain it for future users. An extension of this is the way we use technology to abuse our 'commons' despite the fact that the modern 'commons' includes fresh air, water and other finite natural resources.

The theorists identified the continuing problem of the conflict of interests – economic, political and environmental. Governments would only act when it really suited them. Boer's 'ecological human' had no doubt arrived, but if he or she was unemployed or hungry there was the danger of slipping back to the old persona. Similarly, governments facing economic contraction or recession would place the environment further down the political agenda.

There was no doubt that there had been an increased awareness of the need to protect the environment. Our environment was not just a community, country or continent – it included our planet and universe. A paradigm shift had occurred but could this environmental ethic expand throughout the world? This required a commitment of governments, industry and communities to the concept of *sustainable development*, which attempted to bring together a wide variety of economic, political and environmental considerations.

7.3 The origins and development of environmental law

Environmental and pollution regulations have evolved in a way different to many other areas of law because of the impact of philosophy, enforcement and the types of sanctions used to punish violators.

Pollution and environmental legislation has evolved over many centuries. Whilst the development of this body of law differs from many other branches of our common law system, it is possible to see similarities with other areas of law, often referred to as 'welfare' law, having a response to the social issues of the day.

Bates (1995) identifies pollution regulation as early as 1273 when Edward I prohibited the use of coal because the fumes were regarded as damaging to human health. In 1388 there was statutory prohibition of throwing refuse into rivers close to human inhabitants. During 1531, the *Bill of Sewers* empowered the Crown to issue commissions to keep sewers and ditches cleaned and to avoid flooding and the erosion of the coast. By the late 19th century the *Public Health Act* 1875 and the *Rivers (Prevention of Pollution) Act* 1876 had been passed by the English Houses of Parliament. It was not until 1952 that air pollution was dealt with by Parliament following the smog-related death of many people in London.

Over the centuries it is possible to see governments' response to economic development and the challenges of industrial pollution as being largely a reaction to concerns about public health. With increasing economic growth and industrial development after World War Two, attention was drawn to the interrelationship of the food chain and likely impacts on human health.

During the 1960s and early 1970s, a number of events occurred that affected public interest in the environment.

1. The 'space race' gave people the opportunity to view the planet and, to many, provided the impetus to look at environmental issues on a global basis.
2. There was an increasing awareness of environmental concerns, and legislation was enacted to protect humans from various types of pollution. Rachel Carson (1965), through careful research, alerted the public to the impact of chemical pesticides on the environment. The media and special interest groups were able to publicize the negative impacts of industrial development on the public. With this growing community awareness came the first attempt to develop environmental legislation that recognized the necessity to incorporate the public interest, and to empower citizens to enforce legislation under the *National Environmental Policy Act* 1969 (US) often referred to as NEPA. This required a full assessment to be made of the environmental impact of proposed developments of federal agencies and provided for an Office of Environmental Quality to advise the President on 'environmentally significant proposals'.
3. Active consumer groups were effective in emphasizing the environmental costs of economic growth. Following simultaneous movements to address conservation and environmental issues in the United States and the United Kingdom during the 1960s, environmental issues became a social concern in many other parts of the world. Bates (1995) describes the environmental movement as 'one of the great social revolutions of the century'.

Rather than a gradual development of law with hundreds of years of precedents, pollution and environmental regulations have developed largely as a response to an increasing environmental consciousness. Cunningham (1982) suggests that governments have responded to the intensity of environmental disasters and the pressure of skilled interest groups. Given the rapid pace of media communications, these newsworthy stories have been quickly brought to the attention of the public. Political parties have recognized that environmental policies must now form part of their party platform because the environment has become a vote winner, particularly in a time of economic prosperity. This has particularly been the situation in Australia as the balance of power in the parliaments has recently been held by parties or independents with a strong commitment to the environment.

Furthermore, scientific evidence has been able to prove links between pollution and the impact on human health. Hard statistical data, which are essential within a regulatory framework, have been applied where, in previous times, emotional displays were rejected.

It can be seen that pollution and environmental law developed in a similar way to a number of other areas of law because of a response to public demands at the time. A number of writers have commented on the interesting coincidence that the 'welfare offence' law and environmental law grew in the United States and the United Kingdom at a similar time, so it could be suggested that this was due to the needs of those societies as they industrialized and prospered. As social conditions change, legislatures must be responsive to the public will and must carry out their legislative task. Courts bound by the rules of precedent do not have this flexibility and are often slower to respond, except where they are able to apply the principles of equity. Courts have, however, been able to develop a framework for judicial analysis of legislation.

More recently, emissions of pollutants into the environment that impact on the health of people have become described as 'toxic' torts. Cashman (1992) highlights the difficulties the common law has to personal 'environmental' injuries because of proof of causation and the lack of precedents to establish a course of action. With scientific developments moving at an increasingly rapid pace the impacts of these new technologies and 'wonder drugs' is often not fully understood for many years. Insurers in the United States are introducing a no-fault scheme, which could be extended beyond accidental injuries to include toxic illness.

This type of action probably has the greatest common law potential when seeking a remedy for an environmental or pollution injury-related wrong given the recent limitation of strict liability in Britain over water pollution (Haydon and McDonald, 1993) and also in Australia (Iles, 1994).

7.4 Sustainable development

It is easy to see why confusion has developed over the evolution of pollution and environmental legislation. This area of law is very much a response to the philosophical and environmental ethic of the day, as well as to demands for economic growth. Governments must respond to local and national demands, as well as abiding by election promises with respect to the environment. In many countries, legislators have a responsibility to prepare legislation that encapsulates its obligations with respect to international agreements and conventions. This has led to a layering of legislation that flows from the international to local government level, which provides huge challenges, particularly for those involved in the construction industry.

With this increasingly ecocentric approach has come a wider appreciation of the distinct parts of the environment that should be protected. As a result, a proliferation of legislation has developed over the last 30 years that has often led to confusion over legal responsibilities and complex methods of administration. This legislation has related to new awareness and knowledge. The Club of Rome's report, *The Limits to Growth,* challenged the traditional assumption that the natural environment provided an unlimited resource base for population and economic growth and could assimilate increasing quantities of waste. In 1972, the First United Nations Conference on the Environment focused on the link between environmental problems and economic development. One of the outcomes was an international agreement on the responsibility to protect the environment.

The World Conservation Strategy identified the need to maintain the planet as our living resource base and aimed to advance sustainable development.

The Brundtland Report (World Commission on Environment and Development, 1987) presented an optimistic view of improved ways of meeting the needs of future generations. However, while the report put great faith in the possible improvement of technology and social organization, there was no detailed blueprint.

The developing world has an unrelenting commitment to growth based on western methods. Essential elements of sustainable development are firstly, establishing legislative force; and secondly, empowering administrative structures. The major criticism of sustainable development is that it is merely a complex extension of the environmental impact statement. Whilst there may be some truth in this, it is important to remember the dynamic nature of the issue. It is simply inappropriate to establish cumbersome, inflexible bureaucratic structures to develop policy and ultimately legislation that may be of little value in the future.

The Second United Nations Conference on the Environment and Development (also known as the Earth Summit) in June 1992 was attended by 170 countries. The key outcome was recognition of the need for international agreements to plan for environmental protection. These objectives were outlined in the *Rio Declaration* and *Agenda 21.*

Given the ecological crisis the world faces, it is encouraging to see ongoing world participation in the Earth Summit II, the Osaka conference, Law of the Sea, Antarctic Treaties and the Ramsar[1] Convention. The 'commons' approach is seen as essential to the ongoing survival of the planet. As a result, a substantial body of international law continues to develop. This comes from:

1. treaties and conventions which require ratification by party nations
2. declarations which provide the basis for future treaties
3. the International Court of Justice.

Clearly, difficulties can arise in the policing of these agreements. Most environmental problems are found within the global commons. The challenge is that most environmental degradation occurs within national borders and yet can have a serious impact on neighbouring countries. A further difficulty is that governments are mainly concerned with their nation's economic objectives and this is understandable. They are often reluctant to ratify agreements because of concern over the economic and political consequences of inappropriate environmental management practices. This is particularly obvious in developing nations who do not feel they should forgo opportunities for development. A good example of this is the CFC and ozone layer debate. Countries in the developed world have also been challenged to reconsider their international environmental responsibilities. This was most recently observed in Osaka when Australia was reluctant to reduce greenhouse gas emissions and was only able to limit its original commitment because of widespread environmental sinks or forests. Companies who pollute as part of their production process can obtain environmental credits by acquiring and maintaining forests able to process polluting gases.

However, the use of international trade sanctions and censure has proven to be a powerful punitive mechanism. Furthermore, the realization that failure to achieve sustainable development may have serious, irreversible impacts on human survival will no doubt be the strongest incentive.

The impacts on the construction industry have been particularly significant. The trickle-down effect of international agreements to local building codes has been substantial. The paradigm shift has led to a change in management attitudes to energy and resource efficiency. Think tanks such as the Rocky Mountains Institute have developed new design strategies for the construction industry including ecodesign, and industrial ecology.

The ecodesign approach to resource management includes consideration of the prevention of pollution through ecologically appropriate materials, fabrication techniques and construction management processes. Closed loop techniques, where waste output from one process becomes a resource for another, have also been a valuable initiative.

Industrial ecology includes a life cycle perspective and examines whole business sectors to find environmental efficiencies that create economic opportunities.

Many local building approval authorities have included the principles of sustainability by requiring developers to consider their selection of power technologies, the choice of energy systems for buildings and the selection of building materials with low embodied energy. These initiatives are embodied in legislation that has largely been a response to the huge attitudinal changes that have occurred at the international level. The challenge now is to strengthen and remain aware of our individual and corporate responsibilities to the environment.

7.5 Conclusions

The realization at the global level that environmental considerations must be factored in to any type of development has rapidly evolved over the last 20 years. The development of the environmental ethic has had major impacts on government decision making from local government to the international level. Environmental awareness has been 'mainstreamed' in political party policies and in the education of young children. Increasingly, people are indicating their concerns over the future of the environment, as environmental issues need to be addressed if our species is to continue.

This has had two outcomes for the construction industry. First, the construction professional must be aware of environmental policies and legislation in the area they are responsible for developing. There are numerous examples of development proposals that have stalled because the applicant has failed to consider all the environmental requirements of the consent authority. Furthermore, after a development has commenced, objectors may be successful in obtaining a judicial review of a development consent and therefore stop the development from continuing, as long as they are successfully able to raise environmental concerns. The costs in delaying the project can amount to millions of dollars. This may ultimately lead to the client seeking compensation by way of a claim for negligence from the person responsible for the oversight.

A second outcome of the growing environmental awareness is that the actual design of the building, the materials used and the energy efficiency of both the structure and the fittings all need to be carefully considered. As costs will continue to rise for energy and water, purchasers are becoming more discriminating when considering the ongoing expenses of the building they occupy. Furthermore, we are seeing the growth of a generation that has an expectation that structures will be environmentally friendly.

The value considerations are therefore extensive. On the micro level, value can be added to the individual structure by using ecologically sustainable construction principles. However, the multiplier effect of the impacts of the construction on communities and the planet is enormous and commitment to environmental policies and legislation is imperative.

Endnote

1. On 13 February 1971, 18 countries came together at the Iranian city of Ramsar to sign the *Convention on Wetlands of International Importance Especially as Waterfowl Habitat*. Australia was the first signatory to the agreement and, in 1996, hosted a Conference in Brisbane where 92 contracting countries met. Currently, there are 105 contracting countries. At the time of the convention wetlands were generally regarded as swampy areas infested by insects and were therefore seen as being of little use. Much wetland was destroyed and often filled. Wetlands were defined in the Convention as:

   > . . . areas of marsh, fen, peatland or water, whether natural or artificial, permanent or temporary, with water that is static or flowing, fresh, brackish or salt, including areas of marine water the depth of which at low tide does not exceed six meters.

   The first listed Ramsar site in Australia was the Coburg Peninsula in Northern Territory. There are now over 770 sites. The Convention requires contracting parties to list sites and also to formulate and develop plans to promote the conservation of wetlands. The full list of Ramsar sites can be downloaded from: http://w3.iprolink.ch/iucnlib/themes/ramsar/index.html

References and bibliography

Bates, G. (1995) *Environmental Law in Australia*. Fourth edition (Sydney: Butterworths).

Boer, B. (1984) Social Ecology and Environmental Law. *Environmental and Planning Law Journal*, **1**, 233.

Carson, R. (1965) *Silent Spring* (London: Penguin).

Cashman, P. (1992) Torts. In Bonyhady T. (Ed.) *Environmental Protection and Legal Change* (Sydney: Federation Press).

Cunningham, N. (1982) *Pollution, Social Interest and the Law* (UK: Martin Robertson Press).

Dobson, A. (1990) *Green Political Thought: An Introduction* (London: Unwin Hyman).

Hardin, G. (1983) The tragedy of the commons. In O'Riordan, T. and Turner, R. (Eds) *The Nature of the Environmental Idea: An Annotated Reader in Environmental Planning and Management* (London: Pergamon Press).

Haydon, J. and McDonald, J. (1993) House of Lords rejects strict nuisance for water pollution. *Australian Environmental Law News*, November 4, 27.

Iles, A. (1994) A continuing retreat from strict liability at common law. *Australian Environmental Law News*, June, 26.

O'Riordan, T. and Turner, R. (1983) *The Nature of the Environmental Idea: An Annotated Reader in Environmental Planning and Management* (London: Pergamon Press).

Tribe, L.H. (1974) Ways not to think about plastic trees: new foundations for environmental law. *Yale Law Journal*, **83**, 1315.

World Commission on Environment and Development (1987) *Our Common Future* (Oxford University Press).

PART 2

Analytical techniques

PART 2

8

The capital cost fallacy

Craig Langston†

Editorial comment

The notion of comparing costs over time as a basis for decision making, rather than simply comparing initial or capital costs, is not new. In the construction industry, however, it has generally not received the attention it deserves. The basic concept is simple and the benefits to building owners are clear, yet the cost-effectiveness of reduced costs over time which can be gained by building owners are often ignored in favour of the more easily identified and assessed savings which appear to accrue from reduced capital expenditure.

This may occur in response to a number of factors, for example:

- when a building is constructed for sale or lease and the principal therefore passes on the burden of operating costs (and maintenance costs if the building is sold) to the buyer or tenant
- when a project is for a public client such as a government department or agency and, for reasons of accountability, the best value for money outcome may be perceived as the one which provides 'more building' in return for the public money invested in it, or where budgets are tightly controlled and based on annual appropriations, thus making it impossible for extra funds to be provided initially even though the extra expenditure may produce greater savings in future years
- when clients trading in a volatile marketplace deliberately base their business planning on short time horizons, e.g. only two or three years – this will often sharply reduce the savings that can be identified as many savings will not become apparent until such time as major refurbishment or maintenance is required, or until savings from reduced operating costs accumulate to a point where they outweigh the additional initial costs that actually produce those savings.

The relationship between life-costs and value for money is obvious: a client may pay more initially for their building but over time that extra investment generates savings that continue to accrue over the building's life. While there may appear to be limited scope for the implementation of life-cost planning during the pre-design phase, there can be a clear commitment to the life-cost approach to cost management before design starts. In addition, life-cost planning techniques can provide valuable assistance when assessing alternative design strategies during pre-design

† University of Technology, Sydney, Australia

decision-making, e.g. when the design team and client are choosing between full air conditioning, mixed mode or natural ventilation systems.

At a more fundamental level, life-cost planning may be used when a client is trying to determine whether the construction of a new building is actually the most appropriate solution to their problem, or whether some other solution, such as leasing a facility or even outsourcing the functions or activities which a new building would accommodate, might be the most cost-effective solution in the longer term.

The following chapter looks at the basics of life-costing and discounting and examines some of the problems, actual or perceived, which are commonly associated with life-cost studies.

8.1 Introduction

Although the concept of considering all the costs of a project may seem common sense, in current practice decisions are made largely, and unfortunately, on the basis of capital costs alone. This results in value for money being viewed in a narrow context and limits the possibility of achieving efficiencies in the employment of resources within society. While value assessment must additionally include factors such as functionality, aesthetics, environmental impact and financial return, operating costs remain a significant part of any asset decision. The design professions must be conscious of the importance of operating costs and seek out solutions in their practice that will not ultimately become financial burdens for their clients or others in the community. The ability to deliver value for money on projects whilst concentrating solely on capital cost issues remains a contemporary fallacy.

Past analyses of design solutions for building projects have concentrated on initial capital costs, often to the extent that the effects of subsequent operating costs are completely ignored. However, even in cases where a wider view of cost has been adopted, the discounting process has commonly disadvantaged future expenditure so heavily as to make performance after the short term irrelevant to the outcome, resulting in projects which display low capital and high operating costs being given favour. Thus, design solutions that aim to avoid repetitive maintenance, reduce waste, save non-renewable energy resources or protect the environment through selection of better quality materials and systems, usually having a higher capital cost, are often rejected on the basis of the discounting process.

8.2 Towards a whole-of-life approach

8.2.1 Life-cost studies

Life-cost studies (life cycle cost assessment) comprise both measurement and comparison activities. Measurement relates to the total cost of a particular design solution and involves the establishment of expected performance targets and their monitoring over time. Comparison relates to a number of alternative design solutions and, in addition to total cost (and in many cases differential revenue), must consider the relative timing of receipts and payments. Most of the literature concentrates on the latter activity and hence gives the impression that discounting plays an absolute role in life-cost studies. But this conclusion is disputed, and it is recommended that measurement activities are important and their presentation in terms of real value is more meaningful than the use of discounted value.

Research into life-cost studies may have uncovered and resolved many of the conceptual problems and misunderstandings that have haunted the technique for some time, but it has also highlighted the potential application that exists in the assessment of sustainable development. Life-cost plans (expressed in present-day dollars) can be used to support essential facilities management processes over the life of the project. Budgeting, cash flow forecasting, data collection and design feedback are direct and tangible outcomes from the quantification of life-costs. These are measurement activities and do not involve discounting. The life-cost plan becomes a framework that supports knowledge acquisition and encourages the adoption of a total cost attitude in project design.

Although the capital cost fallacy remains, there is an increasing interest in life-cost considerations. When the technique became popular in the 1960s in the UK it was because there was general concern at the expense attached to the maintenance of an ageing building stock and the impact this would have on the growth and prosperity of the country in the years ahead. The technique was revived again in the mid-1970s when the oil crisis created doubts over the ability of building owners to operate energy-intensive buildings into the future. Today the technique is back on the agenda under an environmental banner. Sustainable development is a global issue and one that may turn out to be a more significant driving force for enabling life-cost considerations to become a routine part of professional practice.

8.2.2 Sustainable development

The philosophy of sustainable development borrows freely from the science of environmental economics in several major respects. A basic component of environmental economics concerns the way in which economics and the environment interact. Recognition of the fact that the economy is not separate from the environment is fundamental to any understanding of sustainable development. The two are interdependent because the way humans manage the economy impacts on environmental quality, and in turn the environment directly influences the future performance of the economy.

The risks of treating economic management and environmental quality as if they are separate, non-interacting elements have now become apparent. The world could not have continued to use chlorofluorocarbons (CFCs) indiscriminately. That use was, and still is, adversely affecting the planet's natural ozone layer. Furthermore, damage to the ozone layer affects human health and economic productivity. Few would argue now that we can perpetually postpone taking action to contain the emission of greenhouse gases (GHGs). Our use of fossil fuels is driven by the goals of economic change and that process will affect the global climate. In turn, global warming and sea-level rise will affect the performance of economies.

Definitions of sustainable development abound since what constitutes development for one person may be neither development nor progress for another. Development is essentially a value word: it embodies personal aspirations, ideals and concepts of what constitutes a benefit for society. The most popular definition is the one given in the Brundtland Report (1987):

> development that meets the needs of the present without compromising the ability of future generations to meet their own needs.

This definition is about the present generation's stewardship of resources. This means that for an economic activity to be sustainable it must neither degrade nor deplete natural resources, nor have serious impacts on the global environment inherited by future generations. For example, if greenhouse gases build up, ozone is depleted, soil is degraded, natural resources are exhausted and water and air are polluted; the present generation clearly has prejudiced the ability of future generations to support themselves.

Development implies change, and should by definition lead to an improvement in the quality of life of individuals. Development encompasses not only growth, but general utility and well-being, and involves the transformation of natural resources into productive output. Sustainable development in practice represents a balance (or compromise) between economic progress and environmental conservation in much the same way as value for money on construction projects is a balance between maximum functionality and minimum life-cost. The economy and the environment necessarily interact, and so it is not appropriate to focus on one and ignore the other.

Development is undeniably associated with construction and the built environment. Natural resources are consumed by the modification of land, the manufacture of materials and systems, the construction process, energy requirements and the waste products that result from operation, occupation and renewal. Building projects are a major contributor to both economic growth and environmental degradation and hence are intimately concerned with sustainable development concepts.

Having said that, there is probably no such thing as sustainable development for the general run of construction projects. But that is not to say that consideration of sustainability is a waste of effort. On the contrary, every project (new or existing) can be enhanced by consideration of whole-of-life methodologies, particularly during pre-design. While most projects will consume more resources than they create, projects that are closer to sustainable ideals will increasingly deliver benefits to their owners and users and to society as a whole. Therefore, if design can encompass assessment and decision-making processes that address sustainability goals, it is likely that over the long term the construction industry will be able to demonstrate a significant contribution to global resource efficiency.

8.2.3 Overcoming past problems

The economic assessment of projects using a whole-of-life approach has been plagued with difficulties ever since the concept was first practised. Even today, after much research and development has led to most problems being laid to rest, there is still a persistent recounting of these difficulties in industry. The most commonly quoted difficulties comprise forecasting accuracy, lack of historical data, professional accountability, technology changes, capital versus operating budgets, and the underpinning theory of discounting.

It is true that forecasting the future is always likely to lead to variations in results between initial predictions and subsequent reality. But this does not add weight to the argument that a whole-of-life approach to economic assessment is flawed. The solution to the problem lies in the routine use of risk analysis, whereby the uncertainty of future events can be evaluated against the robustness of assumptions to add confidence levels to final decisions. The bottom line is that uncertainty may be high, particularly where long time horizons are involved, but the risk of making a poor decision between various proposed courses of action may be quite low.

There is a lack of historical data about maintenance, energy usage, operating costs and life expectancies of materials and systems. Some useful resources do exist, but the ability to create a comprehensive database is problematic. The point is that even if a comprehensive database existed now, it would need to be used with great caution, since historical information is highly context dependent. For example, operating costs of buildings will differ due to the variable durability of materials and systems and the effects of their interaction with each other, the environment and the users of the building. Different occupancy profiles, hours of operation, location, environmental factors, management policy, quality of the original construction, premature damage and building age will all affect the integrity of the data being collected. Long time frames are involved in the establishment of data, and new methods and systems are always being added. Rather than focus on historical data, estimating operating costs from first principles is recommended, and context-specific databases will be formed from the routine comparison of predicted and actual performance.

Professional consultants, such as quantity surveyors and engineers, will not normally be accountable for the accuracy of their work in this area, since it may take many years to test whether their predictions are valid. But whole-of-life planning should not be considered as an estimate of actual cost, rather as a framework within which effective cost management processes can occur. Elemental targets and individual work items act as benchmarks for design and subsequent operation. Where cost overruns occur, the cost management process can focus on the reasons why and on implementing strategies which will ensure that overall cost limits are not exceeded. Professional accountability, therefore, should be about issues of methodology, capability and process, not fortune-telling skills.

Changes in technology and social fashion occur regularly. It would be unrealistic to think that original design decisions will continue to represent good value for money until their life expectancy is reached. The cost management process must be able to adapt and re-evaluate new options as they appear in the marketplace, and this must be seen as another routine activity with any whole-of-life approach. Therefore, systems must be used that support the concept of continuous cost management from concept through to demolition or disposal (cradle to grave). Technology changes are not problems, merely new pieces of information.

A classic difficulty is the artificial barriers that have been created, particularly in the public sector, between capital and operating budgets. Any whole-of-life approach relies fundamentally on the ability to spend more money initially in order to make overall savings, and if this cannot occur then the entire process becomes pointless. Therefore, it is vital that clients remove these internal divisions and treat project budgets as inclusive of initial and recurrent cash flows. There is plenty of evidence of success in this area, even in the public sector, but more policy emphasis is required.

Discounting is generally regarded as a controversial process. This is not because there are unresolved theoretical issues that require further research, but rather that existing research has not been adequately disseminated to industry. One major misconception, for example, is that discounting leads to bias against future generations and therefore is not compatible with sustainable development goals. This assertion is incorrect provided the rate of discount is based on the weighted cost of capital to be used in the project. In the past, high discount rates have led to poor decisions.

There are other minor difficulties which sometimes appear in the literature, but the only real barrier these days to widespread implementation of a whole-of-life approach throughout the construction industry is the attitude of individual stakeholders.

8.2.4 Pre-design issues

As already explained, the measurement of development costs is commonly undertaken on a capital cost basis. Projects are analysed in respect of their likely construction costs, to which might be added land purchase, professional fees, furnishings and cost escalation to the end of the construction period. Budgets similarly address initial costs, and the planning and control processes they foreshadow normally do not extend beyond hand-over. Developers are often not interested in long-term costs that will be borne by one or more future owners, and as yet there is little government regulation that makes either presentation or consideration of long-term costs commonplace.

The need to look further into a project's life than merely its design or construction is trivially obvious, since operating costs usually outweigh expenses involved in acquisition. A whole-of-life approach takes into account both capital and operating costs, so that more effective decisions can be made, and thus satisfies the primary objective of clients and agencies – that of achieving value for money. It delivers the following tangible benefits:

- assesses alternative courses of action
- investigates the sustainable deployment of limited natural resources
- identifies cost-effective designs
- provides a balance between capital and operating budgets
- calculates operating cost cash flows over the life of the project
- enables the financing of future commitments
- enables continuance of cost management throughout the stages of design, construction and occupation

- facilitates meaningful data collection and feedback from existing buildings
- provides information for financial planning and analysis of future expenditure
- evaluates the burden being placed on future generations to maintain buildings
- monitors the success of design choices against estimated targets and highlights possible areas of improvement.

These advantages build on those already implicit in the rigour of the cost management process. Nevertheless, the challenge ahead is to ensure that short-term financial considerations do not prevent the pursuit of long-term benefits.

The ability to balance capital and operating costs is crucial to the effective use of any whole-of-life methodology. The concept is simple and, for most materials and systems, holds true: spending more money initially will lead to lower operating costs over the life of the project. Quality assurance assumes greater importance as poor workmanship may lead to premature failure of expensive and otherwise long-lasting components. Figure 8.1 illustrates the expected relationship.

While this concept is often applied to issues of maintenance, repair and replacement, a similar relationship can be expected with energy. However, in some cases substantial savings can be delivered by using low technology solutions that are inexpensive to construct and operate. The greatest value for money savings may well lie in passive heating and cooling systems and the increased use of natural light and ventilation.

During the pre-design stage it is important to establish a realistic framework for financial management and to investigate options that have the promise of delivering maximum value for money. The focus therefore is on investment

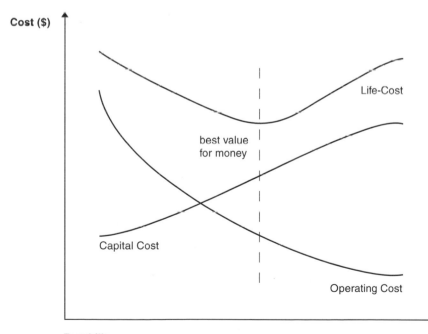

Fig. 8.1 Capital versus operating costs.

analysis and early cost advice that can direct and influence later design without constraining architectural creativity or innovation. This requires the application of detailed knowledge derived from past experience and investigation in a form that can assist new conceptual designs.

Figure 8.2 illustrates the key financial tools within the overall economic management process. Investment analysis is the economic decision-making activity during the pre-design stage. Life-cost planning applies to the project's design development and life-cost analysis applies to both construction and subsequent occupancy. These three linked activities rely on a whole-of-life approach to project evaluation that takes proper account of operating costs over time. Each is now discussed in further detail.

8.3 Investment analysis

8.3.1 Economic assessment

Cost–benefit analysis (CBA) is the practical embodiment of discounted cash flow analysis. There are two types of CBA: economic and social. Economic analysis involves real cash flows that affect the investor. Social analysis involves real and theoretical cash flows that affect the overall welfare of society.

8.3.2 Economic cost–benefit analysis

Economic CBA is a technique for assessing the return on capital employed in an investment project over its economic life, with a view to prioritizing alternative courses of action that exceed established profitability thresholds. It uses discounted cash flow analysis to make judgements about the timing of cash inflows and outflows and envisaged rates of return.

Most experts agree that discounting is fundamental to the correct evaluation of projects involving differential timing in the payment and receipt of cash. Accounting rate of return and simple payback methods, which do not consider changing time value, are quite inadequate substitutes and may produce misleading advice.

The two most common capital budgeting tools used as selection criterion in CBA are net present value (NPV) and internal rate of return (IRR). Both rely on the existence of costs and benefits over a number of years, and lead to the identification and ranking of projects that are financially acceptable for possible selection.

NPV is the sum of the discounted values of all future cash inflows and outflows. If the NPV is positive at the end of the economic life, then the investment will produce a profit with reference to the discount rate selected. If the NPV is negative, then the investment will produce a loss. Mutually exclusive projects should be selected on the basis of the magnitude of the NPV, provided it is positive. IRR is defined as the discount rate that leads to an NPV of zero. Depending on the cash flow pattern, some projects may exhibit multiple IRR while others may have no rate at all.

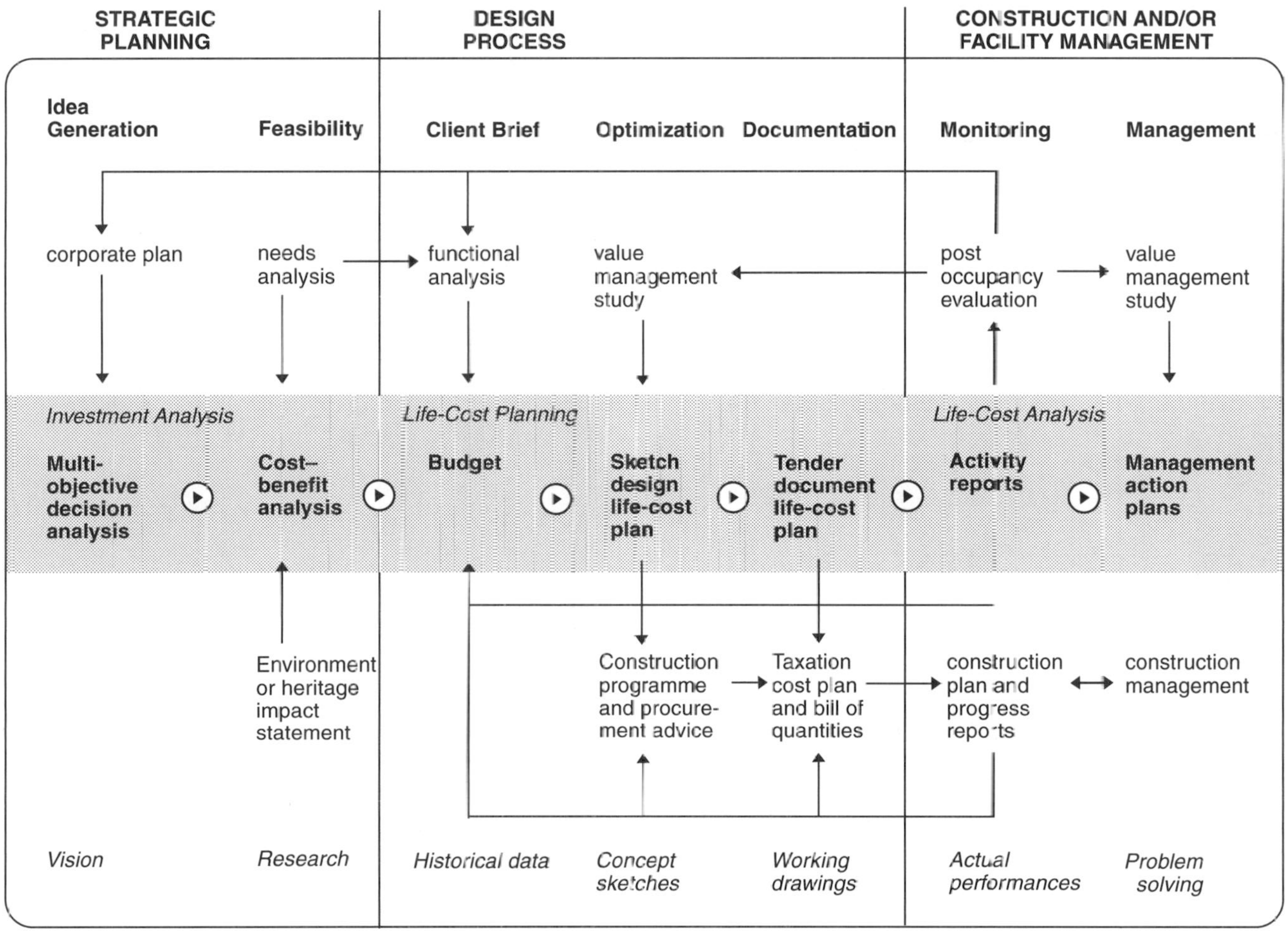

STRATEGIC PLANNING | **DESIGN PROCESS** | **CONSTRUCTION AND/OR FACILITY MANAGEMENT**

| Idea Generation | Feasibility | Client Brief | Optimization | Documentation | Monitoring | Management |

corporate plan

needs analysis

functional analysis

value management study

post occupancy evaluation

value management study

Investment Analysis

Life-Cost Planning

Life-Cost Analysis

Multi-objective decision analysis (▲)

Cost-benefit analysis (▲)

Budget (▲)

Sketch design life-cost plan (▲)

Tender document life-cost plan (▲)

Activity reports (▲)

Management action plans (▲)

Environment or heritage impact statement

Construction programme and procurement advice

Taxation cost plan and bill of quantities

construction plan and progress reports

construction management

Vision | *Research* | *Historical data* | *Concept sketches* | *Working drawings* | *Actual performances* | *Problem solving*

Fig. 8.2 Economic management of projects.

Table 8.1 and Figure 8.3 are examples of a discounted cash flow for a hypothetical project having an expected economic life of 10 years. The discount rate used is 10% per annum. The NPV at this rate is $57 992 and the IRR is determined by iteration or trial and error at 20.63%. The project can be seen to be acceptable at any discount rate that is less than the IRR. No capitalized reversion at the end of 10 years has been allowed.

Table 8.1 Discounted cash flow analysis

Year n	Cost C	Benefit B	Net Benefit $NB = B - C$	Discounted Value $DV = NB(1 + d)^{-n}$
0	120 000	0	(120 000)	(120 000)
1	5 000	32 000	27 000	24 545
2	5 000	32 000	27 000	22 314
3	7 500	39 500	32 000	24 042
4	10 000	47 000	37 000	25 271
5	10 000	47 000	37 000	22 974
6	10 000	35 000	25 000	12 112
7	10 000	35 000	25 000	12 829
8	10 000	35 000	25 000	11 663
9	10 000	35 000	25 000	10 602
10	10 000	35 000	25 000	9 639
			NPV =	**57 992**

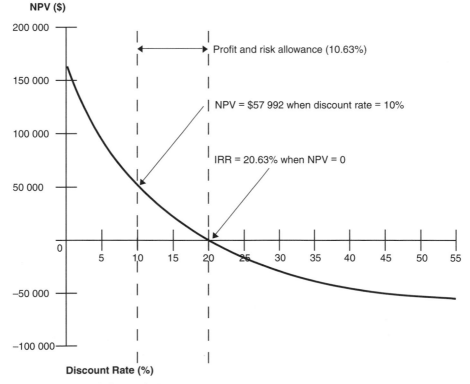

Fig. 8.3 Discounted cash flow analysis.

8.3.3 Social cost–benefit analysis

Social CBA uses the concept of collective utility (aggregate efficiency) to measure the effect of an investment project on the community. Externalities are assessed in monetary terms and are included in the cash flow forecasts, even though in many cases there is no market valuation available as a source for the estimates. Nevertheless, the technique is useful in that it attempts to take both financial and welfare issues into account so that projects delivering the maximum benefit can be identified.

Because social CBA is directed at maximization of collective utility, it is generally employed as an investment technique by government agencies. Furthermore, the amount of research and estimation that the technique demands means that it is feasible for use mainly on large and complex projects, typically those involving infrastructure, such as transport, health care, water irrigation, energy supply and the like.

More frequently, the technique is being called upon to evaluate environmentally-sensitive projects. It has, however, come under attack from many conservationists because it can justify investments that cause damage to the environment provided that this damage is outweighed by an increase in capital wealth. The quantification of externalities clearly has a great impact on the outcome. In addition, distributional effects are often overlooked, as the acceptance criterion is related to overall utility improvement and hence does not recognize whether this benefit is evenly shared. For this reason projects that are likely to involve an uneven distribution of benefits should be the subject of a separate study designed to expose such effects.

Social CBA is the primary tool used in environmental economics, yet the technique still has significant problems when applied to the evaluation of environmentally-sensitive projects. The most significant problem concerns the fundamental concept of aggregate efficiency. In its basic form social CBA does not adequately account for environmental issues. A major reason for this is the Kaldor–Hicks principle that underlies CBA theory, which states that any type of cost to society is acceptable as long as a project generates greater benefits. Therefore, environmental damage is acceptable if benefits, such as increases in capital wealth, are valued more highly.

No CBA principle prescribes that part of the benefits should actually be reinvested in measures to avoid or compensate for environmental damage, and therefore substitution between capital and environmental wealth is perfect. Yet from a sustainability point of view, it would be preferable to limit environmental damage in absolute terms once a particular threshold has been reached regardless of the increases in capital wealth that may be anticipated.

Externalities often pose significant problems in their monetary translation, and shadow pricing may become necessary. These costs and benefits can be significant, and unrealistic assessment can easily distort results. Where externalities are beyond assessment altogether, and hence become intangibles, important aspects of the project may be neglected. Environmental costs and benefits may often be considered intangible, and clearly this fails to address adequately the sustainability question within the boundaries of the technique.

8.3.4 The application of discounting

The application of discounting and the translation of social and environmental costs and benefits into monetary terms are two further aspects of the technique that are the subject of much controversy. The discounting process is extended in social cost–benefit analysis to adjust externalities (and other items that ought to be left as intangible) along with the more conventional cash flows. Utility is actually being discounted. In particular, the impact of discounting on subjective social and environmental considerations infers that matters like time savings, the value of human life, lifestyle improvement and environmental protection are linked to the rate at which interest is accrued or lost in the financial market.

It is questioned whether the traditional application of discounting theory to the assessment of social and environmental goods and services is still appropriate. Discounting should be applied only to those items in the CBA which are tradable; and even for these items, discounting should be used purely for comparative selection purposes. The discount rate comprises project-based, product-based and investor-based attributes, and fluctuations in living standards will directly affect the affordability (and hence value) of future costs and benefits. Discount rates based on the true time value of money do not introduce bias against future generations and hence discounting in this form is compatible with the intergenerational equity principles of sustainable development.

8.4 Life-cost planning

8.4.1 The purpose of cost planning

Cost planning is part of a wider cost management process that commences with the decision to build and concludes with the completion of design documentation. During this period the main objectives are:

- the setting of cost targets, in the form of a budget estimate or feasibility study, as a framework for further investigation and as a basis for comparison
- the identification and analysis of cost-effective options
- the achievement of a balanced and logical distribution of available funds between the various parts of the project
- the control of costs to ensure that funding limits are not exceeded and target objectives are ultimately satisfied
- the frequent communication of cost expectations in a standard and comparable format.

The cost plan is one of the principal documents prepared during the initial stages of the cost management process. Costs, quantities and specification details are itemized by element (or sub-element) and collectively summarized. Measures of efficiency are calculated and used to assess the success of the developing design. The elemental approach aids the interpretation of performance by comparison of individual project attributes with similar attributes in different projects, and forms a useful classification system. Life-cost plans differ from traditional capital cost plans only in the type of costs that are taken into account and how these can be expressed and interpreted. Figure 8.4 illustrates some of the outcomes that can be achieved from this process.

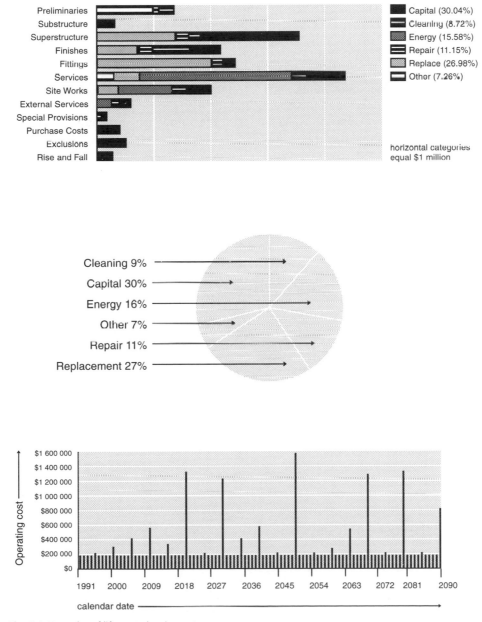

Fig. 8.4 Examples of life-cost planning outcomes.

8.4.2 Types of costs

Life-costs can be divided into various categories that aggregate capital and operating expenses in different ways. The following categories are offered as a preferable division.

- **Capital costs**. Capital costs comprise the initial acquisition of the land and building, and can include:
  (a) *Land cost*. The purchase cost of the land.
  (b) *Construction cost*. The cost of labour, material and plant involved in the creation of the building and other improvements to the land, including all supervision, profit, and rise and fall during the construction period.
  (c) *Purchase cost*. Acquisition costs not directly associated with the finished product, including items such as stamp duty, legal costs, building fees, professional fees, commissioning and the like.
- **Operating costs**. Operating costs comprise the subsequent expenditure required to service the land and building, and can include:
  (a) *Ownership cost*. Regular running costs such as cleaning, rates, electricity and gas charges, insurance, maintenance staffing, security and the like.
  (b) *Maintenance cost*. Annual and intermittent costs associated with the repair of the building including periodic replacement or planned renovation.
  (c) *Occupancy cost*. Costs of staffing, manufacturing, management, supplies and the like that relate to the building's function, including denial-of-use costs.
  (d) *Selling cost*. Expenses associated with ultimate sale, including real estate agent commissions, stamp duty, transfer fees and the like. In some cases demolition or decommissioning costs may be considered.
- **Finance costs**. Finance costs comprise expenditure relating to the interest component of loan repayments, establishment and account fees, holding charges and other liabilities associated directly with borrowed capital.

CBA generally uses a decision criterion of maximum NPV, but comparative life-cost studies use a decision criterion of minimum cost. The two approaches are otherwise quite similar and share a common methodology.

Life-costs are of a significant nature and deserve consideration commensurate with their effect. Their absence from financial decision-making in the past has largely stemmed from the difficulties of applying these techniques and the tendency by professional advisers and building owners to ignore the impact of long-term expenditure.

8.5 Life-cost analysis

8.5.1 The purpose of cost analysis

The collection and interpretation of actual building costs as an input or feedback source for cost planning is the main function of traditional cost analysis. It includes distributing actual costs amongst the appropriate elements (or subelements) upon which cost plans are normally based, and has made possible a degree of control that was previously unknown. Composite rates can be averaged from real projects to provide a valuable guide to the estimation of new projects. This process applies equally to life-costs as it does to traditional capital costs.

Apart from providing feedback for future cost planning activities, analysis of life-costs can form an essential element of overall cost management by highlighting the ways in which potential cost savings in existing buildings might be achieved. For example, it might be better value to replace prematurely an expensive building component with a more efficient alternative than simply to continue with the original decision until its useful life has expired. Prudent control requires that actual and expected performance be constantly compared.

8.5.2 Monitoring and management

The monitoring and management of actual performance is known as life-cost analysis. While life-cost planning encompasses design cost control, life-cost analysis focuses on activities applicable to the construction and occupation of buildings. Together these techniques form the backbone of the cost management process and extend from the decision to build until the cost implications are no longer of concern.

If a life-cost approach is to be effective in reducing the running costs of existing buildings it is necessary that these running costs be continually monitored. Monitoring involves the recording of actual performance of a particular project in a form that facilitates subsequent life-cost planning and management activities. Such performance can be compared against the cost targets and frequency expectations given in the life-cost plan. Areas of cost overrun or poor durability can be explored, potential improvements identified, and better solutions implemented.

8.5.3 Data collection

Data collection is an essential activity in life-cost analysis. Data from different projects will reflect the specific nature of those projects, their locations and occupancy profiles. When measured overall, no two buildings will have identical running costs, nor will the running costs for any specific building be the same from one year to another. Data collection is still a useful exercise, particularly for monitoring activities, but obtained results should be applied to other situations with caution.

8.6 Important pre-design considerations

8.6.1 Discounting

Discounting is a means by which equivalent value is determined. Costs and benefits which arise in different time periods must be brought to a common base so that a proper comparison can be made. This comparison concerns not just the timing of cash inflows and outflows for a given project, but the relationship of cash flows across projects.

Discounting is merely a technique invented to help make judgements between investments that have different timing, of costs expended or benefits received. Cash inflows that are received earlier and cash outflows that are incurred later are generally regarded as more preferable than the reverse. Time preference and capital productivity are two philosophies that can be used as the basis for the choice of discount rate. Discounting is based on the compound interest principle in reverse and can thus be described as a negative exponential. Over long time periods the effect, particularly at high rates of discount, can be such as to make future costs and benefits irrelevant. Figure 8.5 illustrates the effect of a 6% discount rate on the estimated operating cost cash flow for a project over a 100-year life.

Discounting is a commonly misunderstood process and this is one of the principal reasons for the slow adoption of life-cost studies into normal practice in the construction industry. It is important to realize that discounting is an artificial mechanism invented so as to disadvantage future expenditure and income over the present for the purpose of equating the value of money in different time periods. The sum of all the discounted values does not represent real dollars, but is used solely for selection purposes. Therefore, discounting can be correctly applied only when two or more alternatives are compared. Discounting includes the cost of finance needed to fund the project, or the opportunity lost by not investing elsewhere, or both. Therefore, interest is never included as a discrete item of expenditure, and to do so would be double-counting.

While discounting has been described as socially unacceptable, because it diminishes the impact of future cash flows and hence favours present rather than future generations, it nevertheless describes the basic tendency of all thinking people. Ask yourself whether you would prefer to receive $100 now or in one year's time. You would normally say that you would prefer the money now. Also ask yourself whether you would prefer to pay $100 now or in one year's time. You

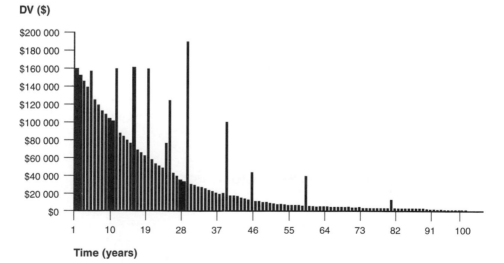

Fig. 8.5 Example of the effect of discounting.

would normally say that you would prefer to pay later. This decision is based, at least in part, on the potential opportunity that money has when you have control over its use. Discounting is a mathematical representation of these otherwise subjective judgements, and enables us to assess complex patterns of expenditure and revenue with objectivity.

Discounting has relevance to CBA, as well as life-cost studies, where a comparison between a number of alternative courses of action is involved. Results are expressed in terms of discounted (present) value and the lowest value is the alternative that is the most cost effective. Table 8.2 illustrates an example of a typical comparison over a 25-year life using a discount rate of 2%. But discounting has no place in the measurement of a single design solution, as would occur in a life-cost plan for a proposed project, and in such cases it is recommended to present all costs in today's terms (real value).

Discount rates based on the weighted cost of capital reflect the true time value of money and do not distort the relationship between present and future events unfairly. This approach is fundamental to comparisons over time because it accounts for the costs of finance, either real or imputed. Underlying the exchange

Table 8.2 Worked example for hypothetical alternative comparison

		Alternative A	Alternative B
Data (real value)	capital cost	$15 000	$40 000
	cleaning cost	$100 pa	$100 pa
	energy cost	$4 000 pa	$2 500 pa
	maintenance cost	$500 pa	$500 pa
	intermittent maintenance cost (additional)	$500/5 years	$500/10 years
	equipment life	10 years	20 years
	replacement cost (less scrap value)	$11 000	$35 500
Comparison (discounted value)	capital cost	$15 000	$40 000
	cleaning cost	$1 952	$1 952
	energy cost	$78 094	$48 809
	maintenance cost	$10 891	$10 172
	replacement cost	$16 427	$23 890
Total life-cost		$122 374	$124 823
	less residual value	($4 572)	($18 286)
	less tax deductions (36% on operating)	($32 737)	($21 936)
	less tax depreciation (10% PC on capital)	($8 358)	($14 030)
Comparative value (lowest is best)		$76 697	$70 571

rate for the time value of money is actually a zero discount rate. This signifies that normally present and future generations are given equal weight.

In investment analysis applications, the difference between the chosen discount rate and the IRR indicates the profit and risk allowance. If the difference is sufficient to cover market expectations given the risk attitude of the investor, then the project may be regarded as acceptable. Obviously this implies a positive NPV.

8.6.2 Study period

The study period, or time horizon, is another important consideration. The study period is commonly taken as equal to the economic life. This is the period of time during which the project makes a positive contribution to the financial position of the owners, both present and future. However, once the owner sells the property to another person, the original owner is no longer concerned with its operating performance or the costs or revenues that it will subsequently incur or generate. From the perspective of the original owner, the study period should logically be the period over which the owner has a financial interest in the property.

The determination of the study period for an investment project highlights the problem of the often conflicting objectives of investor profit and social welfare. While it may be conceptually correct to optimize a project so that society achieves the maximum benefit possible, this strategy may also result in no detailed study ever being commissioned since the investor may be departing on a course that ultimately reduces the expected return. Issues of energy conservation and public sector investment suggest social welfare (economic life) is the most appropriate basis for the study period. Issues of profit maximization and private sector investment suggest investor interest (holding period) is preferable.

The choice of study period is, however, not as critical as might first be thought. In a comparison situation, where discounting is employed, cash flows after 20 to 25 years will generally be rendered negligible to the outcome. Hence, choosing a study period of 100 years, for example, is a waste of time. In a measurement situation where discounting is not employed, the study period can be seen as a continuum along which the property might be bought and sold. In this context the study period is best selected as the period of financial interest of the owner, as this will define the quantification process, but the period can normally be treated as a variable and 'what if?' decisions analysed based on a range of possible values.

Furthermore, government regulation at the time of planning approval application could force projects that are shown to be a cost burden for future generations to undergo redesign until their operation is within maximum allowances. This trend is already happening in various countries, and it may soon be routine to prepare a life-cost plan for all new projects, to undertake annual self-assessment of energy usage and to be well-prepared for periodic external audits and the threat of financial penalties should actual performance exceed legislated limits.

8.6.3 Risk and uncertainty

The forecasting of future events is normally an integral part of the decision-making process. It can also be the subject of considerable uncertainty and therefore requires cognizance of the level of risk exposure. Discounting, CBA and life-cost studies are clearly reliant on appropriate forecasts of future events.

The decision-maker is faced with a fundamental choice: either the uncertainty of the future can be ignored by dealing with only those matters which are known, or tools can be used to help make predictions. It is frequently accepted in the literature that it is preferable to plan even when the accuracy of the plan cannot be guaranteed, for otherwise decisions are made in isolation from the environment in which they will ultimately be judged.

A life cost plan is an expectation of future design performance and hence forecasts events that are inherently uncertain. Even capital costs display a level of uncertainty, but when coupled with costs over a long period of time that are themselves a function of economic factors, obsolescence and operating performance, the entire process may appear an exercise in futility. But the life-cost plan should be seen as a set of targets against which future events can be compared. While the quality of the original prediction is important, a far greater benefit can be realized if areas of poor performance or areas in which additional savings can be made are identified.

CBA and comparative life-cost studies involve further aspects of uncertainty, notably the selection of the discount rate and the study period, but even the degree of uncertainty in these areas can be diminished to some extent since they always apply to a range of solutions and hence can have a corresponding effect on estimated costs and benefits across all alternatives. Figure 8.6 illustrates a sensitivity

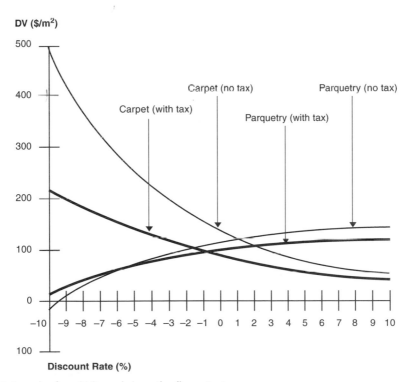

Fig. 8.6 Example of sensitivity analysis on the discount rate.

analysis on the discount rate for a life-cost comparison between carpet and parquetry with and without tax considerations.

Most forecasting techniques will inevitably fail to predict catastrophic events but instead will focus on a range of outcomes that may be reasonably concluded from history. If best, worst and most likely scenarios point to the same decision, then although the uncertainty contained in the assumptions may be high, the risk of making a bad decision is low. Despite its shortcomings, forecasting provides essential information that aids the decision-making process, and therefore it is a vital part of proper analysis.

The adoption of risk analysis techniques enables the uncertainty of future events to be properly assessed. For example, if the discount rate applied to a CBA or life-cost study is likely to vary within a range of values, but at all values the project outcome is favourable, then what is at risk is the level of the benefit not the possibility of a loss. In this case uncertainty remains high but risk of financial loss is low. Investment decisions are not always so simple, and therefore a number of sophisticated risk analysis techniques are available to quantify the impact of various future scenarios. Risk analysis is a key tool in the assessment of value for money during pre-design.

8.6.4 Multi-criteria analysis

Current research is focusing on the use of multi-criteria analysis (MCA) as an alternative to conventional economic evaluation. Problems with the application of shadow pricing and the monetary translation of externalities and intangibles has led to more interest in qualitative assessment. This technique is not dissimilar in concept to the form of value management adopted by the construction industry. A weighted matrix approach may be a more relevant tool to apply than DCF analysis if used effectively in a multi-disciplined team context.

A combination of MCA and traditional economic assessment can lead to the objective calculation of value for money. The weighted score for each competing alternative arising from an MCA study can be divided by the total cost of acquisition and operation as computed over the life of the project. The resultant value is a statement of balance between maximum utility (benefit) and minimum resource input (cost), where the higher the number the better value for money. This may prove a useful method for making informed decisions about complex problems.

It is recognized that sustainable development includes attributes that cannot be simply represented in monetary terms, and hence CBA and life-cost studies must be seen as tools within a much broader and more complex suite of evaluation techniques. Such attributes may include benefit–cost ratio, social benefit (welfare), total energy (embodied and operating) and environmental risk, where each attribute is measured in different units and combined together into a single criterion using MCA.

8.7 Conclusions

The capital cost fallacy still remains one of the biggest obstacles to achieving value for money during the pre-design stage of construction projects. This can only be overcome through continued efforts in education, marketing, research and development, new software solutions, government policy and legislation. Sustainable development goals provide a new incentive for the consideration of whole-of-life implications, and change the traditional narrow focus on initial construction costs. Ultimate success is a matter of speculation, but as limits to growth are approached it is hypothesized that environmental impact will become a matter of the utmost importance.

I suppose we shall just have to wait and see.

Selected bibliography

Ashworth, A. (1994) *Cost Studies of Buildings* (Second Edition) (Longman).

Boardman, A.E. (1996) *Cost–Benefit Analysis: Concepts and Practice* (Prentice Hall).

Bull, J.W. (1992) *Life Cycle Costing for Construction* (Thomson Science and Professional).

Daly, H.E. (1997) *Beyond Growth: The Economics of Sustainable Development* (Beacon Printing).

Dell'Isola, A.J. and Kirk, S.J. (1995) *Life Cycle Costing for Design Professionals* (Second Edition) (McGraw Hill).

Dell'Isola, A.J. and Kirk, S.J. (1995) *Life Cycle Cost Data* (Second Edition) (McGraw-Hill).

Diesendorf, M. and Hamilton, C. (1997) *Human Ecology, Human Economy* (Allen and Unwin).

Fabrycky, W.J. and Blanchard, B.S. (1991) *Life-Cycle Cost and Economic Analysis* (Prentice Hall).

Field, B.C. (1994) *Environmental Economics: An Introduction* (McGraw-Hill).

Flanagan, R. and Norman, G. (1983) *Life Cycle Costing for Construction* (Surveyors Publications).

Flanagan, R. and Norman, G. (1993) *Risk Management and Construction* (Blackwell Science).

Flanagan, R. and Tate, B. (1997) *Cost Management of Building Design* (Blackwell Science).

Flanagan, R., Norman, G., Meadows, J. and Robinson, G. (1989) *Life Cycle Costing: Theory and Practice* (BSP Professional Books).

Hanley, N., Shogren, J.F. and White, B. (1996) *Environmental Economics in Theory and Practice* (Oxford University Press).

Goodstein, E.S. (1998) *Economics and the Environment.* (Second Edition) (Prentice Hall).

Langston, C. (1991) *The Measurement of Life-Costs* (Sydney: NSW Department of Public Works).

Langston, C. (1991) *Guidelines for Life-Cost Planning and Analysis of Buildings* (Sydney: NSW Department of Public Works).

Langston, C. (1994) *The Determination of Equivalent Value in Life-Cost Studies: An Intergenerational Approach.* University of Technology, Sydney: PhD Dissertation.

Langston, C. (1996) Life-cost studies. In: *Environment Design Guide* (Royal Australian Institute of Architects).

Langston, C. (1997) *Sustainable Practices: ESD and the Construction Industry* (Sydney: Envirobook).

Pearce, D.W., Markandya, A. and Barbier, E.B. (1989) *Blueprint for a Green Economy* (Earthscan Publications).

Price, C. (1993) *Time, Discounting and Value* (Blackwell Publishers).

Seeley, I.H. (1997) *Building Economics* (Macmillan Press).

van Pelt, M.J.F. (1993) *Ecological Sustainability and Project Appraisal* (Avebury).

9

Feasibility studies

Tony Collins†

Editorial comment

The financial evaluation of a project is a critical step in the early definition and concept development stage. If clients are to get value from their investment in buildings the decisions made have to be informed by the use of quantified objectives which define the goals that are set. This is the primary function of a feasibility analysis, or economic evaluation.

A preliminary feasibility analysis is used to establish the options available to the client and to define a range of likely outcomes for a project. The business goals of the client can then be linked to the project, which will ensure that the business outcome is satisfactory and value has been delivered.

The analytical processes used in feasibility studies require specialist skills and are often not easily understood by those who make investment decisions. However, the use of quantitative techniques has been steadily increasing over the past decade and many projects now have effective quantitative analysis performed using structured net present value techniques.

For clients, project evaluation and feasibility studies help in two critical decisions. First, the decision to invest (the go/no go decision): if the analysis shows that the project is unlikely, under the circumstances, to be profitable, then the decision to invest is not just about returns but also about avoiding losses. Secondly, the financing mix, or proportions of debt and equity the project will carry, will influence the net return of the investment. In many cases these two decisions are blended together, with an emphasis put on cash flow and the end valuation of the project rather than the refinement and testing of the project's functional and financial characteristics.

Clients seeking value from their projects will use a range of analytical approaches in the evaluation stage. Thorough and detailed feasibility studies are an essential part of this stage of project development.

9.1 Introduction

This chapter looks at value obtained by feasibility analysis from the client/ developer's viewpoint and therefore looks at value gained by the client through the

† ABSA Bank, South Africa (Western Cape)

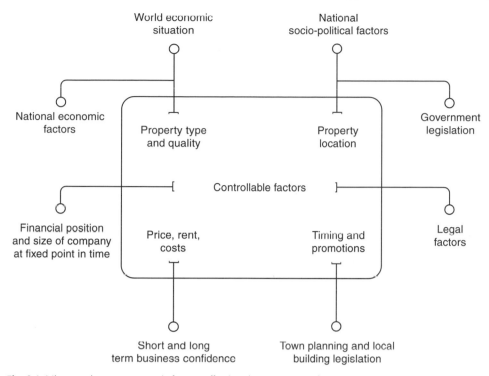

Fig. 9.1 Micro- and macro-economic factors affecting the property market.

early decision-making stage of the development process. The steps in the development process are covered, with check lists to assist with the collection of relevant data since control over variable data inputs allows for flexible decision making, which in turn allows the analyst to work out the best 'value for money' scenario for the client.

In addition, since the client is always faced with many alternatives in the market place, a discounted cash flow model that is sensitive to changes in inputs and outputs is recommended since various decisions can be taken on timing, amount and certainty of the project's cash flow. The client can therefore juggle with market-related data to find the optimum solution, since commercially viable property development should be market driven.

As Fig. 9.1 shows, there are many factors that affect property markets, some of which are beyond the control of those who drive property development. Whether developers will maximize the value which they realize from their activities will depend to a great extent on the way in which they analyse the available data related to these factors and how such analysis affects potential projects.

9.2 A rational approach to investment analysis

Approaches of investors to investment analysis range from being cautious and analytical, to the seat of the pants/gut feeling type management decisions. Furthermore, the investor, according to McKeever (1968, p. 5) 'may be gifted with

Market Study

A study of the market variables which influence the supply of and demand for real estate. (This is the broadest possible of demand-oriented real estate studies).

Marketability Study

A narrow defined study to determine the conditions under which a specific property can be sold. The key conclusions relate to both price and time required to sell.

A Feasibility Study

A study to determine the probability that a specific real estate proposal will meet the objectives of the developer and/or investor.

Highest and Best Use Study

A study to determine that use among other possible and legal alternative uses which results in the highest land value for a specific site.

Fig. 9.2 Forms of analysis.

strong hunches and good horse sense, but he is better equipped when he has definite facts at hand'. Consequently, developers often gloss over important areas in the investment analysis process, as explained by Maritz below.

According to Maritz (1983a, p. 288) the investment analysis process consists of a rational approach to property development, as shown in Figs 9.2 and 9.3. This incorporates a series of steps for the gathering of information, analysing that information and taking a decision on the basis of this analysis. The result is that the characteristics of an investment property are considered objectively and are related directly to a particular investor's needs.

This research process has much in common with the key stages in the development process.

9.3 Stages in the development process

According to Barrett and Blair (1987, p. 8), the development process comprises the following key stages.

Stage 1. Initial planning stage of the project which embraces four phases comprising

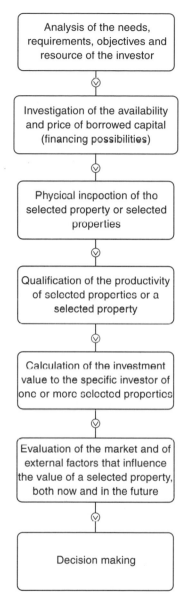

Fig. 9.3 The investment process.

Phase 1. formulation of the developer's objectives
Phase 2. conducting a market analysis
Phase 3. preparation of a financial feasibility study and
Phase 4. taking a decision on whether to continue with the project, shelve it until a later date, or abort it completely.

Stage 2. Acquiring the land – an option can be taken on the land at an earlier stage.

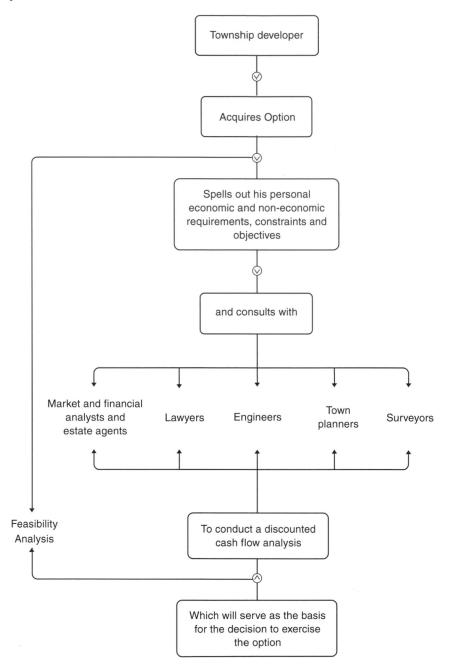

Fig. 9.4 The township development process.

Stage 3. Developing the land.
Stage 4. Constructing the building(s).
Stage 5. Marketing and leasing space and/or units.
Stage 6. Setting up a property management and maintenance system.

Furthermore, in township development, because of the lengthy period it could take to legalize and convert bare land to useable township land, the developer should first acquire an option to purchase before committing his or her capital to the project (Fig. 9.4). The option would hold for the extent of the planning and decision-making period.

It is important to note that some of these stages could run in tandem with other stages. For example, the marketing of a development often commences before any construction work is done. In addition, it should be noted that these stages, according to Graaskamp (1970, p. 11), could differ in sequence according to whether

- the *use is known* and the site is to be determined or
- the *site is known* and the use is to be determined or
- *an investor is looking for involvement in the two above options.*

Other factors include the type of development, e.g. township development, or shopping centre development where tenant mix is of prime importance (Stevens, 1991, p. 7), and the magnitude and complexity of the anticipated project.

9.4 Initial planning stage of the project

The initial planning stage will originate from an idea that a developer has on how to make money and is driven either by (Wurtzebach and Miles, 1980, p. 594)

- an idea in search of a site (Fig. 9.5) or
- a site in search of an idea (Fig. 9.6)

whereas an investor looking for involvement in both options is a possible third source (Graaskamp, 1970, p. 11).

In addition, this stage is categorized by Graaskamp (1970, p. 5) as being part of the feasibility analysis as a whole and could vary time-wise according to the magnitude and complexity of the development project. Furthermore, there is usually an overlap between the various stages and phases in the development process.

9.4.1 Phase 1 – formulation of the developer's needs and objectives

'If you don't know where you are going any road will take you there.' – Unknown source.

Investor needs and objectives

The first step (Barrett and Blair, 1987, p. 8) is to define the developer's goals and objectives. These are determined through discussion prior to establishing uses for sites or finding sites for uses. Furthermore, developers' objectives will differ as the private developer will be driven primarily by the economic principle while

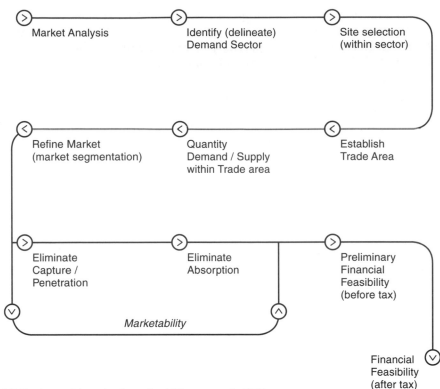

Fig. 9.5 Use known/site to be determined (Messner *et al.*, 1977).

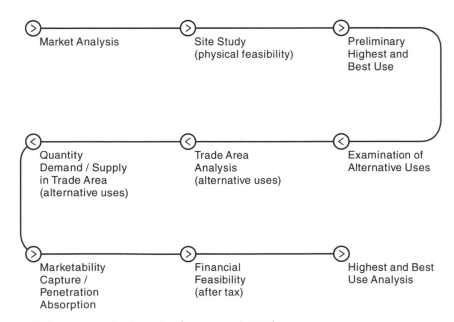

Fig. 9.6 Site known/use to be determined (Messner *et al.*, 1977).

government/state bodies and welfare organizations will consider the social costs versus the social benefits as well. One therefore needs to establish clearly what the client's objectives are at this stage and then assess and define their legal, financial, managerial and technical capabilities for comparison with the scope and objectives set for the project.

The investor begins by conducting an analysis of his or her own needs; on the basis of this (s)he then formulates her/his own criteria for selecting and evaluating investment opportunities. While needs differ between investors, they can (Maritz, 1983a, p. 290) be formulated in terms of the following considerations, namely:

- expected yield or return on capital
- liquidity of the investment
- risk attached to the investment
- income tax considerations
- management requirements of the investment.

Resources of the investor

Maritz (1983a, p. 290) lists the typical questions that should be answered by the developer before (s)he can proceed:

- How much equity capital is available for investment?
- What is the maximum amount that can be invested in a single property?
- Is there sufficient cash to obtain possession and transfer of the specific property?
- Will there, for example, be sufficient money available to pay transfer duties, legal fees and mortgage bond registration costs?
- What other financial assets are there that can possibly be employed to carry the property, either temporarily or permanently?
- What appropriate non-financial resources does the client/developer possess, e.g. know-how, to administer the property?
- To what extent is the investor willing and able to take in partners for the intended investment?

When these questions have been answered the client/developer and his or her professional team have a clear understanding of the resources that can be applied to a potential project as they go through the feasibility process.

Management requirements

Requirements of management involve the investor assessing and defining his or her financial, managerial and technical capabilities for comparison with the scope and objectives set for the project. A further consideration would be the assessment of any community and/or political resistance to the project, particularly regarding the size and possible impact of such resistance. In addition, management systems for controlling maintenance and security are most important, as many of the costs incurred on the inside of the building, such as maintenance, electricity and water costs, are passed on to the clients, who then become more careful about the services they pay for themselves.

Non-financial requirements of the investor

Investors often have special requirements for properties because of company policy, or due to past experience in a particular field. Such requirements, and the possible advantages or benefits that they generate, are often difficult to measure financially. They could include, however,

- special space configurations to promote efficiency and productivity
- special work environments to promote worker well-being
- environmentally friendly approaches to design and construction
- saving trees on the site
- architectural styles, in terms of company image
- tenant selection and tenant mix, in terms of proposed project image.

Investigation of financing possibilities

Once the investor's available equity, resources and investment needs have been identified, it is necessary to determine whether there is adequate additional finance or borrowed capital available. Particular care is needed in a high interest rate market, since an investment property cannot be considered as a proposition by an investor unless the required finance is available, and the financial conditions fall within his or her acceptability limits. Furthermore, the limitations and benefits of creative financing packages should also be evaluated.

In addition, clients should negotiate with various lending institutions, since a good client can usually persuade a lending institution to improve their offer if the client can show that the lender's opposition is prepared to offer a better deal.

Financing the development could take various forms, such as

- loans from institutions, which could diminish the viability of the project in a high interest rate economy, and if the leverage is too high could create a risk due to the low equity capital of the developer
- self-financing by the developer
- syndication by involving more than one investor
- involving one or more major partners such as a financial institution, or a pension fund
- selling units off before they are built where trust funds are used as security for the purchaser
- developing and selling in sections or under sectional titles.

Physical inspection of the selected property

Before taking the final decision to purchase a property, the investor usually conducts a thorough physical inspection to ensure that the property meets his or her objectives and requirements. Apart from the property market and general economic conditions, the most important factors influencing the potential net cash-flow of a property (Maritz, 1983a, p. 300) are the utility or productivity generating attributes, namely:

- the physical nature of the land and improvements
- the location of the property
- the institutional attributes of the property.

An evaluation of these utility-generating attributes is the object of a physical inspection of the property. This enables the investor to reach a conclusion regarding the nature, amount, quality and other relevant aspects of the yield the property can provide in competition with alternative properties.

Land and improvements

The physical appropriateness of the property, or its *in situ* attributes, reflects the services and yield that the property can provide as well as the time period over which they can be provided. Their quality and quantity are also determined by the nature of the improvements.

The contribution of the land is affected by:

- the size and form/shape of the site
- topography
- surface and sub-surface conditions and
- the availability of municipal services.

All of these contributions can influence the productivity (income-producing ability) of the completed development with regard to the improvements in terms of functional, economic and physical efficiency of the property and therefore the income-producing ability of the building is also affected. The ability, therefore, of a building to be rated as an A-grade or B-grade building in terms of facilities offered to clients dictates the level of market rental the building may command in the market place. A-grade and A-grade plus (very plush) buildings are usually new buildings with all facilities such as air conditioning/climate control and undercover parking. B-grade buildings tend to offer similar facilities, but are approximately 10 years older. C-grade buildings are older buildings generally lacking air conditioning and on-site under-cover parking facilities, while D-grade buildings include everything else not falling within grades A, B and C in the market place.

Functional efficiency refers to:

- the fact that the size and permanence (attributes or ability of the building to meet the client's goals and objectives in terms of the holding period the client decides on keeping the building for, e.g. the client requires the building to last for 20 years to perform his functions economically) of the structures determines the quality of services that can be provided within the physical, technical, philosophical and financial constraints of the client
- how serviceable a building is for its existing or proposed use
- the building-to-land coverage ratio
- the appearance of the building
- the adequacy of improvements to land and buildings in promoting the generation of income, for example the rentable floor area to gross area efficiency as a factor of design efficiency.

Economic efficiency refers to the ability of the development to command a market-related rental which will generate a positive cash-flow, whereas *functional efficiency* refers to the ability of the property to fulfil the functions it was designed for and *physical efficiency* refers to the quality of the structure and finishings. Therefore, deterioration of the property may also be referred to in terms of functional, economic and physical obsolescence.

Location

It has been suggested that 'location determines the demand for the services which the property is physically capable of producing and is a function of the *convenience network* and the *exposure network* of the property' (Maritz, 1983a, p. 128).

The convenience network refers to the links that may exist between the property and the basic public the tenants have to deal with. The movement of people and goods gives rise to four types of associated cost, not all of which are readily measurable. These are:

- out-of-pocket transport costs such as bus fare or expenditure on a private motor vehicle
- the time cost of travelling
- terminal costs, e.g. parking money that is paid out when visiting the city centre
- irritation costs, such as irritation caused to the traveller by delays, or traffic congestion.

The exposure network, however, comprises all the positive and negative factors to which the subject property is exposed as a result of its proximity to other properties. No movement is implied in this concept. Exposure along a busy street, or to a railway station or a beautiful view, can be regarded as *positive exposure* if passing trade is important, while *negative exposure* may be the result of location near properties that are undesirable due to poor physical condition or appearance. Negative exposure could also be due to surrounding properties being sources of sound, air or light pollution (e.g. airports). Location close to an airport can, however, be a positive factor for the client who finds it important to be located close to an airport for *convenience*. Location, through the combination of convenience and exposure, is possibly, therefore, the most important productivity-generating attribute of a property as it greatly influences the utility or benefits which it provides for building owners and tenants.

Institutional attributes

The institutional attributes determine the legal use to which property can be put. One of the principal objectives of a physical examination is to ascertain whether the current use for which the property is being used is also the highest and best use. The right of use is limited by public and private legal restrictions such as taxes, and local authority rules, regulations and ordinances and therefore it is essential to ascertain whether the maximum potential of the property in this regard has been exploited.

The major goals are to determine to what extent the property utilizes the total allowable floor space, and/or to what extent it is possible to establish a new use on the property that will increase the value of the land and improvements above their present value.

Return on capital to the investor

The future benefit that would accrue to the investor is referred to as the yield, or return. Four investment approaches in particular can be distinguished in terms of the type of yield or return obtained from possession of property. These are investments for:

- own use
- regular income
- capital growth (primary objective is selling)
- amassing an estate/wealth.

The investor could, however, have different combinations of the above approaches in mind when acquiring an investment property.

Although many investors may say that they require a return expressed as a given amount of dollars per month, most investors require a specific *rate of return*. These yield rates can be calculated on a before or after tax basis, and expressed as

- required rate of return on total capital invested or
- required rate of return on equity or
- an anticipated internal rate of return.

The norm is that investors require a certain before or after tax rate of return on the equity invested in a property. It is therefore assumed that the investor wishes legally to own the subject property for a specific time period, and furthermore, that the benefits that accrue to the investor during that time, such as income stream, capital growth, financing possibilities and tax benefits, will be taken into account when the before or after tax rate of return is calculated.

Risk

Business Risk.

Risk will always be with us. Throughout the decision-making and property development process we are faced with a variety of risks that can be classified into two categories: namely, business risk related to business decisions, and financial risk related to loan capital. The magnitude of perceived risk is very personal and consequently the manner in which risk is dealt with differs according to factors such as:

- an individual's propensity to take a risk
- the financial backing and expertise available
- the education of the individual
- the individual's cultural and religious background
- the individual's personality.

Other factors affecting the manner in which we perceive risk include socio-political and economic factors. The approach to risk would also differ depending on whether the property to be developed is owned, or whether a property is to be purchased for development.

Rational risk takers, according to Greer and Farrel (1988, p. 22),

- carefully specify their investment objectives with respect to required rates of return, timing of returns and acceptable risk levels
- identify and quantify major risks as accurately as possible
- eliminate risks where possible
- transfer risks through insurance policies
- reduce manageable risks and
- base decisions on whether the expected returns justify bearing the remaining risks in the pursuit of the investment objectives.[1]

Income tax consideration

The investor's requirements with regard to tax shelters are most important as they form part of his or her total needs. Examples of tax shelters may include:

- tax rebates on apartment blocks/blocks of flats
- tax rebates in respect of manufacturing buildings
- tax concessions granted to industrialists in designated economic development areas
- capital allowances applicable to hotel buildings
- maintenance and repairs
- refurbishment of hotels and industrial buildings.

However, these tax breaks/shelters/rebates would differ according to the needs of individual countries depending on what sectors of their local economies they wish to stimulate.

Tax consultants who specialize in property tax should be consulted at the design stage, to ensure the client gains maximum advantage from the latest concessions and court case findings.

Legal requirements

At this stage the legal requirements, according to Maritz (1983b, p. 102), would include an investigation of:

- the legal description of the subject property and its ownership
- whether the seller is entitled to sell
- the registered rights and limited real rights of ownership
- mining titles, including other relevant information on possible future mining in the area
- unregistered rights of use which again would differ according to legal practice in different countries. An example is the prescription right in South Africa, where rights are gained by an individual or the general public through continuous uncontested use of another person's property for more than 30 years. This could be a road or a section of a farm for which the unregistered user is then allowed to claim the continued use.

---

[1] See Chapter 16 for a more detailed discussion of risk management

It is essential that the checklists above are worked through carefully and systematically if the client is to maximize value for money.

9.4.2 Phase 2 – conducting a market analysis

The first step after appointing construction and development consultants is to begin with the market analysis and the marketability study, either in-house, or through a specialist market research consulting enterprise. Market analysis, according to Graaskamp (1970, p. 27), can be defined as 'a reduction of aggregate data, such as population, employment, and income totals to factors which are relevant to the site, the merchandising target, or the client.' Barrett and Blair (1987, p. 6) assert that 'the market study is perhaps the most important single element in the planning process.' It is also the most difficult part of the process. Barrett and Blair recommend that it should cover the following:

- indirect economic influences such as public regulations, social impact and environmental concerns
- direct economic influences, covering the competitiveness of the site itself in relation to community and regional markets, through analysis of quantifiable demand factors such as population, employment and incomes
- the operational development programme which is established in terms of the developer's business objective and the total site bulk capacity and zoning
- site holding capacity of the client (how long the client can afford to hold the site and pay interest on the loan and other charges before developing) which could be restrained by the financial objectives of the developer, or the design of the architect.

In addition, some real estate specialists talk about 'ripening time' which refers to the site not being ready for development. The result is that the site may have to be held for a period before ground conditions stabilize, or before local government ordinances/regulations are altered or adapted to allow for the legal use of the site to be altered.

Fortune (18 May 1992, pp. 46–59) sounded a warning against neglecting to survey both the macro- and micro-environments. Injudicious development of the property markets in the early 1990s returned vacancy rates averaging 18.8% in the United States of America, whereas in 1981 the average vacancy rate was 4.8%. In isolated cases the situation was so bad that brand new 50-storey office complexes had individual vacancy rates of between 50 and 100%. The development and insolvency problems experienced at Canary Wharf in London are also mentioned in this *Fortune* article. Many of these problems were caused by world economies going through a recession, during which time the Canadian-based Reichmann Brothers of Olympia and York Developments Ltd had borrowed eleven billion US dollars from 92 lenders and had become insolvent in the process.

It is therefore clear that the property development process does not occur in a vacuum, as both the macro-environment and the micro-environment affect the destiny of property developments. Horne (1978, p. 13) lists these macro- and micro-environmental factors as:

Uncontrollable factors	Controllable factors
World economic situation.	Property type and quality.
National socio-political factors.	Property location.
National economic factors.	Price, rent costs.
Government legislation.	Timing and promotions.
Legal factors.	
Financial position and size of company at a given point in time.	
Short- and long-term business confidence.	
Town planning and local building legislation.	

Environmental impact statements

Purchasing a site for development but without conducting a preliminary environmental impact study is tantamount to playing 'Russian roulette'. In addition, during the initial planning stage, a comprehensive environmental impact statement should be commissioned to ensure there are no extra hidden costs or factors that could prevent the project from going ahead.

9.4.3 Phase 3 – preparation of a financial feasibility study

Defining feasibility

Although there are many definitions of feasibility analysis, Messner *et al.* (1977, p. 14) suggest that 'A Feasibility Study is a study to determine the probability that a specific real estate proposal will meet the objectives of the developer and/or investor.'

This definition will suffice, as it is short and to the point.

Framework for total feasibility analysis

Graaskamp (AIREA, 1977, p. 215) recommends that the following components be included as a framework for total feasibility analysis:

1. Objectives of the enterprise for whom the feasibility should be performed.
2. Market trends to identify opportunity areas consistent with objectives.
3. Market segmentation for merchandizing targets.
4. Legal–political constraints.
5. Aesthetic–ethical constraints.
6. Physical–technical constraints and alternatives.
7. Financial synthesis of proposed enterprise form.

Graaskamp also adds that 'the essence of a feasibility determination lies in correctly defining the objectives which the solution must serve, and the context or standards wherein an acceptable solution must be found.' Only then is the analyst in a position to take a decision on the degree of consensus between external factors affecting the decision to be made and the internal ability to achieve individual objectives in the real estate enterprise.

Why real estate projects fail

James Boykin (1985, p. 88) lists the results of a national survey of American real estate executives who identified nine principal causes of real estate failures as being:

- inaccurate, or overly optimistic market feasibility studies
- poor planning
- financing problems
- location problems
- improper timing
- lack of professional expertise
- construction problems / weak project management
- inadequate cash flow projections
- other project problems such as lack of proper marketing, building to satisfy ego needs, poor or wrong designs, insufficient land or too high prices paid for land, and poor communication and liaison with local authorities.

In this survey, executives indicated that feasibility studies were often inaccurate in that the analysts erred by adopting over-optimistic space absorption rates. Furthermore, market acceptance of space was also misconstrued, while market data were often inadequate and poorly interpreted. A further problem pointed out by Boykin (1985, p. 91) was that the level of the quality of presentation available with modern technology is so impressive that investors and developers neglected to check analysts' calculations. Barrett and Blair (1987, pp. 143–8) support Boykin's findings.

9.4.4 Phase 4 – taking a decision on whether to continue with the project, shelve it until a later date, or abort it completely

Framework for analysing investment information

When all the relevant information necessary for the investment analysis has been collected, the next step is to place a monetary value on the possible current and future services or benefits that the property can produce. Such an analysis should be both quantitative and qualitative.

Qualitative analysis

Qualitative analysis of investment information involves the evaluation of both the micro market and the external macro market as both may influence the value of the property. As previously discussed, the value of a property is a function of its utility-generating factors, which include land and improvements, locational and institutional attributes, and the market, and external economic, social and political factors. All of these factors must be analysed by the investor prior finally to deciding whether to purchase the property or not.

Quantitative analysis

Maritz (1983a, p. 302) states that the net monetary income which can be yielded by an investment property consists of the income stream, or net operating income, obtained from running the property as a going concern: for example, the monthly income accrued from the letting of space in an apartment/flat block, office complex, or shopping centre, and/or the reversion or income obtained from the refinancing or sale of the property when its market value has increased.

In addition, depending on the needs of the investor, calculations can be performed at three levels of sophistication according to Maritz (1983a, p. 302). These three levels include quantification of the

- productivity (income stream and reversion) of the property alone (i.e. prior to the employment of borrowed capital and in the absence of income tax considerations)
- before-tax productivity of the investor's equity (i.e. after the employment of borrowed capital but before the payment of income tax) and
- after-tax productivity of the investor's equity (i.e. after the employment of borrowed capital and the payment of tax).

The components of the discounted cash flow model

Of utmost importance to all investors in this process, according to Greer and Farrel (1988, p. 17), are the amount, timing and certainty of revenues, as the investor is faced with an array of alternatives for which (s)he has to determine relative investment values. This is probably the main reason why discounted cash flow techniques are such a useful tool in the decision-making process.

The components of the discounted cash flow model are:

- the operating income produced by the property as a running concern and
- the return achieved on resale of the building where the market value exceeds development cost.

These two main components are discussed in more detail below.

Operating Income Statement, which represents the cash flow of the prospective property; this should be projected over the holding period of the building. The important items to be calculated in the operating income statement should include, at least:

- annual gross rental income
- effective gross income
- net operating income
- normalized before-tax cash flow and
- normalized after-tax cash flow.

The reason for projection according to Greer and Farrel (1988, p. 111), is that 'The economic desirability of an investment proposition is strictly a function of the amount, timing and certainty of after-tax cash flows.'

The Resale Return to the Investor Table is the next important component as, according to Maritz (1983a, p. 310), there are two principal types of value that must be distinguished in investment analysis, namely the,

- utility or use value of the property for a specific owner/user and
- market value, i.e. the value for which the property can be exchanged in the market place for money, or the price that a willing buyer will pay to a willing seller in the market.

The resale return to the investor calculations are therefore important, in that they supply data for the calculation of various real estate investment ratios/ measures as well as allowing for a comparison between development cost and market value. In addition, the present values of the cash from reversion can be calculated over the holding period and can be compared with the initial equity inputs.

The following items could therefore be calculated over the holding period of the property:

- capitalization rates
- market values
- effective gross proceeds from the sale (reversion)
- before tax net worth
- after tax net worth
- justified present value of equity capital and
- justified present value of the project.

The above model allows the analyst and the client to compare the subject property with a variety of alternative properties that could be developed at procurement stage. In the process, differences in the timing, amount and certainty of receipts over the holding period for different properties are compared, which enables the client and the analyst to print out a variety of alternatives for the subject property and the alternative properties. These alternatives allow the analyst and the client to adjust building and land costs, equity and loan inputs to maximize leverage, and play with interest rates, floor areas and rentals as well as any other factors which may affect the feasibility of the property.

The investment decision

The investment decision taken by the investor is based on both quantitative and qualitative information analysis. The buyer and seller are also required to agree on the price at which a property will change hands, as well as on the conditions of sale. Furthermore, the price agreed on in any transaction is influenced by a knowledge of property and the negotiating abilities of the parties concerned and therefore, according to Maritz (1983a, p. 318), once the investor has:

- analysed his or her needs and resources
- investigated the availability and price of finance
- analysed the utility of the property
- quantified the productivity of the property
- determined his or her investment source and
- evaluated the market and external factors that can influence the value of the property;

only then does (s)he decide on whether or not to purchase the property.

At this stage the investor must also become acquainted with the market value of the property and therefore (s)he needs to take cognizance of the following before making the decision on whether to purchase the property (Maritz, 1983a, p. 320):

- The investment value of the property for the investor sets an upper price limit on what can be paid for the property. Up to this price it will be possible for the investor to satisfy his or her investment needs.
- When the investment value of the property is lower for an investor than its market value, it is improbable that (s)he will purchase the property since his or her investment needs will not be satisfied.
- It is also unlikely that a rational, informed seller will accept less for his or her property than the market value.
- When the investment value of a property to an investor is higher than its market value, his or her investment needs will be more than satisfied. However, (s)he will endeavour to pay no more for the property than that for which it can be purchased on the market.

Final evaluation

With both the qualitative and quantitative decision-making information at hand the investor now has to make his or her final evaluation and decision on the investment. Once the project is proven to be feasible, the formal development plan is completed and the future of the project is decided. Four alternatives exist at this stage:

- abort the project
- dispose of the site including the plans
- shelve the project until more favourable conditions exist, taking ripening time into account, or
- continue with the project, seek official approval, obtain financing, and physically develop the site.

In addition, with all the calculations to be made in the decision-making process, the computer spreadsheet is a good tool to use as it is quick, powerful, easy to use and, with the right hardware and software, can provide a high standard of presentation.

9.5 The big three

Land cost, building cost and income/rentals earned are the big three factors in any development and they tend to affect the profitability of the project most. If the land and building costs are too high they can affect the viability negatively while an increase in rentals (market related) will improve feasibility. It is advisable, therefore, when analysing the project, to start with these three factors.

It is important to note that high percentages of loan capital affect the cash flow of the project, but not the market value, as market value for a commercial building is calculated from net operating income (NOI = gross income − vacancies − operating cost) and capitalization rates. Loan payments are only deducted from NOI when calculating cash flow and therefore do not affect market value. This, of course, is logical as each individual will have a different finance structure that will affect the price that the individual can pay or offer and not the price the property could fetch in the market place.

It should also be noted that while maximization of loan capital is good for return on equity, too much loan capital can affect the cash flow of the project due to the size of the mortgage payment.

9.6 Conclusions

The end goals of individual developers differ according to personal goals and requirements as well as in response to the size, nature, complexity and timing of the development and the quality and type of in-house expertise the developer has. In addition, whether there is a site in search of a use or a use in search of a site, there is a further basic premise/assumption, that all commercial developments must be market driven.

Real estate feasibility analysis can be complex and time consuming owing to the large amount of information required, the many calculations required, and because the value of a real estate project extends throughout the holding period of the owner. This holding period often extends for many years past the first year's capital input and, therefore, the ensuing years bring in income and incur *inter alia*, running costs and maintenance expenditure, while the resale at the end of the ownership period must also be considered.

In spite of the above, the feasibility process should always allow the analyst and the client to compare the subject property with a variety of alternative properties that could be developed at procurement stage. In the process, differences in the timing, amount and certainty of receipts over the holding period for different properties are compared, which enables the client and the analyst to print out a variety of alternatives for the subject property and the alternative properties. The above total approach to feasibility allows the analyst and the client to modify variables such as building costs, land costs, equity inputs and loan inputs to find the optimum level of leverage. They are also in a position then to manipulate interest rates within market constraints as well as floor areas and market rentals which may affect the feasibility of the property.

References and bibliography

AIREA (1977) *Readings in Real Estate Investment Analysis* (Chicago: American Institute of Real Estate Appraisers).

Barrett, G.V. and Blair, J.P. (1987) *How to Conduct and Analyze Real Estate Market and Feasibility Studies* (New York: Van Nostrand).

Boykin, J.H. (1985) Why real estate projects fail. *Real Estate Review*, **15** (1), 88–91.

Fortune (18 May 1992) Victims of the real estate crash. *Fortune*, **125** (10), 46–59.

Graaskamp, J.A. (1970) *A Guide to Feasibility Analysis* (Chicago: Society of Real Estate Appraisers).

Greer, G.E. and Farrel, M.D. (1988) *Investment Analysis for Real Estate Decisions*. Second edition (Chicago: Longman Financial).

Horne, L.G. (1978) *Marketing Property* (Johannesburg: McGraw-Hill Book Company).

Maritz, N. (1983a) *The Study Guide for Estate Agents* (Cape Town: Juta).

Maritz, N.G. (1983b) *Study Guide: Real Estate Investment, Development and Finance* (Pretoria: UNISA).

McKeever, J.R. (1968) *The Community Builder's Handbook.* Anniversary edition (Washington: Urban Land Institute).

Messner, S.D., Boyce, B.N., Trimble, H.G. and Ward, R.L. (1977) *Analyzing Real Estate Opportunities: Market and Feasibility Studies* (Chicago: Realtors National Marketing Institute).

Stevens, A.J. (1991) *An Overview of the Property Development Process* (Cape Town: University of Cape Town) unpublished notes.

Wurtzebach, C.H. and Miles, M.E. (1980) *Modern Real Estate* (New York: John Wiley & Sons).

10

Cost or benefit?

Grace Ding[†]

Editorial comment

Cost–benefit analysis (CBA) is generally associated with the selection of alternative developments or projects that are competing for limited or scarce resources. The most common use of CBA is in the selection of publicly funded projects where many possible uses of funds from government budgets exist but such funds are strictly limited and their use is subject to public scrutiny. The selection of those projects that are to proceed requires the use of some systematic method of appraising the competing alternatives and choosing those which provide the greatest benefits to the community – in other words, choosing the alternatives that will produce the best value for money.

While CBA is usually applied to projects such as highways, bridges and similar infrastructure improvements, it has been included here as its fundamental purpose is to identify projects that will provide maximum value in return for the consumption of resources. These resources may be purely financial, but for many large projects the resources may include both tangible resources such as land and materials, or intangibles such as air quality or a quiet environment.

The basic techniques of cost–benefit analysis can also be applied to the evaluation of individual building projects as there are nearly always alternatives to constructing a new building, such as outsourcing or refurbishment or re-engineering a process, which may be shown to be preferable if properly evaluated. If a new building is the appropriate solution to the client's needs then there will be alternative designs which will need to be assessed if the most cost-effective solution is to be pursued.

There is some obvious overlapping of CBA, feasibility studies and life-cost planning, particularly in the way that costs over time are discounted to allow meaningful comparisons of costs and benefits that emerge at different times over the life of the project once it is completed. CBA takes a broader view of costs and benefits, however, as it includes social costs and benefits, which are not usually part of building life-cost studies or investment-driven feasibility studies.

The following chapter outlines the basic types and techniques of CBA, and in doing so provides a slightly different view of the idea of achieving value for money.

† University of Technology, Sydney, Australia

10.1 Introduction

The products of the construction industry are inseparable from our basic social fabric. The construction process transforms land into housing estates to provide shelter, develops infrastructure to improve communication between people in different localities, builds all manner of industrial and commercial structures and, over time, directly improves living standards and quality of life. Indeed it is one of the most important sectors in our society.

Construction projects differ widely in type and size. They can be as small as simple domestic renovations or as large as a transnational infrastructure such as the Channel Tunnel, requiring the collaboration of several countries. In its widest sense, the construction industry is not limited to building construction but has diversified to include civil engineering and mining projects, shipbuilding, aerospace, transportation and energy generation projects, maintenance of existing facilities, and the implementation of new technologies.

A project is defined as 'a discrete package of investments, policy measures, and institutional and other actions designed to achieve a specific development objective within a designated period' (cited in van Pelt, 1993, p. 41). Indeed, projects can be in the form of physical developments, government policies, community activities or welfare programmes. They are often conceived in response to particular problems experienced by the initiator. The project is intended to bring an undesirable existing condition to a desirable new condition within a stated period of time and within budget limits. Project development involves systematic analysis of development objectives in the order of their priorities so as to allocate resources efficiently. A project may also be defined in the broader sense of any use or saving of resources, such as health, social services and environmental control projects.

A project is regarded as successful if it is completed within the imposed constraints of quality, cost and time and it achieves its designated purposes. This involves the fundamental process of selecting the right project, and constructing it according to specification at minimum total cost and within reasonable time. Project selection often involves choosing the best option from a range of possible ideas.

The purpose of a project is derived from a prescribed set of objectives. The objectives of a private development may be to maximize current profit margins, or to increase profitability, efficiency, yearly turnover or employment. From society's viewpoint, the ultimate goal of a project may be to improve social welfare or quality of life, or provide enjoyment. From an environmental viewpoint, however, more project development means more damage to the natural world and depletion of scarce renewable and non-renewable resources. In this way people tend to go to one of two extremes, either focusing on project development without any consideration of the environment or criticizing almost any kind of new development in society. Nevertheless, going to either extreme is not an ideal circumstance and an effective balance needs to be struck.

Economic growth and environmental protection have a two-way interaction. The environment is the prime supplier of raw materials needed for economic growth. In construction, raw materials such as iron ore, timber, and quarried stone are crucial to the existence of the industry. Today the supply of some basic raw materials is under threat as a result of past overuse or finite limits of supply. External effects such as air and water pollution generated from mining, manufacturing and construction processes can also seriously affect the capacity of the environment to perform as a producer of future raw materials. This in turn hampers economic growth and quality of life in the long run. Economic growth and the natural environment jointly affect the well-being of humankind. The efficient allocation of scarce resources for project development is an important issue for both present and future generations. Decisions taken during project appraisal are of paramount importance if the balance of our social fabric is to be maintained.

It is necessary therefore to determine whether a particular course of action is a net cost or net benefit to society. The most popular technique used for this purpose is *cost–benefit analysis* (CBA). It is a powerful tool that assists decision makers in the process of project selection, taking into account the efficient allocation of resources and the conservation of the environment. The ultimate aim is to maintain environmental quality without giving up further economic development. However, it is not without its problems or its critics.

10.2 Principles of CBA

Maximizing social and economic welfare are the ultimate goals of government, and social and economic welfare concern the efficient allocation of scarce resources. The concept of economic welfare is founded on Pareto's sense of welfare improvement, which states that welfare is improved when at least one person is made better off without anyone being made worse off. Society maximizes welfare by using scarce resources in the most efficient manner possible.

This welfare principle only identifies potential improvement as the outcome may be affected by factors such as equity and social distribution. In practice, no project can properly satisfy the Pareto improvement principle. Consequently, the Kaldor–Hicks principle of potential compensation was developed as a modification to the Pareto argument. Under this principle, economic welfare can be increased provided the gainers compensate the losers, regardless of whether compensation is actually paid. This principle implies that an alternative that exhibits the greatest net benefit will contribute positively to society overall and should be given the opportunity to proceed. This concept underlies the cost–benefit analysis technique (Sinden and Thampapillai, 1995).

Efficient allocation of scarce resources involves choosing between alternative projects and such choice requires the evaluation of options. The evaluation process is so pervasive in economics that project appraisal has become synonymous with cost–benefit analysis (van Pelt, 1993). Gilpin (1995, p. 169) defines CBA as 'the identification and evaluation of all costs and all benefits attributable to a policy, plan, program, or project'. Faced with conflicting objectives and limited resources, it is essential that governments have a basic mechanism for determining whether a given use of resources will improve community welfare and, where alternatives are competing for resource allocation, which alternative will provide the greatest improvement. For private developments, investment return is the important criterion; however, the decision support techniques are similar.

CBA captures the trade-off between the real benefits to society from a given alternative and the real resources that society must give up to obtain the benefits, and uses money as the universal metric. It is a tool that assists decision makers to organize information in a systematic manner so as to choose the best option among competing alternatives. By weighing up the costs and benefits of each alternative in monetary terms and ranking alternatives on the criterion of economic worth, society is able to identify the best allocation of scarce resources and therefore maximize social benefit.

The option with the highest net benefits will normally be selected, whereas other options will be more lowly ranked, and those showing negative benefits will be abandoned altogether. Since social welfare is measured by maximization of net benefits, the technique can overlook issues concerning the fair distribution of benefits. In other words, while more people win than lose, the losers may be made a great deal worse off and this will not matter so much as long as they are in the minority. Compensation is theoretically provided from the winners, but this does not usually happen unless a particular compensation plan is built into the successful proposal. Furthermore, where the environment is the loser then it may be argued that the technique is flawed, for all it has done is to undervalue social costs.

CBA came into practice in the United States as early as 1808 when Albert Gallatin, US Secretary of the Treasury, recommended the comparison of costs and benefits associated with water-related projects. This idea was adopted by the French engineer, Jules Dupuit, in the formulation of cost–benefit analysis described in the 1844 publication 'On the Measurement of the Utility of Public Works'. Since then CBA has become established as the most popular technique for evaluating public projects (Hanley and Spash, 1993) and is now the primary technique used in the field of environmental economics.

10.3 Types of CBA

10.3.1 Economic CBA

The technique of CBA applies to the evaluation of options by the identification of the expected expenditure (costs) and the expected income (benefit) over a specified period of time. Where the analysis involves more than a couple of years, CBA uses a discounted cash flow approach to bring costs and benefits into an equivalent monetary value so that the overall net benefit of the options can be calculated. A single decision criterion is usually applied, known as *net present value* (NPV), and the decision rule is that NPV must be positive and greater than the NPV of any rival option, including the implied opportunity of investing elsewhere.

Economic CBA, also known as financial CBA or financial analysis, is used when evaluating the costs and benefits of a number of projects (not necessarily construction projects) from the perspective of a providing authority or investor. Costs comprise acquisition and recurrent expenditure, while benefits comprise operational income and reversion of the asset at the end of the study period. A simple example of a cost–benefit analysis cash flow for a particular option is shown in Table 10.1.

The preferred process for calculation of equivalent value (discounted net benefit) is to use the following formula:

$$\text{Equivalent value} = \frac{\text{net benefit}}{(1 + \text{discount rate})^{\text{year}}}$$

where net benefit equals benefit minus cost, and discount rate is a real rate of return per annum (also known as the opportunity cost or weighted cost of capital) expressed as a decimal (i.e. 5% = 0.05).

NPV is calculated as the sum of discounted net benefits over the study period. Any positive figure indicates the project is acceptable, but the ratio of discounted benefits to discounted costs, known as the *benefit–cost ratio* (BCR), will indicate the significance of the result. A BCR of one equals an NPV of zero and is the breakeven point. A BCR of two indicates a 2:1 ratio of benefits over costs, which is attractive. Since this is a comparison technique, the higher the NPV (or BCR) the better provided that NPV > 0 or BCR > 1. Year 0 refers to immediate transactions, such as land purchase. In the example (Table 10.1), Year 1 allowed for construction, Years 5, 10, 15 and 20 had increased maintenance costs, and Year 20 had a theoretical sale of the asset equal to $18 000 000.

The decision to proceed with the project with the highest NPV is left until an assessment of the risk has been completed. All evaluations of this type are uncertain,

Table 10.1 Example of a cost–benefit analysis cash flow

Year	Cost $	Benefit $	Net Benefit $	Discounted Net Benefit @ 5% $
0	5 000 000		(5 000 000)	(5 000 000)
1	10 000 000		(10 000 000)	(9 523 810)
2	500 000	2 000 000	1 500 000	1 360 544
3	500 000	2 000 000	1 500 000	1 295 756
4	500 000	2 000 000	1 500 000	1 234 054
5	1 000 000	2 000 000	1 000 000	783 526
6	500 000	2 000 000	1 500 000	1 119 323
7	500 000	2 000 000	1 500 000	1 066 022
8	500 000	2 000 000	1 500 000	1 015 259
9	500 000	2 000 000	1 500 000	966 913
10	4 000 000	2 000 000	(2 000 000)	(1 227 827)
11	500 000	2 000 000	1 500 000	877 019
12	500 000	2 000 000	1 500 000	835 256
13	500 000	2 000 000	1 500 000	795 482
14	500 000	2 000 000	1 500 000	757 602
15	1 000 000	2 000 000	1 000 000	481 017
16	500 000	2 000 000	1 500 000	687 167
17	500 000	2 000 000	1 500 000	654 445
18	500 000	2 000 000	1 500 000	623 281
19	500 000	2 000 000	1 500 000	593 601
20	4 000 000	20 000 000	16 000 000	6 030 232
			NPV =	5 424 864

but the sensitivity of the result to the key variables involved in the calculation is a necessary consideration. The interpretation of the risk assessment will be different depending on whether the investor is risk-seeking, risk-neutral or risk-averse.

10.3.2 Social CBA

Social CBA, also known as economic appraisal or extended CBA, follows exactly the same methodology except for one important difference. The costs and benefits are no longer those that pertain to the providing authority or investor, but rather relate to society generally. This variation in approach leads to the common inclusion of *externalities*, defined as costs and benefits that are beyond the operation and interest of the providing authority or investor. As a direct consequence, this technique is usually applied to public sector projects and, because the assessment of externalities can be quite a difficult process, it is usually applied to large projects where the time involved in evaluation is proportionate to the value of the project.

Social CBA is therefore community-centred rather than investor-centred, and aims to identify those projects that provide the greatest social benefit. Externalities include issues like environmental impact, accident reduction, time savings and employment, and are assessed using shadow pricing mechanisms based on surrogate market valuations. Where externalities defy monetary assessment, they are known as *intangibles* and are considered separate to the financial calculations.

Public sector projects aim to be progressive in that they favour the poor and disadvantaged over the rich and privileged. Where projects have the opposite effect they are termed regressive. Equity issues are considered through separate distributional analysis, and imbalances may be overcome through the employment of direct compensation as an additional cost.

10.4 The CBA model

Although different investors and agencies have their own procedures for the appraisal of projects, a typical cost–benefit approach would normally have the following framework (Fig. 10.1).

CBA is a method of quantitative analysis, which transforms data into manageable information for direct and easy comparison. It starts with the determination of the project scope and objectives. This is an important step as the compilation of clear and precise objectives at the outset significantly enhances the likelihood of identifying innovative solutions. The identification of project constraints is also important as it ensures that the selected alternatives fit within the permissible boundaries.

Project costs and benefits are identified descriptively and are often presented in a balance sheet format. Once all relevant costs and benefits have been determined, they are quantified in monetary terms and allocated to their respective years of the cash flow. The net benefits over the life span of a project are converted to discounted values to enable proper comparison. Risk exposure is often determined by running

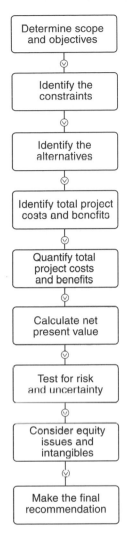

Fig. 10.1 Cost–benefit analysis framework (Department of Finance, 1997).

scenarios on the cash flow for various reasonable assumptions (i.e. best and worst case). Before making a recommendation, judgement issues such as equity distribution and intangibles are noted in the report as a guide to the decision maker.

10.5 Methodological approach

10.5.1 Project planning

Most projects, policies, programmes and activities have objectives that justify their claim for implementation. The first important step in CBA is, therefore, to define the project scope and objectives, which should be clear and unambiguous. Project scope is usually related to the initiator's needs and requirements, and may involve solving a specific problem, exploiting an opportunity or obtaining a benefit. Project implementation is intended to narrow the gap between an existing situation and a more desirable situation. The project scope and objectives are the foundations which support and give rise to the implementation of a development activity. Further alternatives may also be determined and short-listed to guarantee that the best solution is pursued. Sinden and Thampapillai (1995) define such alternatives as specific ways of using particular factors of production to provide various goods and services. They are particular combinations of physical and technical inputs, which can be used to satisfy the different requirements of projects of varying design, location and size.

Project constraints have to be identified simultaneously. It is important to understand whether the project will be technically and commercially fit for the required purpose. Internal and external restrictions may directly affect the success of a project. Internal constraints relate to the scope of the project in the form of technology, resources, effect of other projects, corporate policies and strategies, administrative aspects, financial objectives and marketing, while external constraints include legal and regulatory concerns, as well as environmental impacts. Defining project constraints can help to set the appropriate boundary of the analysis. A project that appears financially viable may be subject to environmental restrictions that render it unsupportable at the stage of implementation.

In a social CBA, the populations of gainers and losers must also be considered. It is important to determine the size of the population whose quality of life will be significantly affected by development of the project. Demographics may be confined to the immediate vicinity of the project or may be at a regional or national level. The proportions of gainers and losers are also significant since a project may highly benefit a small group of people while imposing a small cost on a large group of people. During project planning, compensation or remedial measures may have to be determined and costed to minimize the impact felt by the project losers living in the immediate vicinity.

10.5.2 Project costs and benefits

Projects generate impacts when implemented and these impacts need to be carefully identified and evaluated before a decision to proceed is made. By examining the project objectives thoroughly, costs and benefits may be identified and feasible alternative choices of development selected that satisfy stated goals. The total project costs and benefits of each alternative require evaluation and subsequent comparison so that the analysis reflects the true and comparable outcomes of each alternative. Project selection can then be made on the basis of that evaluation.

Costs and benefits (including externalities in the case of social CBA) are quantified in monetary terms wherever possible and are processed in a discounted cash flow to determine NPV. The accuracy of their quantification depends on the significance of the project, data availability, cost of obtaining missing data and the clarity of project objectives. At this stage all calculations are performed under varying levels of uncertainty. Therefore, it may be possible to attach probabilities to uncertain events and calculate an expected value (Hanley and Spash, 1993).

Project costs comprise the total cost of resources consumed by a project over its expected life. They can be broadly divided into development and operation costs. The development costs refer to the expenditure for the construction of a project and include land and other property costs, infrastructure expenditure, plant and equipment, energy, materials and labour. Operation costs begin once the project is finished on-site and run to the end of its life span. Operation costs include the energy consumed during the operational period, routine maintenance and repairs, major repairs and regular cleaning. Selling costs should be included in the final year.

Indirect costs, such as theoretical environmental damage, should also be included. Environmental damage includes the negative impact of a project during its construction and over its operational life, such as air/noise pollution, stormwater run-off, deforestation and the like.

Project benefits comprise the total benefit related to the positive impacts associated with project implementation. Direct benefits may be revenue produced from the project (as distinct from income earned by selling goods and/or services produced by the project) or in the form of periodic revenue received by renting the property in the marketplace. It may also include increased productivity or employment opportunities, which can be generated by project implementation. Indirect benefits may be a better living environment, improved leisure facilities or better traffic arrangements.

The measurement of indirect costs and benefits in monetary terms is rarely complete, and so may end up being undervalued in the final decision or completely ignored. The complex nature of the environment and the non-market characteristics of environmental goods and services are the major hindrance in this respect. The unpriced impacts (or intangibles) are still an important ingredient in the final decision, and should be included descriptively in a CBA report so that decision makers can determine the weight to be given to these issues in their policy determination.

10.5.3 Discounting cost and benefit flows

Once all the relevant priced costs and benefits are identified and distributed in the cash flow, it is necessary to convert them into equivalent present value. Discounting is the mechanism used to achieve this conversion. Costs and benefits in future years are disadvantaged against early years to account for the opportunity cost of money related to either borrowing of capital or use of equity reserves. Costs and benefits in later years have a lower equivalent value than those arising in early years, and this disparity conforms to a negative exponential curve.

Once the net benefits for each alternative have been discounted for each year, the sum (NPV) can be used for comparison. Alternatives that show discounted benefits exceeding discounted costs (or in other words a positive NPV) are considered to represent an efficient shift in resource allocation. Projects with positive NPV will be accepted for consideration whilst projects with negative NPV will be rejected, and the project with the greatest NPV will be the most preferred. The choice of discount rate is important but not necessarily significant to the ranking of project alternatives.

Other criteria can also be used as supplementary tools in the consideration of project alternatives. One common criterion is *internal rate of return* (IRR). The IRR is the discount rate which would produce an NPV equal to zero. If the IRR is greater than the chosen discount rate, the project is considered profitable. Conversely if the IRR is smaller than the chosen discount rate, the project is unprofitable and should be rejected. This method is useful for people accustomed to 'rate of return' concepts but may rank projects differently to the NPV approach. It favours projects with low capital requirements and high returns in early years but is biased against projects with high returns in the long run. NPV is the primary decision criterion, but the difference between the chosen discount rate and the IRR can be used to assess the level of profit and risk contingency relative to the breakeven point.

10.5.4 Testing for risk and uncertainty

The available data needed to make the necessary forecasts of project costs and benefits often involve uncertainty or are incomplete because some elements are difficult to predict and suitable data may be unavailable at certain stages. The sensitivity of particular key variables affects the relative desirability of the alternatives. The outcome may change the accept–reject decision, or the ranking of alternatives. For instance, it is difficult to forecast when major repairs will be required for building projects. Therefore, the existence of this uncertainty often jeopardizes the reliability of a CBA outcome.

Sensitivity analysis is often used to test the robustness of results under different scenarios. It helps to analyse the economic structure of a project in such a way as to identify those variables that have more or less influence upon economic desirability. This sort of test gauges the effect which changes in assumptions have on the ranking and comparison of alternatives. It helps to derive a range of values within which an alternative is economically desirable and the certainty levels that can be expected, and requires considerable judgement and experience. Key variables typically include discount rate, project life span, physical quantities and quality of inputs and outputs, and investment and operation costs (OECD, 1994). Initially, alternatives are ranked on the basis of the original NPV; if the recalculated NPV does not alter the accept–reject decision or the ranking of alternatives, then the original alternative is insensitive to the changes and therefore has a low risk level. On the other hand, if project viability or the ranking of alternatives is affected, then the risk attached to the project selection is much higher.

There are more sophisticated methods of risk analysis that can be applied to NPV calculations, but the ultimate purpose is to assess the likelihood of choosing a project that might go bad. High returns are often associated with risky projects, but this situation is often accepted by investors who are risk-seeking. In the case of social projects, however, governments are often risk-neutral or risk-averse and so would be looking for projects that reflect stability and a reasonable level of confidence that they are in the public good.

10.6 Limitations of CBA

It is the ultimate goal of CBA to assist those evaluating proposed projects to choose the best option among alternatives and to promote resource efficiency and social welfare. The methodology of CBA, however, exhibits some conceptual and practical difficulties, which means that its effectiveness as a decision tool may be questioned.

The general approach of CBA is to value all project costs and benefits in monetary terms based on market prices. This limits the scope of any analysis to consideration of only those factors that are the subject of market transactions. Other factors that may be relevant to a social CBA include all public goods in the global commons and all environmental impacts. Most often these costs (or benefits) are neither recorded nor incorporated in the project cash flow. The Department of Finance (1997, p. 82) states that 'the use of the money yardstick for measuring costs and benefits lends a false accuracy to the result of a cost–benefit analysis'. This means that social CBA results may not reflect the true benefits of a project if intangible values are present. Double counting of benefits is also a common pitfall.

Likewise the concepts of welfare and the compensation principle suffer from the total ignorance of equity issues that contrast with the goal of CBA to promote social welfare. Under these principles no concern is given to who gains and who loses from a project; as long as the total benefits outweigh the total costs, the option with the greatest positive net benefit wins. A project that generates net benefits to the rich has an opportunity equal to that of a project that generates net benefits to the poor. However, a project that yields benefits to the poor has a desirable distributional impact. The equity issue is generally not a criterion for consideration within the CBA framework and largely weakens the effectiveness of CBA when applied in social contexts.

The outcomes of CBA may be manipulated to suit private purposes and may be less useful where political decisions dominate. These outcomes may be influenced in various ways, e.g. by the adoption of particular values, shadow prices or discount rates in order to produce a pre-determined result. This is actually in direct conflict with the welfare objective of CBA as it leads to support of less desirable alternatives rather than promoting options that are more resource efficient and socially desirable.

Market failure is another limitation of CBA. One of the fundamentals of CBA is the assessment of project costs and benefits at local market prices and, as a result, the outcome of CBA relies heavily on prevailing market valuations. However, markets may be distorted by a number of factors, such as government intervention, interest rates and foreign exchange rates. In such cases the use of market prices will not reflect true project costs and benefits and thus may lead to inefficient allocation of resources.

The use of discounted cash flow analysis as part of the methodology is the cause for some disquiet. Although the discounting philosophy is conceptually correct, its application to social and environmental issues is arguable and may lead to an undervaluing of these costs and benefits. The choice of discount rate is slightly controversial, particularly where high rates are chosen that rapidly disadvantage future cash flows and quickly make them irrelevant to the decision.

While CBA is an important tool in decision making and its systematic arrangement of information enhances the decision-making process, the limitations of CBA are serious and cannot be neglected. The CBA framework can be supplemented, however, by the adoption of other techniques, such as effectiveness analysis and multiple criteria decision making, which may help to produce a better evaluation tool.

A modern variation of traditional CBA is *cost effectiveness analysis* (CEA). Useful for public sector projects, this technique assumes a common level of social benefit and therefore focuses on measuring the costs. As the benefits are equal they cancel out and can be essentially ignored. The difficulty in estimating externalities is also largely obviated by this approach.

10.7 Multiple criteria decision making

In daily life people are continually faced with decisions. The process of decision making frequently involves identifying, comparing and ranking alternatives based on a number of criteria. People often go through this process without even thinking much about it. However, for big decisions, where a lot of money is involved, there is a tendency to simplify the objectives of the 'project' into a single decision criterion. Everything is converted into dollars, at least where possible, and the decision is based on the alternative with the highest monetary value. This is the essence of traditional CBA.

Project appraisal techniques are often employed by decision makers to structure the complex array of data relevant to a project into a manageable form and to provide an objective and consistent basis for choosing the best solution for a given situation. In CBA much effort has been put into assessing the input costs and output benefits by means of a market approach. With the increasing awareness of possible negative external effects and the importance of distributional issues in economic development, the usefulness of CBA in this respect is increasingly controversial. Consequently, in the past decade a good deal of attention has been paid to multi-dimensional evaluation approaches (Nijkamp *et al.*, 1990). One such approach is known as *multiple criteria decision making* (MCDM).

The identification of value for money on construction projects is clearly related to monetary return. But other issues are also relevant, particularly for social infrastructure projects, and some are becoming increasingly significant. For example, issues such as welfare enhancement and resource efficiency are vital to the assessment of environmental impact in the wider social context. Since no single criterion can adequately address all the issues involved in complex decisions of this type, a multi-criteria approach to decision making offers considerable advantages.

Traditional CBA uses price as the main tool to evaluate projects based on market transactions. However, over the past decade criticisms of CBA have been many, and relate mainly to attempts at putting the underlying welfare economic theory into practice. It is often difficult or even impossible to improve social welfare in a society if the natural environment continues to be abused and depleted. Indeed, within the CBA framework environmental assets are often ignored or underestimated as there are often considerable difficulties in measuring all relevant impacts of a project in monetary units (Abelson, 1996).

Ecologically sustainable development (ESD) is now a constant focus for the mass media and a matter for widespread public concern. Consequently, intangibles and externalities have become major issues in project development. The presence of externalities, risks and spillovers generated by project development often preclude the meaningful and adequate use of a market-based methodology. When the analysis turns to assessment of environmental quality or loss of biodiversity, it is rarely possible to find a single variable whose direct measurement will provide a valid indicator of the severity of these effects. Although many efforts have been made to arrive at values for intangibles and externalities, it is almost impossible in practice to place anything more sophisticated than arbitrary numerical values on such effects. The requirement for incorporating environmental issues into the project appraisal process is becoming increasingly apparent, and as it does, the application of market prices to these factors becomes more and more questionable.

Various alternative methods have been researched and suggested to replace CBA completely. Such techniques identify environmental effects but do not require that these effects be monetarized, since they are difficult or even impossible to measure in this manner. Cost-effectiveness analysis and environmental impact assessment are prominent examples. Others have suggested supplementing CBA with a technique that can measure environmental costs in other than monetary terms (Nijkamp et al., 1990; van Pelt, 1993; Hanley, 1992; Abelson, 1996). MCDM is now widely accepted as an aid to decision choice when dealing with environmentally sensitive projects.

MCDM is a technique designed to value two or more criteria and it is particularly useful for those environmental impacts that cannot be easily quantified in terms of normal market transactions. MCDM transfers the focus from measuring criteria with prices to applying weights and scores to those impacts and so determines a preferred outcome. Total scores are used to rank project alternatives and thus arrive at a balanced decision. MCDM is a more flexible methodological approach as it can deal with quantitative, qualitative or mixed data, whilst CBA is limited to quantitative data, and it does not impose any limitation on the number and nature of criteria. Therefore, MCDM is a more realistic methodology in dealing with the increasingly complex nature of building development. The debate on conventional versus modern evaluation analysis tends to regard CBA and MCDM as complementary tools rather than as competitive tools (Nijkamp et al., 1990; van Pelt, 1993).

10.8 Enhancing traditional CBA

A multi-criteria approach to project appraisal has advantages. While considering conventional monetary issues it can also include social and environmental issues using non-monetary measures and still arrive at an objective conclusion. Traditional CBA can be enhanced to embrace the wider issues of sustainability through the creation of a comparative index made up of key project attributes.

Ding (1998) identifies these attributes as benefit–cost ratio (BCR), total energy (TE), social benefit (SB) and environmental risk (ER). Both benefit–cost ratio and total energy are concerned with the resource input in project development. This is particularly important today as the supply of natural resources is under serious threat. BCR reflects the effective allocation of scarce resources by the measurement of total project costs and benefits discounted by time. This is the ratio of the discounted value of benefits to the discounted value of costs, and includes only those cash flows that are normally part of an economic CBA. The greater the ratio the more profitable the proposal. Total energy consumption includes both embodied energy and operating energy consumption over the project life span. Energy is measured in physical units, e.g. GJ per m^2. The lower the total energy consumption the better.

The other two attributes focus on welfare maximization. It is difficult and sometimes impossible to deal with environmental goods in terms of economic transactions. *Social benefit* refers to the positive contribution of a project in terms of improving living standards, such as time savings and accident reduction arising over the operational life of a project. These non-market goods should be valued beyond an economic framework and a weighted criteria approach can be used to assess social issues across alternatives. High scores indicate that social benefit is significant. Environmental risk focuses on the judgement of long-term impact on the environment. Often, during the project evaluation stage, the information available to make the necessary forecasts of project costs and benefits involves uncertainty and is incomplete because some elements are difficult to predict at the outset. In addition, the complexity of ecosystems enhances the degree of uncertainty. Environmental risk is the percentage probability expressed to reflect the level of environmental uncertainty. The lower the risk level the better.

Such an approach leads to the development of a sustainability index. The attributes are each expressed in different units that are best suited to their quantitative assessment. When combined together they can indicate the relative sustainability of competing investment options. Attributes to be maximized are divided by attributes to be minimized, so the higher the index the more sustainable a project is by comparison with its alternatives. The following formulae may be used:

$$\text{welfare maximization } (WM) = \frac{SB}{ER}$$

$$\text{resource efficiency } (RE) \quad = \frac{TE}{BCR}$$

Therefore

$$\text{sustainability index } (SI) = \frac{WM}{RE}$$

$$= \frac{SB \times BCR}{ER \times TE}$$

where SB = social benefit (value score)

BCR = benefit–cost ratio (ratio:1)

ER = environmental risk (% probability)

TE = total energy (GJ/m^2)

Conventional project appraisal techniques measure net social gain to determine project performance whilst the sustainability index measures the relative ranking of projects from a sustainable development viewpoint. The index is not a measure of net social gain, and so there is still a role for traditional CBA, but often projects are necessary for other reasons and the task of the evaluation process is one of available choice.

10.9 Conclusions

Cost–benefit analysis is a systematic and consistent method of project appraisal widely used by developers, investors, governments and international funding agencies. All project developments, policies and programmes will have different approaches or proposals in order to achieve the same objectives. Projects need to be properly evaluated before a decision is made to proceed. The approach used in project appraisal, therefore, becomes important in choosing the best option among alternatives.

Economic and social CBA are tools used to assist decision makers to compare alternatives by applying economic theory to help make choices. The main theme of CBA is to monetarize and weight the total flow of costs of proceeding with a project against the total flow of benefits obtained from it, relative to other options. Alternatives with a net positive benefit are acceptable whereas other alternatives showing negative outcomes should be abandoned. The higher the NPV the better, given a reasonable BCR and an acceptable level of profit and risk contingency.

But the technique is not without problems and, for public projects where externalities and intangibles are common, the calculated outcomes may be highly questionable. A considerable advantage lies in the rigour of the technique itself and the ability to evaluate different scenarios using a range of variables that are significant to the analysis. In a sense the greatest benefit of CBA is its ability to allow for social and environmental issues objectively, and yet this is also its greatest weakness.

Building development involves complex decisions and the increased significance of external effects has further complicated the situation. Society is not just concerned with economic growth and development but is also conscious of the long-term impacts on living standards for both present and future generations. Certainly, sustainable development is an important issue in project decisions. The engagement of a conventional single-dimensional evaluation technique such as CBA in assisting decision making is no longer relevant and a much more complicated model needs to be developed to handle multi-dimensional arrays of data. The development of a sustainability index is a way to address multiple criteria in relation to project decision making. The use of a sustainability index will greatly enhance the assessment of external effects generated by construction activity, realize sustainable development goals and thereby make a positive contribution to the identification of optimum design solutions.

References and bibliography

Abarchar, A. (1984) *Project Decision Making in the Public Sector* (Lexington).

Abelson, P.W. (1996) *Project Appraisal and Valuation of the Environment: General Principles and Six Case-Studies in Developing Countries* (Macmillan).

Burns, M.L., Patterson, D. and LaFrance, L. (1993) Cost-benefit analysis: an application in two elementary schools. *International Journal of Educational Management*, **7** (6), 18–27.

Dasgupta, A.K. and Pearce, D.W. (1972) *Cost Benefit Analysis: Theory & Practice* (Macmillan).

Department of Finance (1997) *Handbook of Cost-Benefit Analysis* (Canberra: Australian Government Publishing Service).

Ding, G.K.C. (1998) The influence of MCDM in the assessment of sustainability in construction. In: *Proceedings of 14th International Conference on Multiple Criteria Decision Making*, Charlottesville, June.

Dmytrenko, A.L. (1997) Cost benefit analysis. *Records Management Quarterly*, **31** (1), 16–20.

Doeleman, J.A. (1985) Historical perspective and environmental cost-benefit analysis. *Futures*, **17**, 149–63.

Gilpin, A. (1995) *Environmental Impact Assessment (EIA): Cutting Edge for the Twenty-first Century* (Cambridge University Press).

Hanley, N. (1992) Are there environmental limits to cost benefit analysis? *Environmental and Resource Economics*, **2**, 33–59.

Hanley, N. and Spash, C.L. (1993) *Cost Benefit Analysis & the Environment* (Edward Elgar).

Harlow, K.C. and Windsor, D. (1988) Integration of cost-benefit and financial analysis in project evaluation. *Public Administration Review*, **48** (5), 918–28.

Hueting, R. (1991) The use of the discount rate in a cost-benefit analysis for different uses of a humid tropical forest area. *Journal of Ecological Economics*, **3** (1), 43–57.

Johnnson, P.O. (1993) *Cost Benefit Analysis of Environmental Change* (Cambridge University Press).

Joubert, A.R., Leiman, A., de Klerk, H.M., Katau, S. and Aggenbach, J.C. (1997) Fynbos (fine ash) vegetation and the supply of water: a comparison of multi-criteria decision analysis and cost-benefit analysis. *Ecological Economics*, **22** (2), 123–40.

Kirby, J., O'Keefe, P. and Timberlake, L. (1995) *The Earthscan Reader in Sustainable Development* (Earthscan Publications).

Langston, C. and Ding, G. (1997) Sustainable development. In Langston, C. (ed.) *Sustainable Practices: ESD and the Construction Industry* (Envirobook), 21–8.

Martin, F. (1993) Sustainability, the discount rate, and intergenerational effects within a regional framework. *Annals of Regional Science,* **28**, 107–23.

Mishan, E.J. (1972) *Elements of Cost-Benefit Analysis* (George Allen and Unwin Ltd).

Mustafa, H. (1994) Conflict of multiple interest in cost-benefit analysis. *International Journal of Public Sector Management*, **7** (3), 16–26.

Nijkamp, P., Rictvcld, P. and Voogd, H. (1990) *Multicriteria Evaluation in Physical Planning* (North-Holland).

NSW Treasury Technical Paper (1990) *NSW Government Guidelines for Economic Appraisal* (Australian Government Publishing Service).

OECD (1994) *Project and Policy Appraisal: Integrating Economics and Environment* (Organization for Economic Co-operation and Development).

Perkins, F. (1994) *Practical Cost Benefit Analysis* (Macmillan).

Sagoff, M. (1988) *The Economy of the Earth* (Cambridge University Press).

Schofield, J.A. (1989) *Cost Benefit Analysis in Urban & Regional Planning* (Allen & Unwin).

Sinden, J.A. and Thampapillai, D.J. (1995) *Introduction to Benefit-Cost Analysis* (Longman).

Spence, R. and Mulligan, H. (1995) Sustainable development and the construction industry. *Habitat International*, **19** (3), 279–92.

Starr, M.K. and Zeleny, M. (1977) *Multiple Criteria Decision Making* (North-Holland).

Tabucanon, M.T. (1988) *Multiple Criteria Decision Making* (Elsevier).

van Pelt, M.J.F. (1993) *Ecological Sustainability and Project Appraisal* (Avebury).

World Commission on Environment and Development (1987) *Our Common Future* (Oxford University Press).

Zeleny, M. (1982) *Multiple Criteria Decision Making* (McGraw-Hill).

11

Functional use analysis

Peter Smith†

Editorial comment

The ultimate success of any building may be judged according to how well it satisfies the functional requirements of the owner or occupant. Those requirements may be as simple as the provision of secure storage space, with protection from climate, intruders and pests, or may be as complex as the myriad interdependent functions of a hospital. Questions of security, circulation, movement of materials, internal and external communications, and indoor environmental control must all be addressed as the relationships between spaces, functions and activities are considered and balanced during the design process.

The physical nature of a building will be determined to some degree by the function(s) which it is expected to fulfil. The 18th-century architect Horatio Greenough was the first to suggest that 'form follows function', although later architects, notably Louis Sullivan, have used similar phrases to describe the relationship between the purpose and physical form of buildings. It may not always be the function that is accommodated within the building, however, which is the determinant of form. A good example of this is the Sydney Opera House: the 'sails' of the unique roof, while performing the normal functions expected of a roof (watertightness, security and so on), have an equally significant function as they help to make the building a symbol of the city and its harbour, and an internationally recognized icon. This sort of functionality, however, is apparent in many buildings of less significance than Sydney Opera House. Functionality may be manifested in an opulent hotel lobby, or a grand entrance to a corporate headquarters building; in both cases an important consideration for the client may be the physical display of prestige to potential clients.

Any determination of the value of a finished building must give appropriate weight to all the functional requirements that the client expects the designers to satisfy. Consequently, any attempt to maximize the value of a building requires a systematic analysis of all of those requirements, and the interdependencies and conflicts which may exist between them. Such analysis provides a sound basis upon which the design process can proceed. This chapter examines the links between functionality and value in building, and introduces some basic techniques for a structured approach to functional analysis that can be undertaken before design

† University of Technology, Sydney, Australia

commences, with the aim of producing buildings that better satisfy the needs of those who pay for them.

11.1 Introduction

Functional use analysis (FUA) during the pre-design stage of a project involves an analysis of client requirements and the functional requirements of the built facility for the purposes of setting parameters for both the building design and performance. Accordingly, this analysis goes to the very core of good design practice. It provides the springboard for the design process.

Despite its fundamental nature, the architect has traditionally carried out this analysis alone with little input from other consultants. It has become increasingly evident that FUA, in terms of optimizing value for money, has not been generally effective due to misconceptions about what a client really wants and which building functions add value. The traditional lack of a team approach to this important area is a further reason. The inability of clients, and their advisers, to establish exactly what their real objectives are and then to reflect that in their designs compounds all this.

For most commercial and public sector building owners and users, the building is really a means to an end for their business purposes. These business purposes usually equate to improving business performance and the 'bottom line'; profit levels. For these building types, the building needs to be recognized as a 'dynamic business facility' rather than a dormant product, with the design process focusing on optimizing business operations within the facility. Corporations and governments around the globe are becoming less concerned with the actual construction process and building; they essentially need 'space' that will optimize and add value to their core operations.

FUA focused on producing designs that reduce operational costs and improve worker, equipment and business performance can provide the catalyst for value optimization for both owners and users. The real value of a facility lies in its ability to enhance the satisfaction and performance of the users and, where applicable, financial returns. For non-income producing buildings, a focus on improving user performance and/or satisfaction with the building will also optimize value.

After all, a building costs money but only its function has value.

11.2 The functional use analysis process

FUA is now increasingly incorporated as an integral part of the value management (VM) process during design and construction, with an emphasis placed on the relationship between function, cost and value. Whilst always an integral part of the design process, FUA has been developed as a formal process –work credited to Lawrence D. Miles. He founded the technique of *value analysis* in the 1940s and incorporated FUA as a vital element of this technique. The process had its early application in the manufacturing sector and it was not until later that it began to be applied in the construction industry.

To be most effective, FUA should be applied from the earliest conceptual stages of the project. It is at these stages that the major decisions for a project are made and these ultimately have a dominant influence on the value of the project. The application of FUA from the outset will facilitate a more accurate feasibility study and the development of a more appropriate client brief which, together, will shape the future direction of the design and construction process. FUA should then be applied throughout the process as the design detail unfolds and construction proceeds. The purpose of this chapter is, however, to describe FUA during the pre-design stages.

FUA at the pre-design stage of a project will establish and define the overall functional requirements which, in turn, will provide the rudder to steer the project team through the design and construction process and optimize the functionality of the detailed components.

11.2.1 General principles

Applied from the outset, FUA will identify the key functional requirements of a project. For commercial and public sector projects, these requirements will largely be the provision of space designed to maximize user performance and, hence, business performance and, where applicable, monetary returns. Accordingly, this is where the pre-design and design focus should lie: on productivity and business performance enhancing design. However, this has traditionally not occurred. Design teams are often not fully aware of the full range of requirements that the client expects the building to fulfil; a lack of awareness which is usually a result of a failure by clients to pay design fees that allow in-depth analysis of their requirements. The above considerations are typically equated with the project feasibility analysis when, in reality, they should also underpin the design process. The actual building may well be only a means to an end.

Project functional definition needs to occur before any design/technical alternatives and solutions can be explored. Seeley (1996) subdivides functions into primary and secondary functions. Primary functions are those fundamental to the project without which the project would fail or be seriously adversely affected. Secondary functions relate to technical solutions for the primary functions. Norton (1992) contends that, for FUA purposes, most secondary functions have zero use value but if they are essential to the primary function they should be deemed essential secondary functions and thus allocated a value. For example, if the primary function for a building is to enhance worker and business productivity, a brick wall may provide a secondary function of providing enclosure but have no real value. However, if this wall was redesigned to incorporate more window area to reap the benefits of natural daylighting on worker productivity, then this design feature would add value.

Traditional FUA approaches allocate cost and worth to each function. The cost relates to construction costs whilst worth relates to the lowest possible cost for which the function can be performed. A general 'rule-of-thumb' has been that if the cost:worth ratio is 2 or less the component will likely be adopted for its cost reduction benefits (Seeley, 1996).

However, value also needs to be incorporated here. The cost/worth analysis focuses on minimizing initial costs but the real value will often lie in solutions that improve user performance, irrespective of whether they incur higher or lower initial capital costs.

11.2.2 Design team approach

As with value management, FUA should incorporate a multi-disciplinary design team approach. The modern-day designer needs to consider a vast number of design variables ranging from client and legislative requirements to environmental considerations and cannot be expected to have the breadth of knowledge and expertise that this requires. The design team must also be able to operate effectively and produce optimum results within the 'real world' time and cost constraints imposed by clients.

11.2.3 General approaches

There is no uniform approach adopted for FUA; applications for this technique vary widely amongst designers and design teams. It is not the intention of this chapter to give an in-depth analysis of these approaches but rather to give an overview of the main and most commonly used approaches and then focus attention on the areas that are lacking in terms of FUA.

As previously mentioned, the development of FUA began with Miles in the 1940s. He developed the method of expressing function as a verb and a direct object. This was further developed by Charles W. Bytheway in the 1960s when he introduced the idea of applying several questions to individual functions in order to identify the primary function. This led to the creation of the function analysis system technique (FAST) which provides a method for systemizing functions by applying a series of questions. This provided a catalyst for further development and FAST is still recognized as one of the main FUA techniques, and is described in further detail below. Around the same time, Arthur E. Mudge developed a function chart that identifies unnecessary functions, overlapping functions and functions that require high costs, to facilitate objective design evaluation in order to reduce costs without adversely affecting the essential role of the product (Akiyama, 1991).

In terms of the property industry, the Building Performance Research Unit at Strathclyde University (Ferry and Brandon, 1991) categorizes FUA during the design process in three stages.

1. Analysis (ascertaining what is required).
2. Synthesis (information from the analysis is used to explore solutions).
3. Appraisal (the solution is represented in some form and then measured and evaluated).

Kelly and Male (1993) subdivide FUA into four phases.

1. Task (the client has a problem or need, and needs to determine whether the construction of a new building or the refurbishment of an existing one is the best solution to this problem/need).
2. Spaces (if a new or refurbished building is the solution the design team determines the client's requirements including space requirements, prepares a brief and undertakes spatial analysis in the form of sketches).
3. Elements (the building assumes a structural form through the conceptual design of the main building elements).
4. Components (the elements become part of the built form and components are chosen to satisfy the requirements of the elements in terms of surrounding and servicing space).

Norton (1992), however, believes that FUA is more complex because primary functions may be considered at different levels. Seeley (1996) provides the example that, whilst a building's primary function may be for a developer to create a profit, another primary function on a lower level is to enclose space. This results in different functional hierarchies from which the function can be considered.

11.2.4 Establishing primary functions

The determination of the client's main functional requirements is not an easy task. Most clients will list 'value for money' as one of their principal requirements but whether that occurs or not will depend, in the first instance, on whether clients or their advisers are able to navigate through the maze and identify what they really want and then have that adequately defined in a brief. Many clients need careful and expert guidance in this area as inexperience and/or misconceptions may result in briefs that do not reflect what the client *really* needs.

Ranking primary functions in order of importance is the normal approach. These functions are, however, combined with detailed technical requirements that can stifle the process. Attention needs to be placed initially on the overall functional requirements of the building; these will then guide the functional use analyses of the detailed building components at a later stage.

11.2.5 FAST diagrams

FAST is a technique used to establish and organize hierarchical functions in order to achieve project objectives. It enables a problem to be broken down into manageable components and then analysed in a balanced manner. The relationship of item/component importance, value and cost can be readily determined and focus placed on value-adding items. The following section is adapted from Seeley (1996, pp. 284–95) with the examples provided by Davis Langdon Management. The FAST system incorporates a decision-tree style using a 'how–why?' approach. An example is given in Fig. 11.1.

The main procedures for compiling FAST diagrams include:

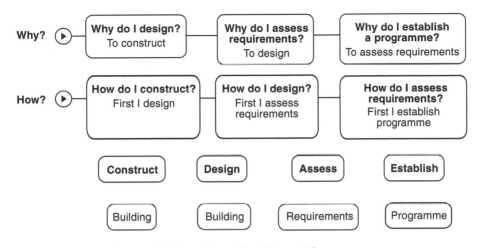

Fig. 11.1 FAST conceptual approach (adapted from Kelly, 1992, p. 284).

1. Identifying the key functions of the project.
2. Compiling a FAST diagram as shown in Fig. 11.1 by working from left to right. The analysis resembles a decision tree by answering questions 'Why?' when reading from right to left and 'How?' when reading from left to right.
3. Divide/subdivide functions/elements into components to an appropriate level of detail.
4. Number each item.

Figure 11.2 provides a conceptual example of this process.

11.2.6 Case study – bank processing centre, northern England

This case study, drawn from Davis Langdon Management, provides an example of the FAST technique. The primary requirements of the client for this project were for the design and construction of a high quality building that was efficient, user friendly and with the lowest possible operating costs. This was all required within an 18-month time frame. Figure 11.3 shows a FAST diagram for the main functional requirements of the project and their estimated costs. The most expensive items, providing accommodation and the provision of the internal environment, are the most likely to receive detailed examination.

Detailed FAST diagrams were then developed for each functional element. Figure 11.4 provides an example of the analysis for one of these elements, the 'provide acceptable working environment' function. This also shows how the FAST approach can be extended to incorporate evaluation of the impact of building design on worker productivity. The components are determined and priced. All the alternatives are considered with the emphasis on meeting the specified performance standards at lower cost, improving standards for the same cost and increasing value for money.

Carried out as an integral part of a formal value management process, the FAST

Function (1.1)	Element (1.1.1)	Component (1.1.1.1)
		Component (1.1.1.2)
	Element (1.1.2)	Component (1.1.2.1)
		Component (1.1.2.2)
	Element (1.1.3)	Component (1.1.3.1)
		Component (1.1.3.2)

PROCESS (1.0)

Function (1.2)	Element (1.2.1)	Component (1.2.1.1)
		Component (1.2.1.2)
	Element (1.2.2)	Component (1.2.2.1)
		Component (1.2.2.2)
	Element (1.2.3)	Component (1.2.3.1)
		Component (1.2.3.2)

Function (1.3)	Element (1.3.1)	Component (1.3.1.1)
		Component (1.3.1.2)
	Element (1.3.2)	Component (1.3.2.1)
		Component (1.3.2.2)
	Element (1.3.3)	Component (1.3.3.1)
		Component (1.3.3.2)

Fig. 11.2 FAST functional unit breakdown (adapted from Kelly, 1992, p. 285).

		Functions	£
	1.1	Prepare Site	905 000
	1.2	Provide Accommodation	1 776 000
	1.3	Provide Internal Environment	2 034 000
	1.4	Provide Acceptable Working Environment	640 000
	1.5	Provide Welfare Facilities	866 000
PROCESS (1.0)	1.6	Satisfy Safety Requirements	381 000
Rationalize Operations and	1.7	Satisfy Statutory Regulations	455 000
Achieve Operating Costs	1.8	Provide Flexibility	324 000
	1.9	Provide for Expansion	120 000
	1.10	Reduce Costs	216 000
	1.11	Enhance Quality/Prestige	120 000
	1.12	Secure Operations	470 000
	1.13	Provide External Environment	671 000
	1.14	Provide Temporary Facilities	350 000
		Total Cost	**£9 328 000**

Fig. 11.3 Determination of functions and costs (adapted from Kelly, 1992, pp. 290–1).

FUNCTIONAL ELEMENT (1.4)
Provide Acceptable Working Environment.

	Functional Elements	£			Functional Components	£
1.4.1	Provide External View	0				
1.4.2	Provide Internal Daylight	423 000		1.4.2.1	External Growth Curtain Walling	135 000
				1.4.2.2	Roof Glazing	288 000
1.4.3	Accommodate Disabled	39 000		1.4.3.1	Ramps	7 000
				1.4.3.2	Toilets	32 000
				1.4.3.3	Parking	see item 1.12
1.4.4	Additional Building Height	40 000				
1.4.5	Up lighting to Roof Space	26 000				
1.4.6	Upgrade Exposed M&E Services	17 000				
1.4.7	Acoustic Insulation to Roof	13 000				
1.4.8	Decorate Exposed M&E Services	14 000				
1.4.9	Tubular Roof Steel Work	68 000				
	Functional Cost	**£640 000**				

Fig. 11.4 FAST analysis for 'provision of acceptable working environment' functional element (adapted from Kelly, 1992, pp. 290–1).

analysis yielded cost savings of £296 100 on the original budget of £9 328 000, which represented an overall saving of over 3%. The cost of the study was £8000 which represented a cost–benefit ratio of 1:37.

11.2.7 Spatial analysis

Determining the amount of space required for a building and optimizing the spatial arrangements for the various functions of the building are a rudimentary part of the design process. For most buildings, space is the most influential determinant of end cost and, in all likelihood, value for money. Accordingly, an array of techniques is used by designers to optimize spatial requirements. The client and end-users should also be involved in carrying out such analyses.

Whilst it is common for the client to give parameters for the amount of space required (usually in the form of floor area or functional units) the design team needs firstly to determine the appropriateness of these parameters and then seek to arrive at optimal spatial solutions. The design team may be able to reduce space requirements (and thus building costs) by devising more efficient functional use of the various spaces. An example of this can be found in current trends in redesigning office work space where large individual offices are giving way to open plan collaborative team work spaces. The end result is that office floor space requirements are being reduced by up to 60% (Ferguson, 1996). Detailed inform-ation about end-user operations and requirements is a pre-requisite for this type of analysis but this can only be obtained through allocating time and funds for proper research by the design team. Unfortunately, the necessary time and funds are not readily available in the current climate of tight design fee competition and fast tracking of projects. Minimizing circulation space and unnecessary end-user travel within a building are usually basic design objectives.

Spatial analysis requires the design team to focus on end-user activities and operations. As mentioned, this has always been an important part of the design process; prior to designing a building the designer needs to know what the building will be used for and what the end users will actually be doing in the building. How well such analysis is done will depend on how well the designer understands what the end users will be doing. For example, with commercial projects, designers will ideally need an intimate knowledge of the organizational and work practices of the firms that will occupy the building. This enables the designer to understand end-user operations, workflow, circulation, proximity of functionally related spaces, security, maintenance, furniture and equipment layout, services, movement patterns and supervision and control and other operational requirements (White, 1986). Unfortunately, as pointed out earlier, this analysis normally does not receive the time and attention that it deserves largely due to time and fee constraints. It is here that the greatest impact on value for money and user satisfaction can be made.

The principal objectives of spatial analysis are to minimize the time it takes end-users to move between functionally related spaces and to eliminate unnecessary movement, not only of people, but also of materials and equipment. In commercial buildings, time spent moving between spaces is unproductive. Proper achievement of these objectives requires detailed analysis of end-user operations and activities, procedures and products of each activity, the relationship of the activities, analysis of the key steps in the operation, the conditions and resources needed to perform the activities, the duration and scheduling of each step and the characteristics and needs of the end-users (White, 1986).

11.2.8 Spatial analysis techniques

Three common spatial analysis techniques are the *matrix diagram*, the *bubble diagram* and the *zoning diagram*. Figure 11.5 illustrates the concept and the relationship between these diagrams.

The matrix diagram is a two-dimensional grid which examines and ranks the importance of the relationship of functional spaces. A building's functional spaces are listed on both the top and side of the matrix and importance indicators are used to rank the relationships between them. The client and end-users have an important role to play here.

The bubble diagram takes the matrix analysis further by transforming it into a more tangible graphical form. Each functional space is represented by a 'bubble' with lines connecting the spaces that need to be adjacent to each other. Varying line thicknesses or colours can be used to indicate the relative importance of the space relationships.

Zoning diagrams can be used to superimpose functional zones over the bubble diagrams. A number of simulations can be conducted to sort the functional spaces into functional zones. Sorting criteria may, for example, be based on grouping functional spaces into public zones, working zones and amenities zones.

These diagrams can then be transformed into initial sketch designs. Further simulation is required to optimize spatial relationships and then attention needs to be directed to the actual building form that will house these spaces.

Matrix Diagram

	Reception	Administration	Enquiries/Sales	Manager's Office	Meeting Room	Kitchen	Toilets	Display Room	Loading Dock
Reception		●	●						
Administration	●		●	●	●	○	○		
Enquiries/Sales		●				○	○		
Manager's Office		●			●	○	○		
Meeting Room		●		●		○	○		
Kitchen		○	○	○					
Toilets		○	○	○	○			○	
Display Room							○		●
Loading Dock								●	

● Critical
○ Desirable
(Blank) Not Critical

Bubble Diagram

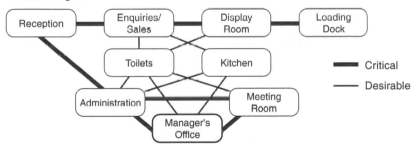

■■■ Critical
— Desirable

Zoning Diagram

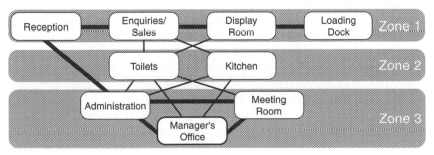

Fig. 11.5 Spatial analysis diagrams.

11.3 Functional considerations for value optimization

11.3.1 Where does real value lie?

Chapter 20 describes how the real value of commercial buildings will often depend on how the building can enhance business and end-user performance and thus increase profit levels. For these types of building users, the actual building is only a means to an end for their primary motivation: making more money. In terms of life costs, research findings are cited in Chapter 20 that show how actual user or worker salaries can account for 75–92% of total facility life costs and render other capital and operational costs almost insignificant. The work environment, which is created by buildings, has a big influence on worker productivity and further studies have shown that building design that can improve worker productivity by as little as 1% can offset a building's entire annual energy and maintenance costs. The Lockheed example in Chapter 20 shows how productivity enhancing design features did increase productivity levels and business performance to the point where the competitive edge was provided in winning a billion-dollar contract; the profit from which paid the entire cost of the company's new building. To Lockheed, this was where real value lay.

For non-income producing buildings, real value may lie in user satisfaction and/or user performance. For example, if good building design can improve the learning abilities of students, the recuperative rates of hospital patients and the like, this may be of great value, albeit in a less tangible way.

Consequently, optimizing building value will normally lie in optimizing facility performance and end-user performance. For businesses and governments, this will equate to greater operational efficiencies and productivity, reduced operational costs and, where applicable, increased profits. For other building types the benefits may be less tangible but real nonetheless.

11.3.2 Value in use versus value in exchange

A prime determinant of overall functional use requirements lies in whether the client places greater value in the use of the building or in the exchange (sale) of the building. Many commercial buildings are, in fact, procured for a quick sale and hence the developer, in those circumstances, may not place a great deal of importance on the value of the building during its actual operation. However, building purchasers and users are becoming increasingly aware of the significance of a building's operating costs and their effect on user performance and satisfaction. As a consequence, *value in use* and *value in exchange* are becoming intrinsically linked because the higher the level of potential user satisfaction the greater the value of the building to a future purchaser and/or user. The better informed building users become about these issues the closer these links will become.

11.3.3 Treating commercial buildings as dynamic business facilities

Buildings have traditionally been procured as static products, with the ongoing involvement of the design team during operational stages usually limited or nonexistent. As mentioned earlier, commercial buildings are normally procured for the purposes of facilitating business activity with the users' prime objective being to optimize the performance of their operations. The actual building can have a major influence on this objective.

There is a clear relationship between business performance and the work environment and it is the building design and its physical layout that create this environment. Buildings are an integral part of the business process and can provide users with competitive advantage if appropriately designed. The contribution that building design can make to user productivity and profit levels is still not fully appreciated but users are becoming more educated in this respect.

Additionally, the capital cost of a building may not, in fact, be that significant for major corporations. Atkin (1988) contends that, for some corporations, the capital cost of their building may be less than one day's trading on the international markets. These types of users can realistically afford better quality buildings than the industry is currently able to offer, particularly with respect to performance enhancement and optimizing telecommunications and information technology capabilities.

This leads to the concept of utilizing the building as an investment. Patterson (1996) compares building ownership to the sharemarket by citing the example of a $100 million building. If an organization had a $100 million portfolio of stocks and shares they would normally have a very sophisticated portfolio management team to maximize their potential. The building should not be treated any differently, it is not merely $100 million worth of bricks and mortar. Expert asset/ facility managers can optimize business returns and add value to existing buildings. Patterson cites the example of one facility management firm in the United States that employs over 100 market researchers who analyse everything from market trends to occupancy costs. The traditional building manager focuses on cleaning, maintenance, repairs and ensuring services are operating adequately. Innovative facility managers will do that also but add significant value by applying business acumen and financial administration to not only the building's operations but also the users' operations.

11.3.4 Shifts in traditional thinking

For many designers and clients, this will require a fundamental shift in thinking in terms of building value. For commercial operations, a building will cost money but only its function will have value. The actual functions carried out in a building are where value optimization lies.

This will require FUA to focus on improving user and/or business performance through building design. This is the true primary function of most facilities, a fact that is often overlooked or given scant consideration. The mode of thought needs to shift to building design that enhances end-user and equipment performance.

This must, however, be tempered by the fact that many commercial buildings are not procured for the purposes of owner-occupation and use. Many proprietors or developers have the objective of maximizing profit through maximizing the difference between total capital cost and selling price or leasing prices. In other words, value lies in exchange not in use. Proprietors are typically far less concerned with operational costs and the end-user. They may not even be able to readily identify the type of activities that will be undertaken by future users. Nevertheless, businesses, governments and other building users are becoming more astute and aware of the effect of building design on business/worker performance. They are becoming increasingly selective about the types of buildings they purchase or lease and, as the concept of occupancy costs becomes more widely understood, buildings that exhibit user performance enhancing design features are likely to have a higher market value and greatly improved sale or leasing prospects. Exchange value can thus be increased through greater use value. These benefits can be used as a marketing tool to provide a proprietor with a marketing edge.

11.3.5 Re-evaluation of traditional design approaches

Traditional design approaches need to change for the above to be achieved. First, a design team approach is essential, as the vast number of issues and variables cannot be adequately addressed by a single designer. The design team needs to include architects, engineers, services designers, cost management experts, contractors, project managers, asset/facilities managers, end-users and even organizational and sociological experts.

Ferguson (1996, p. 137) agrees that the architectural process needs to be reformed. She suggests that designers need 'to know how to use management data and internal culture so that you can link the design of the organisation with the design of the workplace'. She adds that, whilst management approaches are becoming increasingly innovative, building design remains largely conservative and misdirected. Lovins and Browning (1992, p. 22) go further: 'the lack of real design optimization has cost the United States up to US$1 trillion worth of unnecessary space-conditioning equipment and its power supplies. A similar story could be told for virtually any other aspect of design'.

Current trends towards minimizing time and money spent on front-end design issues and fast-tracking projects have created an environment that is not conducive to adding value. More appropriately directed attention to front-end design issues, and FUA and value analysis techniques in particular, provide far greater scope for adding value to the end facility for the owner and/or users. Proprietors themselves need to lift their heads out of the bunker and look at the overall picture rather than the narrow, and largely misplaced, focus on initial capital costs.

Traditional design fee structures, which are based on a proportion of initial capital costs, and the current trends towards minimizing design costs actually reward ill-conceived, poorly thought-out and inefficient designs and inhibit, or make impossible, efficient 'value-adding' design that focuses on end-user performance. Design fees based on operational savings and two-stage fees based on initial costs with attractive financial incentives for reducing energy and other operational costs are beginning to emerge in many countries. This provides real motivation for the design team to incorporate detailed analysis of occupancy, and functional and operational variables in the design process; in other words, motivation to produce designs that focus on end performance rather than initial cost. Productivity enhancing design has the potential to offset entire design fee costs many times over; annual user salaries in the built facility can be hundreds, even thousands, of times the initial design fee.

11.3.6 Performance enhancing design features

In Chapter 20 a number of design features that can improve user performance are identified. Whilst the actual financial benefits of incorporating these design features are difficult, if not impossible, to quantify accurately, the case studies strongly suggest that the benefits will be real with the only intangible being the actual extent of the benefit. Many of the design features can, in fact, be incorporated for little, if any, extra cost. Where extra costs are incurred, the payback periods will often be counted in months rather than years if improved business performance is incorporated into the equation.

The following design areas are identified as having the greatest influence on user performance and thus having the greatest value-adding features:

- visual acuity
- thermal comfort
- air quality
- acoustic/noise insulation
- ergonomic factors (workstation/furniture design, etc.)
- spatial arrangements and design.

In addition, building design can be used to optimize machinery and equipment performance by closely linking enclosure design with the operational requirements of such equipment.

The services component of a building has a major impact on many of these areas so clearly services designers need to be included in the design team approach to FUA. The cost of the services component of buildings is increasing dramatically, accounting for well over half of total capital costs for a growing number of buildings (Atkin, 1988).

11.3.7 Intelligent buildings

Productivity and user performance will also be enhanced with buildings that maximize telecommunications and IT capabilities both now and into the future. The need for 'intelligent buildings' has never been so great and will continue to escalate. Intelligent buildings are increasingly required by big business where being part of a world market demands considerable interorganization communication, which their buildings must provide. The financial consequences of poor communication technology due to building and design constraints can be disastrous for many organizations. Cabling requirements, both now and in the future, need to be fully incorporated into a building's design. Built-in obsolescence of IT services will occur in shorter time frames. IT is revolutionizing the way in which buildings are

used and design needs to anticipate changing demands brought about by technological advances (Atkin, 1988).

Computing and telecommunications technology is developing at a pace that
far outstrips the ability of both new and existing buildings to take full
advantage of this technology. Most existing buildings are unable to cope with
current, let alone future, technology requirements, and substantial refurbishment is often required. This is constrained and, in many cases, made impossible by inadequate floor-to-floor heights, floor plans, insufficient services
space and other design problems. Rapidly changing work practices have been
a direct result of this technological change with firms continually re-
engineering the way that they do things to maintain competitive advantage.
Downsizing, 'hot-desking', high performance collaborative work teams and
telecommuting are just some of the developments that are redefining how
organizations use building space. Designers need to listen to users and analyse
the marketplace.

Worthington (1988) describes the direct physical consequences as being more
wires, larger ducts, larger services spaces and a greater diversity of standards to
reflect the wide range of functions. Space will tend to be increasingly allocated to
tasks rather than people. Whilst the services component of most buildings will
continue to rise both in terms of cost and space, the actual floor space requirements, particularly for facilities such as office buildings, will actually decline due to
changing work practices. More emphasis is now being placed on teamwork and
shared spaces rather than large, individually allocated spaces, bringing with it a
reduction in total floor space needs.

For example, Harrison (1996) cites international research on office space
utilization which found occupancy rates of only 40–50% during standard working
hours. Compounding this low level of usage was the finding that, for
approximately half the time workers were observed in the office, they carried out
work at locations away from their assigned workstations. Occupancy rates fell to
negligible levels outside of standard office hours. Accordingly, with the cost of
space usually at a premium, organizations are striving to reduce floor area
requirements per person. Simultaneously, they are focusing on improving their
employees' productivity through better designed furniture and workstations,
utilizing information technology for data/information storing and providing more
functional work spaces tailored to suit a range of activities. More flexible working
hours are also being utilized to ensure a greater use of space around the clock
rather than from 'nine-to-five'.

A lot of debate exists, though, on what constitutes an intelligent building.
DEGW International and Teknibank (cited in Harrison, 1996, pp. 45–6) define an
intelligent building as one that 'provides a responsive, effective and supportive
intelligent environment within which the organisation can achieve its business
objectives'. They further refined this definition to 'the efficient use of buildings,
space and business systems to support organisations in the effective operating of
the business'. Accordingly, an intelligent building does not necessarily have to
incorporate high levels of technology.

Harrison (1996) segregated the key issues for intelligent building design into four
facility elements:

- site issues
- shell issues
- skin issues
- building services and technologies issues.

These issues were compiled primarily for office buildings but they may be applicable, in whole or in part, for other types of commercial buildings. Nevertheless, they provide an insight into the concept of intelligent buildings.

Site issues include:

- telecommunications infrastructure (to ensure at least two separate routes for telecommunications and allow clear lines for satellite/microwave reception)
- local amenities (general amenities should be within a five-minute walk of the building)
- access (close proximity to public transport, e.g. less than 500 m to rail station)
- car parking (if sufficient parking not possible, parking stations should be within 200 m of site)
- site security (secure, well lit and entry control)
- aspect (surrounding buildings not within 15 m of each facade to allow reasonable views and daylighting).

Shell issues include:

- thermal strategy (effect of building shape/orientation on thermal environment)
- structural grid (structural column grid should be a multiple of the internal planning grid to tie in with ceiling, partition and other building components)
- planning grid (grids that suit internal layout requirements, e.g. flexible planning grids are often 1.5 m which provide 3 m offices)
- floor size and shape (these affect internal communications and circulation routes; irregular or complicated shapes brings penalties in terms of usage; square or rectangular spaces provide the most usable internal spaces)
- space efficiency (landlord efficiency indicates proportion of net lettable floor area whilst tenant efficiency relates to the proportion and quality of rentable space)
- floor depth and sectional height (sectional height is a key dimension for determining building adaptability as it relates to the building's servicing strategy with relation to such factors as air conditioning, cable distribution method and IT servicing; insufficient heights impose severe limitations, particularly in terms of services, whilst excessive heights are costly in capital and operating terms and reduce the number of floors that can be built)
- communications infrastructure (services risers should have adequate space provisions, at least two separate riser locations in the event of fire or other problems, and a separate riser for voice and data communications)
- staff and visitor access (well located, clearly visible and well signposted)
- goods access (separate goods vehicle entrance with easy street access and turning/waiting space)
- exterior/interior maintainability (low maintenance internal and external finishes).

Skin issues include:

- services strategy (the building skin is no longer just an external barrier, it is becoming an integral part of a building's servicing strategy)
- solar control (direct solar radiation is often the most significant external source of energy, which contributes to cooling loads; the use of shading, shutters, louvres and the like can reduce these problems).

Building services and technologies issues include:

- occupant requirements (building services and technologies can only be judged on how well they meet the occupants' requirements)
- shorter life spans (typically much shorter than most other building elements; provision needs to be made for replacement and regular maintenance/repairs)
- heating/ventilation/air-conditioning (HVAC) zoning and control (more flexibility in HVAC controls)
- small power (power for IT, audio-visual and office equipment and personal lighting)
- backup power provision (particularly where IT is critical for functioning)
- cable distribution system (structured cabling systems that are readily adaptable)
- communication systems (wide area communications are essential; digital links, provision for satellite/microwave communications, computer integrated telephony, cordless telephony, remote access to data networks and television and image distribution cabling systems)
- lighting systems (lighting that reduces glare/reflection problems, provision for personal settings)
- building automation systems (moving away from large, cumbersome and centralized control systems to smaller, more flexible systems with individual controllers on plant items)
- space management systems (utilizing CAD and facility management information systems)
- business systems (office/functional automation)
- access control and security (security systems appropriate to occupants' needs).

The intelligent building needs to respond to these and many other requirements throughout the life of different building elements and the building itself. Addressed properly through techniques such as FUA, facilities can be designed with a reasonable degree of confidence that the end product will satisfy current and future user requirements and will enhance user functionality and performance.

11.3.8 Flexible building design

Loose-fit adaptable design is the most effective means of addressing the above questions and other obsolescence factors. Rapid technological and concomitant societal changes have increased the importance of obsolescence considerations in the design process. Whilst always a factor, the way that buildings are used, and their requirements, are changing so quickly that a building can become obsolete in a matter of a few years.

The economic adaptability of a building is now the main determinant of building life. Adaptability is often constrained by inadequate structural design and inefficient layout, which can make demolition and then construction of a new building a better financial option than refurbishment. 'Long life, loose-fit' design is an increasingly important consideration for the design team. Buildings can no longer afford to be static permanent structures when the organizations using them and their activities are constantly changing.

11.3.9 Ergonomic considerations

Ergonomics is the study of the efficiency of people in their working environment and involves not only human factors but also machine and equipment design that can enhance human comfort and performance. Ergonomics clearly has an important part to play in the FUA process for building design but has traditionally received little attention in a formal sense other than in furniture and equipment design. The relationship between ergonomics and building design has generally not received the attention that it undoubtedly warrants. Whilst there is a considerable amount of literature devoted to ergonomics, very little has been associated directly with building design.

Harper (1990) argues that the main concern in building projects lies in the relationships between person, space and artefact and that focus during the design stages should be placed on building occupancy and total efficiency through ergonomic studies. He describes efficiency in terms of:

1. safe shelter
2. cost/value
3. operations aimed at successful subsequent building performance
4. user performance and satisfaction and
5. public acceptance.

Ergonomic studies relate directly to areas 3 and 4 and, in turn, can have a significant influence on cost/value.

Ergonomics encompasses sociological, psychological and physiological human factors and these have a place in building design when the aim is optimization of user performance and satisfaction. Although probably premature considerations, given the current development of ergonomic applications in building design, they nevertheless warrant attention. Sociologists and psychologists may well provide a further value-adding component during FUA.

11.3.10 User participation

This leads to the actual users of buildings, a sorely neglected element in the traditional design process. The real experts in terms of building use, what buildings need to do and operational problems, are the actual building users, yet they rarely have a role in the formal design process. Users continually assess the suitability of a building for their activities and the more they use the building the more familiar they become with its benefits and shortcomings. This knowledge rarely finds its way back to the designer due largely to the lack of *post-occupancy evaluation* (POE) studies by designers and an aversion to the inclusion of users in the design team. The blame for inadequate POE studies does not lie with designers but with clients themselves who have traditionally not extended design fees to cover such activities. Most clients do not see any financial advantage in user feedback. Additionally, some designers may even resent such studies due to the exposure to negative criticism.

Kernohan *et al.* (1996) contend that users and providers live in separate cultures; the providers have technical/design expertise that users depend on but users have a wealth of knowledge that designers could greatly benefit from. The belief that the designer (or the design team) is the expert and the user has little to offer still holds true for most projects. Users may, in fact, resent becoming involved because of the misconception that the designer 'knows all' and, after all, that is what they are being paid to do. Kernohan describes the two cultures as having different (and often opposing) values, rarely making contact and, when they do, often experiencing conflict. One side tends to avoid expressing their discontent whilst the other avoids acknowledging it.

Wherever possible, the FUA process should include intended end-users. This should not be restricted to senior management but include a broad spectrum of users, particularly at the operative 'groundlevel'. For many buildings, there may well be a variety of intended user organizations that would need to be accounted for. Consideration must also be given to the fact that users will change many times during the life of the building. The concerns of immediate users are paramount but subsequent users should not be ignored.

11.4 Case example – Albury Hospital, Australia

The following example, drawn from Service (1998), provides an insight into some of the issues described thus far, particularly with respect to occupancy and productivity costs. The example also introduces the benefits of performance-based contracts in optimizing FUA approaches.

Traditional procurement practices are increasingly being re-evaluated by proprietors the world over. The NSW Department of Health, the largest procurer of hospital facilities in Australia, falls into this category. Owing to substantial capital and operational cost increases for their facilities during the 1980s, the Department was forced to review its project delivery systems (mostly prescriptive lump sum arrangements). A particular problem was quickly identified; capital and operational costs were controlled by a cost plan developed for each particular hospital. Each cost plan, however, was based on the costs of recently completed facilities. Thus, facility budgets were based on the excessive costs of recently completed facilities. This created a cost spiral effect as new designs were developed within excessive cost budget parameters.

Another major finding was that hospital staff salaries and other labour services accounted for over 70% of total life costs. Capital, maintenance, energy and other building operational costs were found to comprise only 12% of total life costs. The greatest scope for reducing costs lay in the actual functional operation of their facilities.

The Department decided to move to innovative performance-based contracts, utilizing an integrated 'design, construct and maintain' team approach. Albury Hospital was chosen as the first facility to be procured using this approach. Focus was placed on outcomes rather than prescriptive requirements with the outcomes hinging largely on reducing recurrent costs, functionality, ease of use and buildability. Accordingly, FUA figured prominently in the pre-design and ensuing design stages. A team approach utilizing hospital administrators, health planners, architects, engineers, services consultants, contractors, cost experts and other key consultants, was used to examine functional requirements closely. As staff costs easily comprised the greatest life-cost element, the analysis centred on staff hours per patient. Service (1998, p. 11) describes an early realization of the group:

> Hospitals can normally run quite economically with a full patient load but when the load drops off, for instance around the weekend, the cost escalates exponentially. Applying the principle of balancing staff numbers to patient numbers it becomes obvious that, if certain changes are made in the initial concept design of the hospital, significant recurrent cost savings, often equating to the yearly capitalised value of the hospital, flow to the operational cost structure of the facility.

The following were some of the main hospital design principles that the group found could significantly reduce operating costs and, hence, increase value (Service, 1998, p. 11):

- the use of the swing-bed principle to allow the progressive temporary closure of beds during non-peak periods without a reduction in optimum staff/patient ratios
- the reduction of travel time between facilities within the hospital by attention to the grouping or clustering of closely related clinical functions (i.e. clinical services grouped in functional clusters)
- elimination of the 'belts and braces' approach to engineering services, including more efficient allowances for engineering facilities and the ability to shut down services progressively when they are not required
- a focused attention on buildability aspects of hospital construction including the application of commercial building techniques
- the application of outsourcing for certain 'back of house' functions such as laundry
- a realistic view of the standard of design and construction incorporated into hospital facilities – functional facilities, not architectural monuments.

Other important changes resulting in reduced operating costs included design focus on total life costs rather than initial capital costs, changing engineering standards to emphasize performance during design life and a shift from high-rise to low-rise construction. All of these measures led to a substantial reduction in total hospital floor area, which resulted in not only reduced operating costs but also reduced initial capital costs; thereby debunking the myth that operational savings can only be achieved through increased capital costs.

The end result for Albury Hospital was a capital cost 40% below department forecasts, significant productivity increases and dramatically reduced recurrent costs. The delivery team are also responsible for the long-term maintenance of the hospital, which provided an added incentive to address properly long-term operational cost factors during the design process. The knowledge and expertise acquired is now utilized and is being further refined on further hospital procurement and refurbishment resulting in substantial ongoing savings for the public hospital sector of NSW.

11.5 Conclusions

FUA is not new; it has always been an integral part of the design process. Nevertheless, there is tremendous room for the development of this approach to a more professional value-adding level. Formal design team approaches to FUA within a value management framework provide perhaps the greatest scope for optimizing value. Clients and design teams need, however, to be able to determine where real value lies and reflect that in functionality analyses. If the primary objectives of users is to make money, the primary functionality of their building design should reflect that. Ultimately, optimizing user satisfaction and/or performance is usually where real value lies.

References and bibliography

Akiyama, K. (1991) *Function Analysis – Systematic Improvement of Quality and Performance* (Massachusetts: Productivity Press Inc.).

Atkin, B. (ed.) (1988) Progress towards intelligent building. In: *Intelligent Buildings – Applications of IT and Building Automation to High Technology Construction Projects* (London: Kogan Page Ltd).

Ferguson, A. (1996) Time for the office building to lift its productivity too. In *Business Review Weekly*, October.

Ferry, D. and Brandon, P. (1991) *Cost Planning of Buildings*, 6th Edition (Oxford: BSP Professional).

Harper, D. (1990) *Buildings – The Process and the Product*, 2nd Edition (Chartered Institute of Building).

Harrison, A. (1996) Intelligent buildings in South-East Asia. In: *From Envisioning to Implementation* (Property Council of Australia Congress), September.

Kelly, J. and Male, S. (1993) *Value Management in Design and Construction* (UK: E & FN Spon).

Kernohan, D., Gray, J. and Daish, J. (1996) *User Participation in Building Design and Management* (Oxford: Architectural Press).

Lovins, A. and Browning, W. (1992) Green architecture: vaulting the barrier. *Architectural Record*, December.

Norton, B. (1992) Value added. *Chartered Quantity Surveyor*, 21–23 June.

Patterson, M. (1996) Investing success. *Buildings Online The Magazine*, October.

Seeley, I. (1996) *Building Economics: Appraisal and Control of Building Design, Cost and Efficiency*, 4th Edition (London: Macmillan).

Service, B. (1998) NSW Commission leads to performance-based contracts. *Chartered Building Professional*, August.

White, E.T. (1986) *Space Adjacency Analysis* (Architectural Media Ltd).

Worthington, J. (1988) Retaining flexibility for future occupancy changes. In: *Intelligent Buildings – Applications of IT and Building Automation to High Technology Construction Projects*, Atkin, B. (ed.) (London: Kogan Page Ltd).

12

Cost modelling

Allan Ashworth†

Editorial comment

Modelling construction is concerned with predicting outcomes: how much will the project cost (budgets and estimates), how long will it take to build (construction programming), what will be the timing and magnitude of the client's expenditures (cash flow forecasting) and so on. Through comparative modelling of construction costs and the testing of various alternative models, clients seek to increase the value they will gain from their investment.

The use of cost modelling tools allows the client, for example, to balance borrowings against predicted expenditure, thus minimizing interest paid on those borrowings, i.e. by delaying borrowing until staged payments are due, the period of each loan is minimized.

Comparing alternative design solutions allows clients to make informed decisions about design parameters such as materials, components, plant, building form and spatial layout.

The term *cost modelling* covers a range of techniques or tools which provide the basis for comprehensive cost management of building projects. Cost models, as the name suggests, are a numerical or mathematical representation of a project based on predicted construction costs. These models can, however, be utilized by clients and their consultants to provide much more than simple predictions of project cost and time. Advanced cost modelling techniques can help designers optimize relationships between project attributes such as floor area and external wall area, such that internal spatial requirements are satisfied with minimum expenditure on the building elements which enclose the necessary floor area. In any such case the purpose of cost modelling is to provide the client with improved value for money, not merely to predict a final project cost.

As with many of the techniques described in this book, for cost modelling to provide real benefits for clients it must be implemented from the earliest stages of the project. The common practice of 'costing a design' has little to do with cost modelling or cost planning, and has a relatively small chance of improving value for money. In fact it often leads to nothing more than punitive cost cutting aimed at lowering predicted costs to meet budgets. This often leads to the adoption of unsatisfactory, sub-optimal design solutions that have as their only positive feature the fact that they meet pre-determined cost limits.

† Liverpool John Moores University, UK; UNITEC, New Zealand

The implications for clients are clear: better value for money can be achieved through the use of a variety of cost modelling tools but if the benefits of cost modelling are to be maximized then it should be implemented from the start of the procurement process and the resultant models should be continually refined and expanded as the design proceeds. This enables the cost consultant truly to 'manage' cost, not merely to monitor and report on costs, with clients gaining maximum return on their investments.

12.1 Introduction

Cost modelling is the symbolic representation of some observable system that exists or is proposed, and which can be described in terms of its significant cost features for the purposes of display, analysis, comparison or control (Fortune and Hinks, 1996). Cost modelling, as a term, is used when referring both to forecasting construction costs for clients, and when referring to estimating resource costs for contractors.

Calculation of the costs of a proposed building or civil engineering project has traditionally been done by applying appropriate unit rates to measured quantities and descriptions of various components of the proposed scheme. At the design stage of a project the measurements and descriptions may represent little more than the spatial requirements provided by the client. For a contractor, these may be sufficiently detailed to describe the various components and processes of the project. In either case, quantifying the work can be reasonably precise but the judgements involved in allocating correct prices to these quantities and descriptions is very variable; hence two distinct models have evolved.

12.2 Types of cost models

The designer's cost model uses models of previously completed buildings to which estimates of future costs are applied. The design model may be based upon an analysis of work-in-place, such as that represented by the different rules or standard methods of measurement. At a different level, the model may be based upon a number of building elements or functional units of a building. Models produced in the earliest stages of a project often consist of single quantification to which all-in price rates are then allocated.

The constructor's or production model, which is prepared as part of the tendering process, is compiled prior to construction works commencing on site. In this case the aim is to construct a model of the *processes* involved in construction rather than of the *components* of the finished structure. These models use operations or activities, often in considerable detail, as a basis for modelling the costs involved.

Mathematical models have been developed which seek to identify variables that best describe cost. These mathematical models are, to some extent, characterized by a reduction in the number of priceable units in the model. These are thought by some to be an oversimplified representation with an insufficient number of cost centres being identified. However, Sleep (1970), as long ago as the late 1960s, claimed that such models were able to provide better forecasts of cost at much earlier dates than the traditional models.

12.3 Purpose of models

It is most important to distinguish clearly the purpose of cost modelling. Identification of the purpose will influence the structure of the model and the level of quantification required. Cost models may be provided for several different purposes and whilst there may be some overlapping, the characteristics of each model will be different.

Design optimization models are mainly concerned with securing value for money in building design. These are frequently used as a part of the overall design economics and cost planning process. Until the design-driven economic consequences of construction are more fully understood it is not possible to advise a client properly. This understanding comes from two sources: a general understanding of the principles of design economics, and a particular understanding of the project that is under consideration. The strength of these models relies on the cost adviser's ability to analyse and compare them with previously completed projects and with other alternative proposals under consideration.

Tender prediction models are used to forecast the likely tender sum that will be accepted by a client from a contractor. In addition to identifying the probable costs of the project and the model's imperfect predictability, these models must also take into consideration the contractor's own estimating variability and those factors that affect market price. Due to such considerations, the predictive models are less reliable than the design type models and some inaccuracy in forecasting costs is therefore to be expected.

Cashflow models are prepared on behalf of both the client and the contractor. These models indicate both the funds that are likely to be required by the client in order to pay the contractor for the work in progress, and the time when such payments will be required. They take into account the overall contract period and the contractor's likely method of construction. It is generally more difficult to forecast the cost of civil engineering projects since their costs are process determined, whereas the generally accepted view is that building costs are quantity related.

Life cycle costing models are concerned with the whole life of a project and are thus not restricted to design and construction alone but also include use and occupation by the client. The values attributed to life cycle cost models are of much less importance than their comparison against a number of different design alternatives. The ranking of projects, based upon the value of the model and the differences between one model and another, is of much greater significance than any predicted final costs. The use of techniques such as sensitivity analysis can assist in refining such models and in selecting the most appropriate solution from a set of alternatives.

Resource based models have also been developed to assist contractors in their own estimating and forecasting process. Whilst costs are incurred as a result of utilizing resources, design cost models are not normally constructed at this level of detail. In fact those who are involved in design cost modelling generally do not have access to appropriate resource data. These models are employed to improve estimating accuracy. Such models also have the inherent advantage that, by reducing the contractor's own variability in estimating and tendering, they will, in time, improve the performance of the client's predictive models.

12.4 Classification of cost models

Design cost modelling uses many different techniques, as shown in Table 12.1. Many of these techniques have become known as single price methods, even though in some cases they use a limited number of cost descriptors or variables. The choice of a method will depend upon many different factors, such as the user's familiarity with a given method, and confidence in the results expected and achieved. All of the methods require access to sources of reliable information and cost data if acceptable results are to be achieved.

Table 12.1 Cost modelling of building design (Ashworth, 1986, 1999; Fortune and Lees, 1996)

Method	Description
Traditional models	
Conference	A consensus view of the team
Financial methods	Cost limits determined by the client
Unit	Used on projects having standard units
Superficial	Total floor area of the project
Superficial-perimeter	A combination of floor area and building's perimeter
Cube	The volume of the project
Storey enclosure	A combination of weighted floor, wall and roof areas
Approximate quantities	An analysis of the major items of work
Elemental estimating	Used in conjunction with cost planning
Bills of quantities	Analysis prepared in accordance with detailed rules of measurement
Statistical models	
Regression analysis	Derived from the statistical analysis of variables
Causal models	Based upon algebraic expression of physical dimensions
Risk models	Monte Carlo simulation
Knowledge based models	Systems such as Elsie (Brandon, 1992)
Resource based models	Normally a contractor's method, using schedules of labour, plant and materials
Life-costs models	Whole life analysis of buildings

12.5 Trends in cost modelling

Table 12.2 suggests some of the trends in cost modelling that have occurred during the latter part of the 20th century.

The emphasis throughout has been on improving the quality of advice given to clients in order to allow them to make better decisions. Coupled with these developments has been the increase in use and application of information technology. The rapid retrieval of data and the ease by which models could be updated have allowed such improved advice to be provided. The trends throughout the above periods of time have swung between a heavy reliance on the importance of experience and judgement to a rationale that construction costs can all be represented in simple (or complex) formulae. There is now a genuine belief that effective cost modelling is a combination of each of these aspects.

Today, the focus is more on providing design and construction solutions that balance economic choices against the specific needs of clients. Value for money is achieved through a process of adding value to the project. Life cycle costing, or whole life cost associated with the project, is an important factor that should be incorporated in these economic choices.

Table 12.2 Trends in cost modelling (Ashworth, 1999)

1940	Forecasting contractors' tenders, deterministic methods
1960	Cost planning
1965	Value for money
1970	Mathematical modelling
1975	Probabilistic methods
1975	Accuracy in estimating
1980	Simulation
1980	Life cycle costing
1985	Value analysis and value management
1990	Expert systems
1995	Added value

12.6 Early price design cost models

One of the first questions asked by a client who wants a building or structure erected is, 'How much will it cost to build?' In the case of a wise or experienced client, the answer to this question will be followed with, 'And how accurate is this figure?' The purpose of an early price estimate is to provide an indication of the probable cost of construction. This is obviously an important factor in the client's decision to build. This model of building costs will provide the basis for the client's forecasting, budgeting and controlling of construction costs. During the design and development phase, this model of cost will be revised and updated many times in order to keep the client and the designers fully informed of the cost implications of design decisions.

Clients of the construction industry typically approach a design team or contractor with one or more of the following three scenarios in mind:

- an idea of what they require in terms of spatial layouts and quality
- when they would like to take occupation of their completed project
- how much they wish to spend initially.

They will also have a broader set of requirements for the project that they will identify within the brief that is offered to the design team.

In many cases, when a cost model is prepared even sketch drawings will not exist for the proposed project. The design may be in its formative stage, where the cost modeller may have limited information apart from the basic units of accommodation required. Design information based upon the geometry of the project and specification data may not exist. However, a model of the probable costs involved will be requested by the client in order that the project can then proceed. Raftery (1991) has identified the central problem that needs to be addressed by all cost models: it is that of translating the client's brief into an aesthetically pleasing and functional design and converting this into some form of economic measure. The design may be constrained due to the nature of the site, or by perceived complexities of construction, or demands for early completion of the project. Considerations of this type need to be included in the early price estimate or cost model of the project.

The models available to the design team are described in Table 12.1. During the early stages of a project not all of these models will be suitable for early price modelling. Some practitioners attempt to analyse projects by breaking them down into their constituent parts or elemental components before any drawn information is available. However, such analyses have been shown to be of limited value in practice. As the design develops it is usual to cost plan the project, using an elemental model as a basis for comparison and evaluation. Only in those cases where the project is some sort of a repeat design will such a model be beneficial in the longer term.

All cost modelling relies upon ready access to reliable cost information. There are many sources available that provide access to such information (Ashworth, 1999). The best source of data is that with which the user is most familiar. Published cost information, which provides a good secondary source of data, is often lacking in the supporting data characteristics that are inherent in primary data sources. During early design cost modelling, the best data that are available relate to the total costs of previously completed buildings. These costs may be expressed as a lump sum, which provides little other than a scale of the cost expected, or the cost may be expressed in terms of functional units of accommodation or costs per metre squared of floor area. Without more extensive design data the more detailed cost information is of only limited use in early stage cost models.

If the predicted cost is acceptable then more detailed cost models are used to provide a framework for cost planning. Eventually the contractor estimates the costs in order to calculate the tender sum. If successful, then this sum is incorporated in the contract documents as the contract sum and whatever model is adopted is then used as a basis for adjusting the final cost that is owed by the client to the contractor.

12.7 Accuracy of models

Some of the differences between estimates or models of cost can be attributed to

- changes in design and specification
- changes in client's requirements
- the introduction of new technologies
- inflationary factors.

Innovatory solutions, in terms of design and/or technology, which often require considerable modification during construction, are extremely difficult to cost model.

The typical levels of cost estimating accuracy that are achieved in practice are shown in Table 12.3. These can be applied to the full range of construction project types.

Contractors' estimating is based upon the measurement of a large number of work items and analysis of their unit costs, based upon previously recorded site performance data. The measured items are usually limited to the most cost-significant items. In theory, the site performance data, including labour outputs and material and plant constants, are retrieved from work done on active construction sites. However, research (Ashworth, 1999) has shown that this theoretical concept is flawed in practice due to the poor and inappropriate recording systems used by contractors, and the lack of confidence that estimators have in feedback data from individual construction sites. The time taken to undertake the different construction operations is also highly variable. The difficulty of capturing these data in a meaningful form, which allows re-use, is a complex task beyond the profitable occupation of most contractors. A comparison of similar items priced by different contractors reveals differences or discrepancies by as much as 200%. Even published data on guide prices can vary by as much as 50% (Ashworth and Skitmore, 1982).

It is now usual to offer a range of possible estimates or confidence limits rather than a single lump sum alone. There is a large amount of data available which measure pre-tender estimates against the final price charged for the project. The relationship is often poor but this belies factors that are normally outside the control of the cost modeller. It is always difficult to forecast the future, and errors and inconsistencies in pricing do occur. Estimates by definition will always include some amount of inaccuracy; they can never be a perfect representation of the final outcome. Whilst the accuracy of construction cost forecasts is poor, the difficulty of forecasting costs, prices or values is a problem common to all industries. This is not to suggest, however, that improvements in practice and performance cannot be made.

Table 12.3 Estimate classification and accuracy (Ashworth and Skitmore, 1986)

Estimate or model of costs	Purpose	Accuracy
Order of magnitude	Feasibility studies	+/− 25–40%
Factor estimate	Early stages assessment	+/− 15–25%
Office estimate	Preliminary budget	+/− 10–15%
Definitive estimate	Final budget	+/− 5–10%
Final estimate	Prior to tender	+/− 5%

12.8 Value for money considerations

An integral part of all cost modelling is an attempt to offer to a client improved value for the money invested in design and construction. Traditionally, the early forms of cost models had but a single objective: that of attempting to forecast contractors' tender sums. This was frequently carried out for the client's own budgeting purposes and to obtain formal approval for a scheme from a board or committee. Cost model forecasts were also sometimes done to test a proposal against the constraints of cost limits on design.

The importance of providing early cost advice to clients was to some extent limited to budget forecasts. During the postwar building boom in the United Kingdom, in the early 1960s, it became apparent that much more information could be provided to clients, particularly in the area of value for money. Throughout this period the importance of value for money became a popular theme, in many areas of life as well as in building design.

The traditional cost models could clearly not identify this aspect other than in the very broadest sense, i.e. that more expensive buildings probably offered improved value for money. The developers of elemental cost models claimed, as one of their objectives, that of adding value for money in building design. They claimed this because their models allowed them to examine the individual elements of the building, and their relationship to each other. However, this technique achieved its major impact only when outline drawn information had been prepared. It was argued that to attempt to cost model a building to this level of detail before shape, spatial layout and specification had been suggested was of very limited use to the design team.

However, modelling costs against a number of different variables, even at such an early stage, may demonstrate that value for money is not always being achieved. For example, whilst units of accommodation (e.g. number of patient beds in a hospital, or number of students in a school) have a relationship to spatial layout and/or floor area, the relationship is not necessarily linear. It was found in the postwar school building programme in the United Kingdom that the ratio of building cost to the number of pupils that a school might accommodate varied significantly across the country. The move towards some standardization based upon pupil costs rather than floor areas alone provided the then Department of Education and Science with improved value for money in their projects.

Early design investigations help to focus a design team on achieving value for money. Once a design has been formulated on the drawing board the efforts involved in achieving value for money become much more restricted. Value for money in building design is about seeking to do more for less. It aims to make out of the ordinary, something ordinary. Issues that early price models seek to address include:

- lowest initial costs
- lowest life cycle costs
- balanced distribution of design costs
- greatest value for lowest cost.

12.9 Conclusions

A good understanding of the behaviour of construction costs is required in order to model them adequately. This knowledge has increased considerably during the past 25 years, but is still lacking in many respects. Researchers have too often been content to accept the status quo, or to believe in the perceived wisdom of practitioners who are involved in the quantification and analysis of building projects. There are still many myths that need to be exploded. What is not easily accepted is that, given the nature of cost forecasting, modelling accuracy is perhaps now almost at its limits.

Cost models may be described as traditional or mathematical, and while the older traditional models were usually prepared manually, today both types are usually computer-assisted. In any event, all cost models rely upon adequate historical cost data, but in many cases the adequacy of such data is questionable, particularly when one takes into account the vagaries of tendering.

Mathematical models that represent a numerical analysis of construction costs have been developed but they are flawed because they do not properly reflect the nature of the costs involved, and because they generally incorporate some form of human interaction. Practitioners are likely to be suspicious of such attempts to describe project cost in purely mathematical terms, and they have, therefore, a limited chance of acceptance in practice.

Traditional modelling techniques placed a great deal of emphasis on human judgement, and recent research in the area of cost modelling (Skitmore, 1986; Fortune and Lees, 1996) has continued to emphasize the human orientation of the research agenda. Brandon (1992), in the development of his expert systems, also recognized that progress in cost modelling relied not only on utilizing techniques and information technology, but also in incorporating what was already best practice from those employed in commerce and industry.

Prior to the development of cost modelling techniques, consultants and clients alike were content with forecasts of costs expressed as a single lump sum amount. Early cost models copied this procedure, and consequently were largely deterministic in their nature. It was not until the early 1980s that the uncertainty inherent in design became accepted as a fact of life. Later models, recognizing this, became of a probabilistic type rather than deterministic. Ironically, about 70 years has elapsed since uncertainty had been recognized, accepted and planned for in many other disciplines.

Many of the former models have been dismissed, often because of their supposed complexity, but in truth because they provided results that were no better than those of the traditional models.

Later models have included expert systems. These systems are intended to build models that forecast price, evaluate alternative designs, prepare cash flows, and carry out life cycle cost analysis in a manner akin to the way that professional practitioners do this work. Cost modelling systems of this type appear to have the best chance of success at the present time.

References

Ashworth, A. (1986) *Cost Models – Their History, Development and Appraisal*. Chartered Institute of Building Technical Information Service.

Ashworth, A. (1999) *Cost Studies of Buildings*. Third edition. Addison Wesley Longman.

Ashworth, A. and Skitmore, R.M. (1982) *Accuracy in Estimating*. Occasional paper 27. Chartered Institute of Building.

Ashworth, A. and Skitmore, R.M. (1986) Accuracy in cost engineering estimates. In *Proceedings of the Ninth International Cost Engineer's Congress*, Oslo.

Brandon, P S (1992) *Quantity Surveying Techniques: New Directions*. Blackwell Scientific Publications.

Fortune, C.J. and Hinks, A.J. (1996) The selection of building price forecasting models. In *Proceedings of Economic Management of Innovation, Productivity and Quality in Construction*, Zagreb.

Fortune, C. and Lees, M. (1996) *Performance of New and Traditional Cost Models in Strategic Advice for Clients*. Royal Institution of Chartered Surveyors Research Papers.

Raftery, J. (1991) Models for construction cost and price forecasting. In *Proceedings of the First National Research Conference*. E & F N Spon.

Skitmore, R.M. (1986) *Towards an Expert Building Price Forecasting System*. Royal Institution of Chartered Surveyors Technical Papers.

Sleep, R.P. (1970) *Topographical Cost Models*. Highways Economics Unit, Department of the Environment.

13

Facility quality and performance

Stephen Ballesty†

Editorial comment

In the manufacturing sector, a new product is designed, prototypes are made and tested, the design is modified, new prototypes are made and tested and the process is repeated many times until the design is refined to the point where the product is ready to be mass produced and marketed. Testing of new products often involves extensive field testing, under conditions that simulate as closely as possible the actual conditions in which the product will be used.

Building designers enjoy none of this: instead, each building tends to be a unique solution to a complex array of interdependent problems. These problems include site conditions, climate, legal restrictions, building codes, intended function. Building designers have few opportunities for testing and refining their solutions; rather they have to wait until the single product of their work is completed for it to be judged.

One obvious possibility is to examine previously completed works and try to learn from the successes and failures that are present in most building designs. Given the large number of competing factors which affect building design it is probably impossible to satisfy completely all the requirements of all the interested parties; e.g. clients may be looking for minimum cost, while tenants are looking for high levels of comfort and aesthetics. Compromise is inevitable, yet designers will aim to produce the best building possible given the constraints within which they must work.

While it may seem obvious that designers should consult those who will use or occupy a building and find out exactly what they require, this is not always possible. When it is possible, the designers should give careful consideration to the views of the potential occupants, but building users are seldom building professionals, and will almost certainly be unaware of the majority of conflicting views and requirements that somehow have to be juggled and accommodated. It is up to the designers to consider the views of users but they must be selective in the weight they attach to those views.

An individual's view of the success or failure of a building may be coloured by

† Rider Hunt Terotech, Sydney, Australia

small personal concerns, e.g. the distance to the tearoom from one's office, or the lack of a light fitting in a precise spot, which causes shadows on a work surface. Highly personal judgements based on aesthetics may mean that some individuals will look for functional reasons to support their dislike of a building when their real concern is that they do not like the look of it.

Appraisal of building performance is becoming an established analytical process, and the results of such appraisals can be of assistance to designers as they can learn from previous designs, refining the successes and modifying, and hopefully eliminating, the failures. Unfortunately these appraisals are often not carried out, and if they are, the results are not generally available to those designing new buildings. If such evaluations are done, and designers can and do make use of the information that is gathered, then better designs should result. And better buildings will, by definition, be more valuable.

13.1 Introduction

From a user's perspective a critical link exists between the value of a facility and its quality and performance. Given the relationship between user satisfaction, productivity and return on investment, optimizing facility quality and performance should be a primary objective of building owners, design teams and managers. However, traditional procurement processes can be seen as 'fragmented', with the financiers and designers commonly focusing on asset creation and initial capital expenditure during the construction stage with little, if any, research, assessment and analysis of the performance of designs beyond completion.

Fragmentation of the property cycle results in the concept, design, construction, operation and disposal stages commonly being seen as distinct and separate. The relationship between design and construction may well be appreciated but the same cannot always be said of the relationship between design and operation. Between concept and disposal the time elapsed and the number of parties involved works against continuity and smooth transition. This inhibits and often prevents valuable feedback to the design team in terms of actual performance. Designers often have to rely on anecdotal or *ad hoc* feedback rather than analytical assessment. This information is critical for achieving real long-term value.

Facilities appraisals, premises audits and post-occupancy evaluations are some of the terms used to describe the structured feedback process between the users and providers of facilities. The key objective is normally to gain an insight into facility performance and user satisfaction through the accumulation of data and knowledge that can feed back into the design process. Such assessments of completed facilities will be generically referred to as *post-occupancy evaluation* (POE) in this chapter. It should also be noted that a facility is commonly a building, but need not be so in all cases as the term facility may have broader implications.

More enlightened owners and managers have realized the benefits conferred by POEs and other forms of facility assessment and analysis. This has seen the use of such techniques escalating. There is a fundamental need for designers to be well informed as to how people respond to their design and how they use (and sometimes change) a facility.

Value may well be best incorporated in the design process by actually looking at the construction process the other way round; that is, focusing on end performance criteria and how these might best be achieved and feeding this in at the very earliest design stages.

13.2 Facilities as assets

Facilities are a significant asset and represent, on average, from 32–44% of the non-financial assets of corporate balance sheets. Conversely, buildings also contribute to the cost of business. For example, in the financial sector, premises and office costs can account for approximately 20% of total costs and house a further 50% of costs, i.e. the staff (Rider Hunt, 1998). The ability of a facility to provide accommodation that supports and enhances the performance of the occupants can therefore have a fundamental impact on business productivity and success.

As facilities are usually the second largest item of an organization's annual costs (after the staff), the management and control of all aspects of facilities expenditure is clearly very important. In this respect, it is not enough to rely upon the reactive information systems of the traditional accounting functions that seek to monitor and control what happens, and report and reconcile what has happened. Such systems relate to, and are a response to, external financial reporting requirements. Traditional accounting functions do not primarily address matters of asset condition, performance and those affecting internal operational control. These facets are influenced by time, technology and the organization's strategy and needs as well as facilities management resources and approaches. These issues can, however, be addressed via a POE.

Designers, and their clients, need to evaluate the quality of their building's performance, to learn from many of the mistakes that become apparent and to treat facilities not just as an overhead to be minimized but as a vital asset to assist in enhancing 'bottom-line' profitability. Investment organizations now increasingly regard property as a separate asset class, competing for the investment dollar. However, performance of property is influenced by the characteristics of the property itself and these properties are unique.

Organizations that are inward looking, self-centred and bureaucratic are unlikely to succeed in the emerging globally competitive marketplace. This is equally true of facilities that are not in tune with the market's perceptions and expectations of quality and performance. This can be overcome, at least in part, by the use of appraisal techniques or other objective assessment systems that quantitatively measure performance for comparative purposes.

13.3 What is building performance and how is it measured?

Performance, in business terms, means quality of function/output along with constant process improvement. Modern management theory seeks to target and measure the performance of individuals, work-groups and equipment. Appropriate definition and measurement of performance is clearly necessary. The century-old assertions that 'if you cannot measure it you cannot understand it' (Lord Kelvin) and 'if you cannot measure it you cannot improve it' still hold true today.

Performance is a concept central to the otherwise diverse field of property. The performance specification is relatively familiar to people on the construction side of the fence. Many building materials are specified by their performance characteristics and there are numerous institutions specializing in the appraisal of the performance of buildings and materials in use. Building performance is normally associated with building quality and, traditionally, this has been expressed in terms of physical integrity and durability and the associated capital and revenue costs. However, building owners and managers can tend to equate performance with market value and return rather than any particular aspect of constructed functional utility, sometimes as if there were no correlation whatsoever between the two. This group commonly sees facilities as simply a means to an end.

Accordingly, performance is currently defined and expressed in a diversity of ways by the major interest groups (developers, designers, owners, investors, managers, end users). Perceptions of performance expressed by these groups are often diametrically opposed particularly with respect to groups on the supply side of project procurement (developers, designers, contractors) and those on the demand side (owners, tenants, users, property consultants).

To achieve performance and overcome these two seemingly diametrically opposed perceptions of quality, approaches need to be structured, objective and comprehensive. This predicament is further complicated by changing/evolving work practices, organizational structures, space usage and technologies.

The agendas or perception drivers within the demand side of the quality equation can be dissected as shown in Table 13.1.

However, quality alone or 'quality for quality's sake' does not equal value, as it does not take account of cost or time. In the same way that project controls focus on cost, time and quality, facility controls need to focus on value, risk and function.

The assessment, comparison and judgement of performance does, of course, take place all the time as a necessary part of decision processes. As a problem-solving technique 'value management' is often used as part of the design process. This involves the systematic use of function-based solutions to achieve cost improvements without sacrificing the required performance. In its simplest but most important context, value management focuses on the elimination of redundant performance; that is, the avoidance of expenditure on any item of construction that does not add value to the project or which makes the product achieve more than its required function.

Table 13.1 Perceptions of performance

Owners and managers	v	Tenants and users
Marketability	v	Functionality
Adaptability	v	Flexibility in use
Maintenance	v	Cleanliness
Return on investment	v	Affordability

As yet there exists no accepted common language between building professionals and their clients whereby the performance required of buildings can be described in terms that are exclusive of specification and design parameters. Consider the inadequacy of reporting to an organization that their planned multi-million dollar office upgrade would produce accommodation equivalent to, to use a food analogy, of a 'take-away', 'bistro' or 'à la carte'. Such a broad subjective assessment could easily give rise to inaccuracies and misinterpretations of what the organization would get. The food analogy may give guidance on the price they may pay, but would tell them little about the resultant nutritional value or gastronomic satisfaction. To express appropriate quality and truly achieve value for money, facility design needs to relate to the expenditure required to achieve the desired performance consistently.

Facility performance measures and indicators have been developed and are being further refined to address this requirement. These performance indicators can be used to measure the effectiveness and efficiency achieved relative to a desired function/output and these can be expressed in terms of time, cost and quality or, even better, as a combination. The relationship of these variables to specific business objectives gives an insight into more critical issues of function, risk and value. Close relatives of the performance indicator principle are the concepts of *benchmarking* and *best practice*. The impact of facility performance on productivity and profitability cannot be overstated and, this, in combination with the ability of physical assets to store wealth and create liabilities, make proper measurement and management of performance essential.

Facilities are increasingly being seen by users' groups as servicing people, processes and places (space). The design and management of facilities should focus on optimizing people, processes, assets and the work environment to support the delivery of an organization's objectives. A building's response to accommodating these variables represents its performance.

Consequently, regardless of the methodology adopted, a comprehensive POE is about measurement and performance. The three critical areas of assessment for any facilities are:

1. Physical performance relating to the behaviour of the structure, envelope, services and finishes embracing physical properties such as internal environment (heating, lighting etc.) energy efficiency, cleanliness, maintainability, durability and environmental impact.
2. Functional performance relating to the properties afforded by a building to the benefit (or otherwise) of the occupier. Examples are space (quantity and quality), layout, ergonomics, image, ambience, amenity, movement/communications, security, health and safety, and flexibility.
3. Financial performance involving a combination of capital and revenue expenditure, rate of depreciation, investment value and contribution to productivity profitability/efficiency. Financial performance can be related to the physical and functional performance of the building and the way in which it is used.

These three variables are inextricably linked but the significance of this relationship is often missed by those whose pre-occupation may be with only one particular facet of property or construction. In the end, significant benefits for individuals, organizations and the community are available if optimum efficiency and effective performance are achieved.

13.4 Assessing facility quality and performance

Facility developers and proprietors need to be able to identify and understand the factors that drive facility quality, produce better buildings and achieve and maintain tenant/user satisfaction. Tenants and users will not stay if, for the same rental or even higher rentals, they can get something better which will improve their bottom line. Facility appraisals provide the vehicle to facilitate that identification and understanding.

Businesses and the investment market have become more aware of the differences that quality of building design and management can make to their business activities and thus their balance sheet. If buildings are not designed, managed and maintained appropriately then users/tenants will leave. Without tenants, or with an excessive tenant turnover and gaps in tenancies, the owner's income and ability to maintain, upgrade and finally dispose of their capital asset is seriously impaired.

In recent years, tenants, whatever their main business activity, have become much more demanding and knowledgeable about building operations, technology available and fit-out design. Tenants, even if not property specialists, are also entrepreneurs, and wish to use real estate as an incidental component of their business activities. They expect their premises to help them in their business activities. Staff must be happy and productive and the property must provide a quality business location and environment with services such as lifts, air conditioning, lighting and car parking operating efficiently. To this end, designers need to grapple with the often conflicting strategic interests of building owners and the operational interests of building users.

Modern buildings have such a high degree of sophistication that subjective and 'rule of thumb' procedures for developing a 'feel' for quality are no longer adequate. The perception of a building varies between organizations, levels of management, professions and individual users over time. Indeed 'beauty is in the eye of the beholder' as determined by their current situation. This is, of course, far from satisfactory, which adds further weight to the need for the use of standard assessment methods.

13.4.1 Post-occupancy evaluation (POE)

The process of POE provides the basis for facility appraisal. POE can be undertaken for a variety of purposes. An organization may wish to consult their employees and other users about their workplace to improve user relationships with management, demonstrate concern for users' welfare and needs and to improve operational performance. Including users in the overall management of an organization through the process of consultation provides a channel for communication and mutual understanding. It essentially represents good management practice. Flowing from this is the use of POE to evaluate a facility in terms of its work environment for the purposes of improving this environment to enhance worker productivity, machinery/equipment operation and business performance generally. It can also be used as the basis for upgrading existing facilities by identifying the current state of repair of the premises and its suitability for users.

POE should also be given consideration during *due diligence* analyses, often conducted by specialist consultants on behalf of prospective property purchasers, facility users or existing vendors or landlords. Initial due diligence analysis generally relates to the compliance of the property with statutory requirements and its state of repair as well as its suitability in meeting prospective or current users' requirements. Ongoing due diligence reporting can be used to ensure that the value of the property is continually reviewed, monitored, maintained and, where possible, improved. The overall scope of property/facility due diligence auditing covers four main areas: property market issues, legal aspects, financial accounting and technical reviews.

To cover this last area (technical reviews), the scope of reports can comprise forecast expenditure requirements, building quality assessment (described later), architectural and planning reviews, structural and facade evaluation, engineering services evaluation (including Y2K/Year 2000 compliance), environmental assessment, statutory compliance, taxation depreciation, replacement cost assessments, site/building surveys, space utilization and planning studies, life-cost analyses, operational management reviews and outgoings analysis, lease reviews and tenant satisfaction surveys. In addition to being of immense value to prospective and existing users, the information derived from these reports is invaluable to designers. Such reports provide a 'snap shot' or 'report card' which, in part, provides a basis for designer feedback and/or a framework for a facilities plan with regard to improving new designs or planning maintenance, upgrades or refurbishments to meet user needs better.

There are normally three main groups involved in POEs: initiators, facilitators and stakeholders. An organization's management personnel are usually the initiators of the process. Unless they have the necessary expertise, which is not likely, they need to engage a person or firm (the facilitator) to undertake the evaluation. Facilitators conduct the on-site analysis and surveys to measure and record facility performance. A team approach is the ideal as it can draw upon experts from the quite divergent fields involved in the evaluation. The stakeholders are the end-users who supply the information about the facility and, hence, are the key to the effectiveness of the process. Stakeholders normally comprise occupants, employees, tenants and visitors.

There are commonly three levels of assessment applied to POEs. The level chosen is largely dependent on the balance between the forecast benefits to be derived in relation to financial, time and personnel constraints. Each level involves similar approaches in planning the process, conducting the analysis and interpreting the results.

Indicative

The indicative evaluation is the quickest, cheapest and, hence, most common method used. It involves walk-through evaluations and structured interviews with key personnel. Inspection reports which summarize the observable building performance indicators are based largely on the facilitator's professional judgement as time (and fees) will not permit a more detailed analysis. Other activities include data collection, archival document evaluations and group meetings with end users.

The information gathering for this type of evaluation can normally be done in a matter of days, depending on the size, age and complexity of a facility. Accordingly, the findings are generally limited to identifying the major successes and failures of the facility's overall performance and direct assessments of only the main elements and functions.

Investigative

Investigative POE provides a more in-depth analysis of building performance. It will often be conducted as a result of problems identified in an earlier indicative assessment. This approach can take weeks to months, depending on the depth of investigation and the number of personnel involved. Scheduled in-depth interviews, workshop meetings and end-user survey questionnaires underpin the process. Inspection reports usually include a summary rating against objective criteria by comparing the facility to other similar facilities. Physical measurements and photographic records are also made. Solutions to identified problems are then explored in detail and recommendations made.

Diagnostic

Diagnostic evaluation is the most detailed form of assessment and is carried out over months and even years. Whilst being the most time-consuming and hence expensive method, the greater benefits that can be achieved by the more detailed assessment will usually far outweigh the costs of the evaluation. It involves long-term and cross-sectional evaluation studies with a specialist team of consultants. The team reviews the facility at a detailed performance factor level and undertakes a comparative analysis of the relative strengths and weaknesses of the facility. The analysis culminates in strategic planning advice for the commissioning organization to ensure they obtain the maximum potential value out of the facility.

The diagnostic evaluation involves more detailed and sophisticated data-gathering and analysis techniques. This provides designers with the most reliable and valuable post-occupancy data and information possible.

Whilst POE as described above has been around for over 25 years, the majority of facility owners/users still do not have such evaluations carried out. However, the past decade has seen a marked rise in the level of POE studies. This has been fostered largely by the dramatic growth of facility management as a professional discipline and an increasing awareness of the benefits to be conferred by POE.

In the long term, POE will enable the establishment of facility performance databases and the generation of planning and design criteria for specific facility types. Designers will be able to consider documented design performance successes and failures, avoid repeating failures, build on successes and better incorporate performance enhancing design factors. Increases in the number of evaluation studies are also likely to improve the overall quality of design practice; designers who know that their designs will be subject to detailed scrutiny during operational stages will have a clear incentive to design facilities that represent value in the long term.

13.5 Critical performance indicators

The quality of a building is the degree to which it meets the expectations and requirements of the building users. As previously outlined there are two basic sides to the quality equation, known respectively as the supply-side and demand-side determinants of perceived building quality.

Building providers tend to believe they know what is required and are convinced they meet the true demands of occupants, but occupants frequently think otherwise. Empirical evidence from post-occupancy evaluations completed in many countries in the past 20 years shows that many occupant requirements are not adequately provided for. Knight Frank Hooker (1995) undertook research into tenant demands and revealed the following consistent responses to a crucial question for building owners and managers:

'What would you (as tenant) most like to change about your existing building?'

Air conditioning	Lifts
Parking	Amenities
Location	Entrance
Layout	Everything
Rental level	Windows
Internal layout	Technical services
External appearance	Security
Management	Delivery and goods

From this it follows that the two main components of performance relating to facilities are functional and physical efficiency. Infinitely more important in economic terms is the performance characteristic relating to functional efficiency. In commercial organizations, this may well impact on profitability. This implies that there is a contribution to value but without an adequate model for relating performance to value it may not be realized. If you cannot measure it, you cannot understand it or improve it nor, indeed, value it.

On the other hand, physical performance can be conventionally measured and priced. Hence, a common issue is that the 'easy' problems are more likely to be addressed. The physical efficiency of the building will embrace such items as:

- deterioration over the life cycle
- building services efficiency
- cleanability and maintainability
- energy efficiency
- environmental impact.

The Property Council of Australia (PCA) classifies commercial office buildings by five groupings. In descending order of 'Quality' they are: Premium and Grades A, B, C and D. The classification criteria relate to location, building size, floor

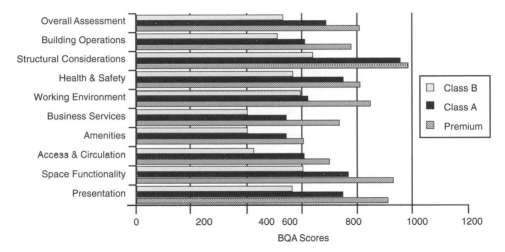

Fig. 13.1 Average BQA scores for office buildings (Sydney CBD).

plate area, age, building services and quality of finishes; these criteria vary between various city locations. The application of PCA classifications remains largely open to subjective assessment but recent enhancements of Technical Services Criteria (Property Australia, 1998b) has seen a move toward objectivity-based assessment. The terminology and criteria are published and widely accepted as a basis for describing the differentiation between competing properties.

Figure 13.1 is the result of Rider Hunt Terotech research undertaken for HSBC James Capel Australia last year for their Property Trust Research – Commercial Property Review. It confirms the generally held perception that quality can be differentiated, as is intended by the PCA classifications. In addition, and perhaps more importantly, it quantitatively reflects the areas and extent of quality difference.

13.5.1 Building Quality Assessment (BQA)

Beyond the PCA classification of buildings there are several methodologies available internationally for the performance evaluation of facilities. One notable method is the Building Quality Assessment (BQA) method, which originated in Australia and New Zealand in 1990. It is now also used in Europe and the UK and is being tested in the United States.

The BQA approach was developed by the Rider Hunt Group, in conjunction with the Centre for Building Performance Research at Victoria University in New Zealand. They sought a means to reduce subjective opinion from facility appraisal. The system was fine-tuned through surveys and questionnaires covering leading property owners, developers, tenants and users. In developing BQA, the approach taken was to identify requirements that are common to most or all users of a building type. BQA emphasizes the requirements of building providers but, as any wise provider will seek to provide a building that meets the needs of existing or future users, it pays close attention to occupant requirements.

It provides an objective base for the measurement of the physical and functional quality of buildings by introducing standards and rules to enable measurement on a consistent and international basis. It is a weighted assessment system with between 129 and 148 standard factors (depending on building type), which both individually and collectively are significant in determining the perception of a building's quality. Each factor is scored out of ten against pre-determined criteria, and is weighted for its relative performance within nine BQA categories for analysis and reporting. The aggregate score out of 1000 is the facility's overall quality rating index.

The nine BQA categories for commercial offices are:

1. Presentation
2. Space Functionality
3. Access and Circulation
4. Amenities
5. Business Services
6. Working Environment
7. Health and Safety
8. Structural Considerations
9. Building Operations.

BQA provides the means for a balanced assessment of building quality and performance, both as a whole and for component parts, against the requirements of a range of users. The standards set are realistic but demanding and are subject to constant review to reflect tenants' and owners' changing needs. The standards are based on best practice. The approach enables owners, investors, users, potential users, facility managers and the like to measure objectively the quality of one or more buildings. It also allows tenant groups and their consultants to compare marketplace options for accommodation. BQA can be used to establish users' needs and the market's quality expectations, and hence benchmark user requirements against standards offered by an organization's current premises or alternative properties.

Since its introduction in 1991, over 100 assessments have been undertaken in Australia, New Zealand, Europe and the UK. The data for all the assessments, all of which have been assessed using a consistent methodology, are returned to Australia for ongoing analysis and feedback. As the number of assessments increases, segmentation of the results is possible by location, city, PCA zones, size of tenancy, type of tenancy and the like. This growing body of information and data is clearly invaluable for the design team.

The major reason for the development of BQA was to drive the improvement of facility quality, equate quality to the business objectives of the organization and thus benchmark the asset against others. In almost every case, the BQA studies have been used to provide a benchmark, which, as defined by the *Collins Dictionary*, is 'a standard in judging quality and value'.

Users of buildings have different requirements, some of which conflict. To be successful, facilities must satisfy these different user requirements at the same time. Therefore, in designing a method for measuring building quality, a basic choice must be made as to whether to assess against all identified requirements, including those that conflict, or against selected requirements, such as those of the providers.

BQA is of particular benefit in that it provides a common basis for measurement, thus enabling a consistent comparison of the quality and performance of buildings of the same functional type.

13.5.2 Other facility evaluation methodologies

Other facility evaluation methodologies have now been developed. These include the Serviceability Tools and Methods (STM) in Canada, the Real Estate Norm (REN) in the Netherlands, and the Intelligent Building concept in Europe and Asia. The number of methodologies already available is testament to the value of improved performance and the need for scientific analysis of value from an occupation viewpoint. It remains crucial, however, that such assessment methods cover performance in all its aspects; that is, functional, physical and financial efficiencies.

In the quest for better methods of measurement for building performance, evaluation of stakeholders and their interaction with the built environment, its occupation and its use, must be considered. However, ultimately the market requires a standard methodology that is sufficiently flexible to reflect individual user needs and provide a basis for consistent comparison. Pressure is increasing for facilities actively to support organizational objectives and not simply provide an enclosure for activities.

13.5.3 The Intelligent Building (IB) concept

Another related contemporary issue is the IB concept, and this chapter would not be complete without at least mentioning it. Buildings can be thought of as either a collection of individual technologies/building characteristics or as a total system that can respond to organizational change over time.

An intelligent building has been defined as any building that provides a responsive, effective and supportive intelligent environment within which the organization can achieve its business objective (DEGW/Teknibank, 1992).

High performance quality buildings need to be 'intelligent'. However, intelligence needs to be more than the inclusion of the latest high-tech gadgetry; it is more about the extent of flexibility and adaptability in accommodating spatial, structural and services configurations and user needs. These aspects are more critical as these will be reflected in physical, functional and financial performance and hence impact on the efficiency and effectiveness of business productivity.

Focus needs to be placed on 'intelligent' design and management of property assets to achieve and maintain maximum utility and tenant satisfaction. Regular benchmarking of facility appraisals against desired or previous performance, and standards/comparisons within and outside the organization, serve as a vital design, planning and monitoring mechanism.

13.6 Conclusions

This chapter is not exhaustive but is representative of the scope of the issues and views involved in the critical area of the assessment and analysis of facility quality and performance.

User expectations have become much more sophisticated and demanding over the past decade. Organizations are required to operate at greater levels of efficiency in shorter time frames and in a more competitive environment. One of the factors that influences an organization's ability to operate effectively is their workspace. Gaining a clearer understanding of quality in relation to workspace, its use and the output it generates is vital. Performance in all areas of business is being measured and analysed in greater depth. Being able to measure building quality and facility performance as variables for analysis and improvement will allow the true contribution of the built environment to the economic prosperity of society to be better expressed and refined.

Duffy and Tannis (1996, p. 14) contend that 'the new gurus talk about vision or the strategic intent of companies, but usually architects and suppliers of physical space are not given much opportunity to link the process of designing [work] space with such strategies'. This requires greater intimacy and sharing of information between design teams and facility owners/managers/users. Combining POE feedback and other facility evaluations tools will enhance communication throughout the property cycle, resulting in greater integration of the concept, design, construction, operation and disposal phases of a facility's life. This will ultimately lead to the delivery of greater 'value for money' for facility owners and users.

References and bibliography

Baird, G., Gray, J., Isaacs, N., Kernohan, D. and McIndoe, G. (1996) *Building Evaluation Technique* (McGraw-Hill).

Ballesty, S.P. (1996) *Post Occupancy Evaluations*. Environmental Economics Series – Master of Construction Economics (Sydney: University of Technology).

Ballesty, S.P. (1997a) Analysing the performance of modern facilities: does 'INTELLIGENCE' equal 'QUALITY'? *International Quality and Productivity Centre – Intelligent Buildings Conference*, Sydney, February.

Ballesty, S.P. (1997b) *Facilities Appraisal Techniques*. Facilities Management and Maintenance, Professional Development Program, University of NSW School of Building.

Ballesty, S.P. (1998) Building quality and its impact on user satisfaction. *International Management Resources*, Property 2000 Conference, Sydney, April.

Bernard Williams Associates (1994) *Facilities Economics* (UK: Staples Printers Rochester Ltd).

Bordass, B. and Leaman, A. (1995) Design for manageability (Unpublished).

Centre for Building Performance (1993) *Building Quality Assessment – Research, Development and Analysis for Office and Retail Buildings* (Wellington: University of Victoria).

DEGW/Teknibank (1992) IB Research Report, cited in Harrison, A. (1996) Intelligent buildings in South-East Asia. In: *From Envisioning to Implementation*. Property Council of Australia Congress, September.

Dell'Isola, A. (1997) *Value Engineering: Practical Application* (R.S. Means Company).

Duffy, F. and Tannis, J. (1996) A vision of the new workplace. In: *From Envisioning to Implementation*. Property Council of Australia Congress, September.

Knight Frank Hooker (1995) BOMA leading edge research – tenant demand. *BOMA Magazine*, September.

NSW Public Works Department's Policy Division (1992) *Total Asset Management System (TAM) – Overview and Guidelines* (Construction Policy Steering Committee).

Pawsey, M.R., Ragusa, S. and Geelings, W. (1995) *Financial Planning Guidelines for Building Asset Management* (Melbourne: National Committee on Rationalised Building).

Property Australia (1998a) New quality grade matrix. *Property Australia Magazine*, **12** (4), Dec. 1997/Jan. 1998.

Property Australia (1998b) Technical services criteria. *Property Australia Magazine*, **12** (4), Dec. 1997/Jan. 1998.

Richardson, R. (1996) *Building Quality Assessment and Property Management*. BOMA Commercial Property Management Course. Sydney: October.

Rider Hunt (1998) *Consultancy Research Studies* (unpublished) (Sydney: Rider Hunt Terotech).

Rijksgebouwenclienst- Directie Huisvestingsbeleid (1994) *Comparative Study – REN/ STM/BQA* (The Netherlands: The Netherlands Government).

Tucker, S.N. and Taylor, R.J. (1990) *Performance Indicators for Building Assets* (Melbourne: National Committee on Rationalised Building).

14

Energy modelling

Paul Strachan and Jon Hand†
Drury Crawley††

Editorial comment

The cost of energy used in buildings has long been treated as an unavoidable expenditure, something more akin to a tax rather than a controllable input to the process of doing business. Because of this perception, until quite recently, little serious thought has been given by designers, developers or operators to managing energy use in buildings in a fashion similar to that in which a manufacturer might control expenses such as material inputs, labour costs, transport or any other component of their expenses.

Superficial attempts have been made in many cases to 'add-on' various energy efficiency measures: typical examples include roof and wall insulation, variable speed pumps and fans and the like, all of which may save a little energy and lower operating costs. There are, however, far greater potential savings available which can never be achieved through this sort of approach, savings which can only be realized if there are some fundamental changes in the design process.

In recent times, a great deal of effort has been directed at the problems of modelling energy use in buildings and in the preparation of computer-based simulation tools. These models and the software tools that are based on them can be used by designers to predict levels of energy consumption and compare potential energy use in alternative designs.

The use of these tools and techniques becomes vital when designers attempt to move away from conventional solutions, particularly when passive ventilation and lighting systems are employed. The success of buildings that employ such systems depends on the designers achieving and maintaining a delicate balance among a multitude of factors including solar penetration, natural air flows, and the thermal mass of the building itself. Balancing all the relevant parameters so that the building will provide a comfortable indoor environment all year round requires a lot of simulation and testing of designs.

From the client's viewpoint, reducing energy use leads to substantial savings in operating costs over a building's life, as energy, whether in the form of electricity, gas or oil, represents a major component of the expense of running any building.

† Energy Systems Research Unit, University of Strathclyde, Glasgow, UK
†† US Department of Energy, Washington DC, USA

Building design is only one of the factors that contributes to the energy profile of any given building, but space heating and cooling and electric lighting are typically major energy users. Energy consumption in both of these areas can be significantly reduced if the building fabric and environmental control systems work in harmony rather than in opposition.

The savings which accrue to building owners and tenants over time are only part of the financial advantages that low-energy design may produce – a better indoor environment can also boost productivity. In short, energy planning and modelling is one part of an overall approach to design that aims to produce better buildings. A commitment by the client at the beginning of the process to allow designers both the necessary time and budget to pursue innovative solutions is essential if optimum solutions are to be realized. Adequate pre-design research is also important, as the designers must fully appreciate client requirements, required building functionality, and site and climatic conditions, before design work commences.

The following chapter describes the philosophy behind some of the planning and modelling techniques now in use, and outlines the sort of procedures that designers can now follow so that their buildings use much less energy than the conventional buildings of the 20th century, dependent as they so often are on full air conditioning and artificial lighting.

14.1 Introduction

It is unfortunately true that the majority of building projects do not include analyses of energy use or internal environmental conditions beyond those required for regulation compliance. Usually national regulations set out maximum permissible elemental U-values, e.g. that an external wall should not exceed $0.35\,W/m^2K$. It is usually also acceptable within the regulations to carry out a more detailed analysis. If it can be shown that a design will consume no more energy than that calculated by the prescriptive method, then the design is acceptable. Indeed, it is often possible to demonstrate considerable savings in capital and operating costs by the use of detailed analysis.

However, energy consumption is only one factor in the energy and environmental performance of buildings. There are many benefits that result from improving comfort (thermal, visual and acoustic), at home, at work and in other buildings, such as hospitals, supermarkets and universities. Analysis of designs can lead to more comfortable, less costly buildings that also have less impact on the environment. Energy efficiency options available to designers are rapidly increasing – for example, use of natural ventilation and daylighting, energy efficient lighting, improved controls for heating and cooling equipment and lighting, advanced glazings, and building integrated photovoltaic modules. Buildings with these features tend to be more responsive to external climate compared with conventional buildings, particularly those with full air-conditioning systems giving constant controlled environments. In such buildings, simplified calculations are no longer sufficient to answer many of the questions posed by designers.

The beneficial effects of attention to energy and environmental performance include the following:

- The working environment is improved. There is evidence that productivity increases in well-designed buildings where there is the option of natural lighting and ventilation and where occupants have control over their environment (see, for example, the case studies in Chapters 19 and 20). This in turn leads to significant long-term financial benefits.
- Life-cycle costs (both capital and operating costs) can be reduced significantly by energy efficient design. For example, the use of natural ventilation can reduce or remove the need for mechanical ventilation plant, giving savings in equipment and the space needed for the systems, as well as reducing recurring costs. Pacific Gas and Electric Company in California found that by aggressively pursuing energy efficiency during design, they were able to reduce significantly the capital cost of the cooling plant – enough that it more than paid back the cost of the energy efficiency improvements (Brohard and Caulfield, 1998).
- Companies with energy efficient and environmentally conscious buildings can use this to promote their 'green' image.
- At a national level, a decrease in energy use reduces reliance on imports and improves trade balances. Buildings can make a significant contribution in this regard. In the UK, for example, over 50% of delivered energy is consumed in buildings, of which more than 60% is for controlling space temperatures. The potential for improvement is large – in the UK, offices (per square metre of internal area) vary by a factor of over 7 in their annual fuel bills (EEO, 1991). In the US, buildings account for 34% of the primary energy and 67% of the electricity use (EIA, 1998). The range of building energy performance is similar to that of the UK.
- Reduction in greenhouse gas (including CO_2) emissions help countries fulfil their obligations under international agreements. In the 1997 Kyoto Agreement, Heads of Government made a commitment to reduce key greenhouse gas emissions in developed countries by at least 5% by 2008–12 (relative to 1990). This would result in 2010 emission levels that are about 29% below what they would have been in the absence of the protocol.

Evidence of direct financial benefits accruing from the inclusion of energy and environmental assessment in building design can be found in the independent analysis of the Energy Design Advice Scheme (EDAS) in the UK. EDAS (McElroy *et al.*, 1997) acts to assist building owners and the design professions to gain access to energy advice and expertise. Assistance varies from brief reviews of the project with energy consultants through to detailed analysis. The independent analysis for the period 1992–8 estimated energy savings from EDAS-assisted completed projects of over UK£16 million annually, with associated reductions in CO_2 emissions of the order of 285 000 tonnes annually. For every UK£1 of government expenditure, energy savings worth UK£30 were achieved.

14.2 Modelling capabilities

Energy and environmental assessment of buildings covers a large range of design issues. Table 14.1 gives some examples of questions that the design team may ask at different stages of the design process. If answers to these questions are not available until after the main features of the building have been decided, then the

Table 14.1 Typical energy/environmental design questions

Site
How do building shape and orientation affect energy consumption?
How do local terrain and surrounding buildings affect the air flow around the proposed building?

Fabric
What benefits would ensue from glazing over the top of an internal courtyard?
For a refurbishment project, what are the relative merits of the options available?

Ventilation
Would cooling the building at night with outside air, together with increasing the thermal mass of the building, be sufficient to avoid air-conditioning plant?
What are the best positions for diffusers to ensure adequate distribution of air in a room?
Will displacement ventilation be able to cope with high levels of internal gain?

Plant Systems
What would be the most energy efficient solution for maintaining close temperature and humidity control in an art gallery?
If office air conditioning is required, what is the most energy efficient system?
What are the demand profiles for a small-scale combined heat and power scheme?
What would be the best heating system for an old church?

Lighting
How can daylight penetration be maximized and glare sources eliminated?
What is the optimum strategy for daylight controlled electric lighting?

General
Does the energy performance of the building and its components as designed comply with building energy regulations?
How does the energy performance of this building compare with others? How energy efficient is the building? How 'green' is the building?

---

scope for influencing the energy cost implications of the design is much reduced. As argued elsewhere in this book, integrated design, in which energy is a consideration even in the early part of the design, is the way forward. Clearly, early decisions on orientation, plan depth and fabric can have major impacts on how well the building performs.

Integrative aspects of design are important, and it is, in general, necessary to consider how design decisions affect lighting, thermal, ventilation, and acoustic performance. Often, an attempt to improve one aspect of performance will conflict with another design goal. For example:

- The amount of artificial lighting displaced in a naturally lit building will have a direct effect on the internal loads for the thermal analysis.
- Increasing window size to improve daylighting may have an adverse effect on summer overheating.
- Using natural ventilation may not be feasible in city centre office buildings where acoustic analysis indicates that sound pressure levels in the building will be too high.

Conflict can only be avoided if the results from different analyses are brought together in design team meetings. Recent developments in software are beginning to address this problem by making it easier to undertake multi-parameter analyses based on a common building model.

Alongside the large range of questions that may be asked, there is a large range of analysis software available, ranging from simple calculators through to detailed simulation tools that can predict the dynamic thermal and lighting performance of a building throughout the year. This is discussed further in the next section.

The term 'computer modelling' in this context indicates the development of a mathematical representation of a building and the associated physical processes (e.g. heat, light, moisture and air flow). 'Simulation' refers to the analysis of this mathematical representation (i.e. model) to predict the building's performance, usually its performance over time (hourly, daily, seasonal or annual variations). Modelling and simulation can provide the detailed information on building performance needed to answer design questions. Not only do they give designers such information as energy consumption and the number of hours in summer when overheating occurs, but they can be used to understand why the building behaves in a particular manner. This allows designers to develop and test innovative solutions, i.e. to ask 'what if' questions.

An indication of the complexity of the energy flows in a building is shown in Fig. 14.1. Owing to the variable climate and the effects of occupants, the energy flows vary in magnitude and direction over different timescales – from short period fluctuations due to switching of lights, for example, through time lags of the order of a few hours for heat transfer through thick walls, to seasonal variations in ground temperature and climate. The number of interactions between energy flows explains why thermal conditions in one room can influence the conditions in adjacent rooms.

So where does energy and environmental analysis fit in the design process? As mentioned above, to be of greatest benefit, it needs to be considered at all stages of the design. Results from modelling can influence all members of the design team: the architect, who may be interested in lighting quality and visualization; the mechanical engineer who may be evaluating different cooling plant options; the cost manager who assesses the value of the building; the project manager who needs to know what equipment is required (and who will want to make sure the modelling is cost-effective); and lastly the client, who wants to be assured of the high quality of the building.

Used properly, modelling can provide a focus around which the design develops and the implications of design decisions are assessed. In the early stages of design, many options are usually studied. The need is for initial access to easy-to-use tools, perhaps based on design charts, which can give an approximate appraisal of the options. Later on, when a few workable options have been identified, detailed analyses can be undertaken to check and refine key aspects of the design.

14.3 Software tools

A large range of modelling tools exists, of varying complexity and functionality. Table 14.2, based on CIBSE (1998), gives a breakdown of categories of tools for thermal, lighting and airflow modelling. Modelling can also be used to study other aspects of building energy and environmental performance: fire and smoke spread,

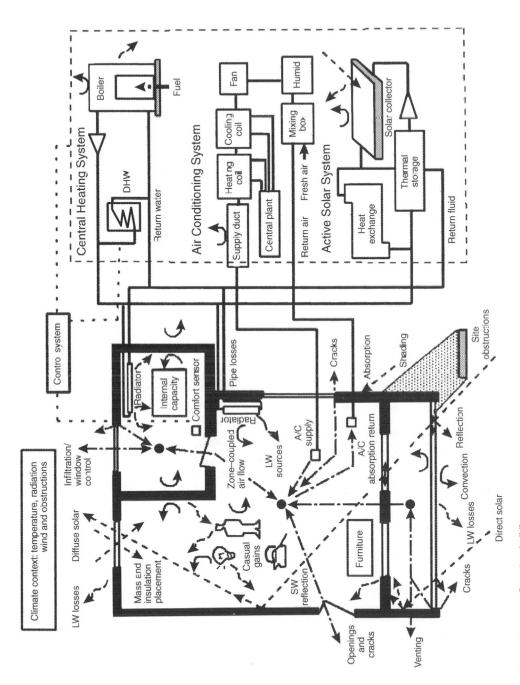

Fig. 14.1 Schematic energy flows in a building.

Table 14.2 Categories of tools

Method	Typical application	Principal limitation
Envelope thermal analysis		
Elemental	Calculate thickness of insulation	Only deals with individual wall/roof constructions, not whole buildings
Steady-state heat loss	Radiator/boiler sizing	Ignores free gains and dynamic effects
Quasi-dynamic analysis	Chiller sizing	All heat gains follow repeating sine wave
Dynamic analysis	Annual heating/cooling loads, overheating risks, control strategies	Fixed time steps (response factor)
Plant systems and controls analysis		
System efficiency	Calculate heating energy from room loads	Cannot deal with air-conditioned or mechanically ventilated buildings
Preconfigured system	System design	Restricted range and configuration of systems
Quasi-dynamic component	System and control optimization	Ignores component dynamics, control lags
Dynamic component	Analysis of control stability	Exacting input data requirements
Combined building and plant simulation		
Annual	Heating energy	Only valid for simple heating systems (e.g. domestic)
Seasonal/bin	Heating and cooling energy	Ignores building dynamics and system/climate interactions
Hourly	Building energy use	Can be difficult to learn and use
Lighting (electric and daylighting) analysis		
Manual	Average illuminance over working plane (electric); daylight factors (daylighting)	Simple room geometry only
Point-by-point	Direct illuminance at a point (electric)	Less accurate where reflected light is important
Daylighting distribution	Distribution of daylight over horizontal plane	Only for standard overcast sky
Lighting simulation	All lighting issues plus accurate visualization	Solution times can be significant
Ventilation and air movement		
Empirical air tightness	Infiltration rates as a function of meteorological conditions	Average values for different types
Simplified theoretical	Time-varying infiltration	Building treated as a single zone
Network	Bulk airflow movement through a building	Assumes each zone has well mixed air
Computational fluid dynamics (CFD)	Detailed air movement in a single space	Long solution times, experience required

pollutant dispersal, acoustics, moisture flow. In terms of building performance analysis, the web site of the US Department of Energy is also a good source for listing available software (US Department of Energy, 1998).

Physical modelling is also possible in wind tunnels for external airflow assessment and under artificial skies for daylighting assessment. This is useful for confirming designs, but accurate models are time-consuming and expensive to construct.

As can be seen from Table 14.2, available modelling tools range from simple calculations, which may be undertaken in a few minutes, through to the creation of detailed computer models that may take several days to set up and simulate. Simplified tools are generally developed to answer a specific set of design questions. Detailed tools can be used to address a greater range of questions and can cope with non-typical buildings. Clearly, selection of the appropriate modelling tool is critical to cost-effective analysis. Some of the factors to be taken into account when choosing suitable software are given in Table 14.3. Not only program capabilities and limitations need to be considered, but also other factors such as level of user experience and cost.

Simplified design tools are often sufficient for energy consumption analysis, particularly where internal conditions do not vary much. There are many design tools available, often customized for the particular characteristics of building stock, heating systems and climatic conditions for each particular country or region. One example of a design tool aimed at the early design stage and for low-energy buildings is given in Section 14.5.2.

The calculation of summertime temperatures can be given as an example of the limitation of simplified analysis. Procedures published by CIBSE (1986) are based on the admittance method, which is a harmonic analysis using a single frequency with a period of 24 hours. The variation in external temperature, solar radiation and internal gains are all approximated to a sine wave with this single frequency.

Table 14.3 Factors involved in selecting modelling software

Functionality of the software

Level of detail and accuracy

Availability of input data

Computer hardware requirements

Program links to other software, e.g. CAD

Ease of use

User support

Availability of experienced users of the software

Cost of purchase and training

Time available for analysis

However, internal gains in many buildings now form an increasingly significant proportion of the heat gain and these gains are not well represented by a sine wave. Also, the use of structures with high levels of thermal mass results in the capability of storing heating and cooling energy, and this is difficult to represent with the simplified calculation techniques.

Because of the complexity mentioned above, ASHRAE now recommends the use of a heat balance method to calculate the complex interactions of internal and fabric loads (ASHRAE, 1997; Pedersen *et al.*, 1998).

14.4 Practical issues

This section deals with some of the practical aspects of using energy and environmental modelling. It must be emphasized that there are no simple step-by-step procedures or manuals which define how a particular building should be modelled. In large measure, experience is gained by working with others who have tackled similar problems. Undoubtedly, an appreciation of environmental conditions in buildings will help in the judgement of the modelling predictions and in the interpretation of the results of the analysis.

There are three important tasks at the start of any modelling study: deriving a specification of what is to be achieved in the study, planning how to undertake the study, and then abstracting the reality to the model.

The first of these involves posing the design question in a form that can be addressed by modelling. Table 14.4 gives some examples.

Table 14.4 Translating design questions to modelling tasks

Design questions	Modelling task
Does this building require air conditioning?	Determine the peak summertime temperatures and their frequency of occurrence with a naturally ventilated scheme
If so, which air-conditioning system will be the most energy efficient?	Compare the degree of temperature and humidity control for various system configurations and evaluate the required capacity and energy consumption
How can daylight penetration be maximized and glare sources eliminated?	Evaluate and compare daylight factors and glare indices for a range of glazing options and shading devices with and without each feature
Will displacement ventilation be able to cope with high levels of internal gain?	Determine the occupied zone thermal comfort levels for a range of loadings and supply air conditions

Once the modelling objectives have been set, it is necessary to plan the modelling study. In this phase, the software to be used and the nature of both the model and the specific assessments to be carried out will be decided. Ideally, there should be regular reporting of interim results to the design team, and quality assurance should be enforced. Phasing of the study is often useful. For example:

- An easy-to-use simplified tool can be used to investigate options for refurbishment, with a more detailed tool used at a later stage of the design for the most favourable option.
- A bulk airflow simulation can provide air change rates as inputs to a thermal simulation.
- Thermal simulations can provide surface temperature boundary conditions for detailed room airflow simulations using computational fluid dynamics (CFD) programs.

With some software it may be possible to run models at different levels of detail at different stages of the design process. Clearly this is more efficient than having to recreate models in different design tools.

Abstracting the physical reality into a model is perhaps the most challenging task, and one in which beginners most often fail. The aim is to keep the model as simple as possible without introducing significant errors that might result from oversimplification. Some points to consider are given in Table 14.5.

Table 14.5 Abstraction from the physical world to the modelling world

---

Which part of the building needs to be modelled?

To what extent can the geometry be simplified without compromising the required accuracy?

Is it acceptable to assume that the air in a room is well mixed, or should account be made of stratification?

What assumptions should be made if the construction materials are unknown?

Is it acceptable to use design air exchange rates or is more detailed analysis required?

How detailed should occupancy profiles be modelled?

It must be remembered that uncertainties always exist in modelling predictions due to assumptions made, program limitations and simple user error. Experience is necessary to assess these errors. Sensitivity analysis is also a useful technique: this involves changing one or more input parameters, for example air ventilation rate, and studying the effect on the output of interest, for example comfort conditions. Such studies can lead to confidence that the modelling predictions are 'robust' and not overly sensitive to the input data and model assumptions.

14.5 Examples

This section shows some examples of the range of analysis that is possible.

14.5.1 Shading analysis

A shading analysis can be useful for:

- allowing the architect to study time-varying patterns of shading on the building
- determining shading factors as inputs to thermal analysis, where the effect of shading can be important for passive solar buildings

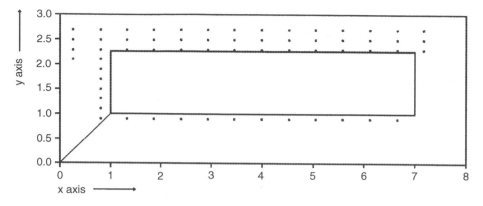

Fig. 14.2 Shading patterns for calculation purposes.

- siting of photovoltaic modules – these should receive no shading, but if it is unavoidable, then the modules must be grouped so that each group is shaded at the same time, otherwise one shaded module within the group can reduce the power output from the whole group.

The program used will depend on whether a photo-realistic image is required. If so, a detailed daylighting program is necessary; if not, then simplified tools are sufficient. Figures 14.2 and 14.3 show the result of both types of analysis. Figure 14.2 shows a gridded wall with a window, the dots indicating grid points that are shaded, allowing a calculation of the percentage of the facade that is shaded, while Fig. 14.3 allows the user to visualize the shading patterns.

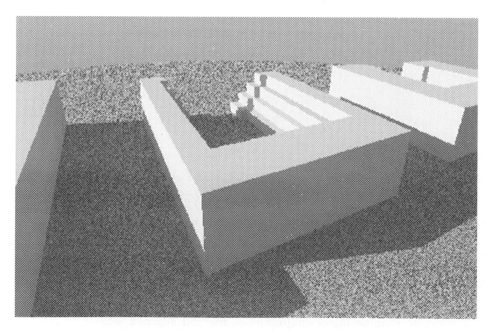

Fig. 14.3 Shading patterns for visualization purposes.

14.5.2 Simplified energy analysis

The example shown here is the LT method (Baker and Steemers, 1994). This method uses a set of graphs (although it can be encoded into spreadsheet form) and is designed to be easy to use. The LT method, L(ighting) and T(hermal), is an energy design tool for non-domestic buildings, aimed at the early stages of design when it is useful to obtain relative estimates of energy consumption for several design options. At this stage, the designer is often concerned with two main issues:

- the form of the building – its plan depth, section and orientation
- the design of the facades; in particular, the area and distribution of glazing.

The method allows the prediction of the annual primary energy per square metre of floor area as a function of:

- local climatic conditions (for several areas of Europe)
- building type (schools, colleges, offices and institutional)
- orientation of the facade
- area of glazing
- obstructions due to adjacent buildings (or parts of the same building)
- optional inclusion of an atrium
- occupancy and vacation patterns
- lighting levels
- internal gains.

The method uses the concept of passive and non-passive zones. Passive zones can be daylit and naturally ventilated and may make use of solar gains for heating in winter, but may also suffer overheating by solar gains in summer. Non-passive zones have to be artificially lit and ventilated, and in many cases cooled to prevent overheating due to internal gains. Figure 14.4 is an example of an LT curve. There are separate curves for cooling, lighting, and heating; a curve giving the combined heating and lighting, and a total energy curve. Data are entered into a worksheet, such as shown in the example of Fig. 14.5. In the upper part of Fig. 14.5, a diagram of the building is shown with the divisions into passive and non-passive zones. The adjacent table gives the areas for each orientation and floor of the building. These areas are entered into Section 2 of the main worksheet, together with the facade glazing areas. Specific energy consumption data, taken from the LT curves of Fig. 14.4, are entered in Section 4 of the worksheet. Urban horizon factors (UHF) are then entered: these take into account shading by adjacent buildings. Multiplying the areas by the specific energy consumption and UHF gives the total energy consumption. Account can also be taken of any buffer spaces, e.g. an atrium. After correction for boiler efficiency, the summary box can be completed, giving the total energy consumption, broken down into figures for lighting, heating, and ventilation and cooling. The associated CO_2 emissions are also shown.

Because of the simple nature of the design tool, there are many in-built assumptions. For example, the wall and roof construction heat losses are those from the Building Regulations, the glazing framing is assumed to be 20%, the glazing type is assumed to be double glazing, fixed occupancy schedules are assumed for the particular building type and the heating set-point is 20°C.

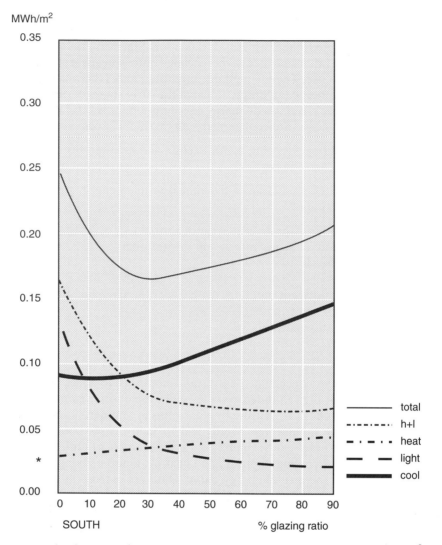

MWh/m^2

SOUTH   % glazing ratio

total
h+l
heat
light
cool

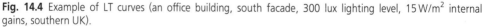

Fig. 14.4 Example of LT curves (an office building, south facade, 300 lux lighting level, 15 W/m^2 internal gains, southern UK).

14.5.3 Detailed integrated study

The example given here is of the Brundtland Centre in Toftlund, Denmark, and is designed to demonstrate the level of complexity and range of issues that can be modelled. However, it must be noted that this study took several weeks of effort.

The study focuses on:

- integrated thermal and daylighting assessments
- the degree to which such assessments enhance design decision support
- how performance information is delivered to the design professions
- the cooperative use of analysis tools.

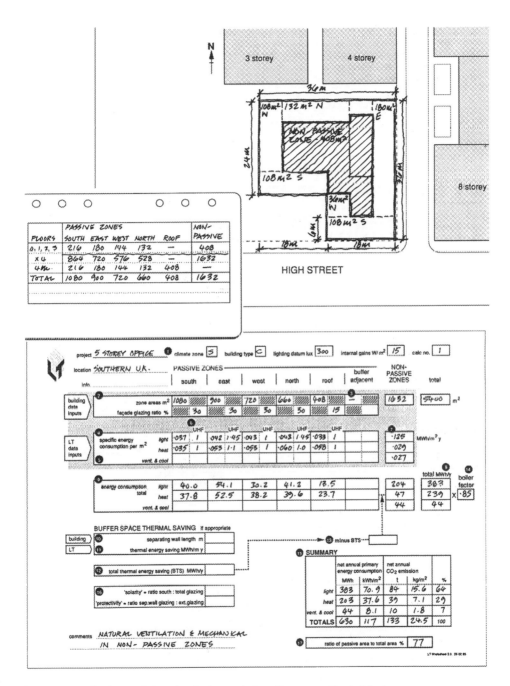

Fig. 14.5 Example of LT worksheet (worked example for a five-storey office building).

Fig. 14.6 Brundtland Centre: principal features.

The project is a design study for a low-energy exhibition centre, as shown in Fig. 14.6. The project exemplifies attempts by the design professions to meld passive and active solar design strategies. It is based on work carried out as part of the two European research programmes – Solar House (Fitzgerald and Lewis, 1993) and Daylight-Europe (Kristensen, 1996).

The design brief for the Brundtland Centre included several goals:

- a high standard of thermal comfort without the use of air-conditioning, and with limited heating costs

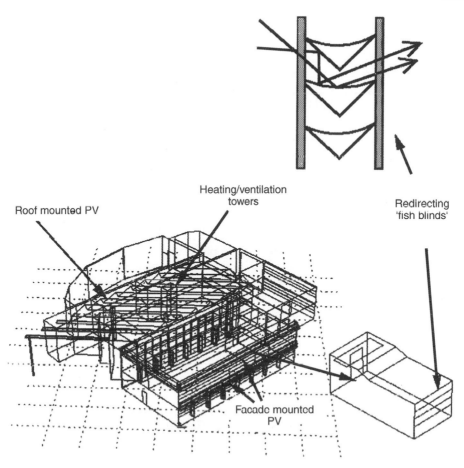

Fig. 14.7 Brundtland Centre: model.

- heating supplied by direct solar gain, radiators in the upper floor, heated floors in the exhibition and atrium spaces and a displacement ventilation system in the lecture theatre
- a high standard of daylighting via borrowed light from the atrium, light directing blind systems and occupancy sensing controls of uplighting and blind systems
- balancing of lighting power demands by the use of facade and roof-mounted photovoltaic modules.

Some aspects of the design involved combinations of novel techniques and technology which demanded detailed assessments of the overall performance of the building and the relationships between design elements. The approach taken in this project was to proceed from the design goals and architectural and operational descriptions to define the metrics of the study (e.g. what comfort, light level and energy use constitutes good performance), the rules for setting up an appropriate simulation model (e.g. where is the detail required, how should building operation be represented) and deciding what assessments to run (e.g. will a typical winter week be sufficient to show the patterns of performance).

The computer model was of sufficient resolution to allow detailed daylighting and glare assessments in monitored offices, the calculation of comfort levels in the exhibition spaces, the determination of overall energy use, and a representation of the ventilation scheme. The resulting model, shown in Fig. 14.7, focused on a typical south-east office, an atrium facing office, the atrium and PV modules, with other parts of the building represented at a level of detail sufficient to define thermal and visual boundary conditions. The thermal analysis was carried out with the ESP-r program (ESRU, 1998); Radiance (Radiance, 1998) was used for the lighting analysis.

The initial task was to reach a consensus on the composition of the model and roughly sketch its layout and digitize critical floor plan coordinates. After creating the model it was tested for errors. Then simulations were run for typical days to see if the resulting performance was reasonable. This process required three working days. Fine-tuning the model continued for several days. A series of assessments were then carried out and the patterns of performance communicated to the design team in the form of integrated performance views, as shown in Fig. 14.8. The figure includes a set of graphs and images focused on specific topics: maximum plant capacity, atmospheric emissions, hourly energy demands for typical days, and thermal and visual comfort. It is possible to compare the performance implications of design decisions by reviewing such presentations.

The analysis showed that the building provided considerable energy savings compared with a conventional building, but that it was subject to overheating if the blind system control failed. It also indicated that on sunny winter days comfort was improved in the atrium when warm air at the roof level was redirected to the lower portion of the atrium.

Another result of the analysis was information on patterns of sunlight penetration in the atrium and glare sources, as shown in Fig. 14.9.

14.6 Future trends

Building energy and environmental modelling software can be used to predict the performance of buildings and to study different design options and 'what if' questions. There is evidence that the use of modelling is increasing, especially in prestige projects and for complex designs (CIBSE, 1998) and, as stated in the introduction, there are good reasons why modelling should become a more important part of the design process. In a survey of energy simulation users conducted in the US in 1995, the Department of Energy found that 10–15% of new building designs were using simulation. In the same survey, users reported that more than 50% of energy modelling was now for existing building retrofit. Such analysis can lead to economic benefits – directly through the reduction in energy use and plant cost, and indirectly through providing a more amenable working environment.

Future trends are likely to focus on better user interfaces, with intelligent defaults to overcome the burden of inputting data. In addition, there is a continuing focus on the integration of different aspects of energy and environmental design performance that will enhance the usefulness of such tools in the

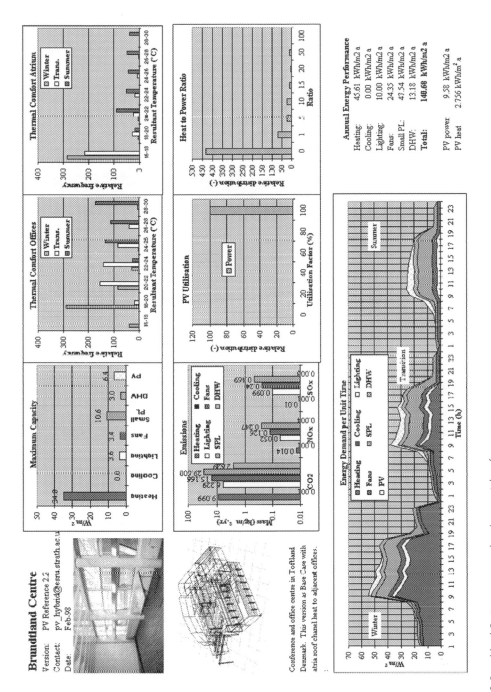

Fig. 14.8 Brundtland Centre: energy and environmental performance.

Fig. 14.9 Visualization study of sunlight patterns.

design process. An additional requirement is for people with the skills to use the modelling software and interpret the output. This is being addressed by the inclusion of modelling in degree courses, and other training as part of Continuing Professional Development.

With these developments, such software should start to move from being used by specialist consultants and more into routine use by designers.

References and bibliography

ASHRAE (1997) *ASHRAE Handbook: Fundamentals* (Atlanta: American Society of Heating, Refrigerating and Air-Conditioning Engineers).

Baker, N.V. and Steemers, K. (1994) *The LT Method 2.0: An Energy Design Tool for Non-Domestic Buildings* (Cambridge Architectural Research Ltd).

Brohard, G. and Caulfield, T.O. (1998) Revisiting the ACT2 commercial pilot, five years later. In: *Proceedings of the 1998 ACEEE Summer Study on Energy Efficiency in Buildings* (Washington DC, American Council for an Energy-Efficient Environment), pp. 3.35–3.46.

CIBSE (1986) *CIBSE Guide A: Design Data* (London: Chartered Institution of Building Services Engineers).

CIBSE (1998) *Building Energy and Environmental Modelling, Applications Manual AM11* (London: Chartered Institution of Building Services Engineers).

EIA (1998) *Annual Energy Outlook* (Washington DC: Energy Information Administration, US Department of Energy).

Energy Efficiency Office (EEO) (1991) *Energy Efficiency in Offices: Energy Consumption Guide 19* (London: IIMSO).

Fitzgerald, E. and Lewis, J.O. (1993) The solar house – design support. In: *Proceedings of the 3rd European Conference on Architecture*. Florence, Italy, pp. 679–82.

Goulding, J.R., Lewis, J.O. and Steemers, T.C. (eds) (1992) *Energy in Architecture – The European Passive Solar Handbook* (London: Batsford).

Kristensen, P.E. (1996) Daylight Europe – Joule Project CT94-0282. In: *Proceedings of the 4th European Conference on Solar Energy in Architecture and Urban Planning*. Berlin.

McElroy, L.B., Hand, J.W. and Strachan, P.A. (1997) Experience from a design advice service using simulation. In: *Proceedings: Building Simulation '97*, Prague.

Pedersen, C.O., Fisher, D.E., Spitler, J.E. and Liesen, R.J. (1998) *Load Calculation Principles* (Atlanta: ASHRAE).

Robinson, D. (1996) Energy model usage in building design: a qualitative assessment. *Building Services Engineering Research and Technology*, **12** (2), 89–95.

UCD-OPET (1995) *Tools and Techniques for the Design and Evaluation of Energy Efficient Buildings*, Thermie Action No. B184 (Brussels: European Commission, Directorate-General for Energy (DGXVII)).

Web sites

Building Environmental Performance Analysis Club (BEPAC):
http://www.bcpac.dmu.ac.uk

ESRU (1998): Energy Systems Research Unit:
http://www.strath.ac.uk/Departments/ESRU

International Building Performance Simulation Association (IBPSA):
http://www.mae.okstate.edu/ibpsa

Radiance (1998): http://radsite.lbl.gov/radiance

US Department of Energy (1998): http://www.eren.doe.gov/buildings/tools_directory

15

Value management

Mark Neasbey, Roy Barton and John Knott†

Editorial comment

The use of value management (VM) is steadily increasing as clients seek better outcomes from their investment in buildings and structures. Some clients are now including the requirement for VM workshops in building contracts, as a way of ensuring optimal solutions. In other cases the project is an outcome of strategic value management processes used in client organizations.

In the quest for better value in building, VM is one of the fundamental tools that can be used. By bringing together the widest possible range of project stakeholders in the VM workshops, where different views and perspectives can be openly debated, many of the problems that typically arise in building projects can be avoided.

There are two particular advantages of VM. The first is the cooperative and inclusive nature of the workshops, which gets people talking to each other and moving in the same direction. The second is the formal process for considering and weighing the options available to the client for a building project. At present, there is no alternative management tool or technique available that can be used for the purposes of VM, or to get the benefits gained from a rigorous VM process.

Although the origins of VM lie in the manufacturing industry of the 1950s it is widely applied in various forms in most industries these days. For the building industry, VM offers a technique to counter the perception of the industry as being essentially unconcerned about the client's business requirements or goals, because VM is essentially about clarifying what these goals are and how they can be met.

In terms of building in value at the pre-design stage, VM is clearly a tool that clients ignore at their own expense. With strong links to other tools outlined in this book, such as functional use analysis and cost modelling for example, VM makes an important contribution to the success of a project. This chapter outlines VM and shows how it is applied to building projects.

† Australian Centre for Value Management, Australia

15.1 Introduction

The purpose of this chapter is to introduce the concept of *value management* (VM), and to describe its application across all stages of the project life cycle. The discussion outlines the VM methodology, the range of application of VM, and briefly describes the way in which a value management study (VMS) may be initiated.

Value management originated in the US during the Second World War, when scarcities forced industry to consider ways of economizing resources or using substitutes. This process was called value engineering or value analysis, and was mainly concerned with the functional analysis of manufactured products. Gains from VM come from examination of function within a system-wide context and from optimizing design solutions to meet project objectives.

The challenge is to determine how the client can achieve best value from an investment in a building and how better buildings can result from a process that is based on good decision-making procedures being put in place before the design work actually commences.

A working definition of VM from the Australian and New Zealand Standards is:

> Value Management is a structured, systematic and analytical process which seeks to achieve value for money by providing all the necessary functions at the lowest total cost consistent with the required levels of quality and performance. (AS/NZS 4183, 1994)

The structure is provided by the methodology, which comprises a five-stage creative problem-solving process known as the 'job plan'. The approach is systematic in that all five stages of the job plan are addressed in sequence. The process involves the identification and the analysis of function, which makes clear the objective or purpose as well as the means of achieving it. This analysis leads to the generation of creative ideas about achieving the function or purpose by alternative means at a lowest total cost whilst achieving specified levels of performance and quality. Figure 15.1 shows the Australian Centre for Value Management Model.

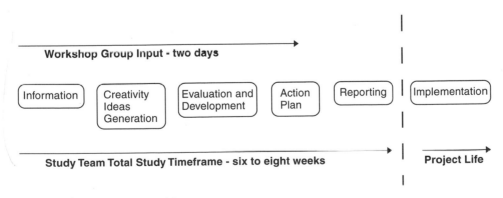

Fig. 15.1 Value management model.

A	B	C
Pre-Workshop Stage • Establish VM timetable • Determine study objectives • Select study facilitation team • Gain stakeholder commitment • Select participants • Arrange venue • Brief participants • Brief presenters • Circulate information	**Workshop Stage** • Information phase • Analysis phase • Creativity phase • Judgement phase • Development phase • Action plan	**Post-Workshop Stage** • Implement action plan • De-brief stakeholders • Distribute study report • Evaluate study performance

Fig. 15.2 The value management study process.

15.2 The value management study

The VM study is that part of the procurement process that commences with the notion to build and ends just prior to the preparation of sketch drawings. A value management study (VMS) is comprised of three separate stages, as shown in Fig. 15.2, each with its own purpose and objectives. The stages are: Pre-workshop; Workshop; and Post-workshop.

15.2.1 The pre-workshop stage

This stage involves the facilitator and the sponsor of the study, conferring to identify all relevant information, to identify the key stakeholders and to organize the workshop. The key activities in this stage are establishing the value management study timetable and workshop objectives, finalizing the value study team including technical specialists, and identifying and gaining the commitment of stakeholders. Nominating, inviting and briefing of participants, arranging the venue and briefing presenters/sponsors for the information phase of the workshop are also done in this stage. Background material is usually sent to the participants as part of the workshop preparation.

15.2.2 The workshop stage

This involves a facilitated workshop comprising key stakeholders. The facilitator leads the group through the five phases of the job plan, which are: Information phase; Analysis phase; Creative phase; Judgement phase; and Development phase.

Phase 1. The information phase

This phase of the workshop is concerned with the sharing, dissemination and clarification of information relating to the project and includes identifying the problem situation, identifying the project givens and project functions. During the project overview, the project's objectives and underlying assumptions will be identified and clarified. The facilitator assists the group to become effective in identifying and addressing the key issues and concerns in a structured way.

Phase 2. The analysis phase

In this phase, the functions of the project are identified and analysed. They may be represented in a hierarchical format and displayed on a function diagram. Key questions asked during the analysis phase are: what does it do and what must it do; what does the function cost (life cycle cost); and what is the value of the function? From the function analysis, functions that appear to be disproportionately expensive are identified.

Phase 3. The creative phase (Figs 15.3 and 15.4)

During this phase, the group explores alternative approaches for the achievement of functions, and generates, in a brainstorming fashion, a number of creative ideas.

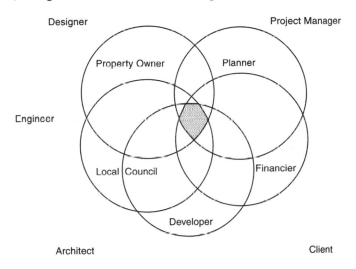

Why use a group approach to creativity? Fundamentally because you can optimize the range of possible solutions. It works because of what is called 'constructive overlap'. Each participant stake holder comes to the process with some particular knowledge, skill or experience, as well as interest in the outcomes. Each is able to add to, overlay or expand the ideas of others in the group – all focused on the proposed facility. Despite their likely different values and objectives they are facilitated to build a range of options for delivering or achieving functions and thus give greater scope to an holistic solution rather than one that is seen as piecemeal or sectoral in terms of benefits.

This is reflected in the diagram above where each 'circle' is intended to represent the different stake holder perspectives.

Fig. 15.3 Constructive overlap.

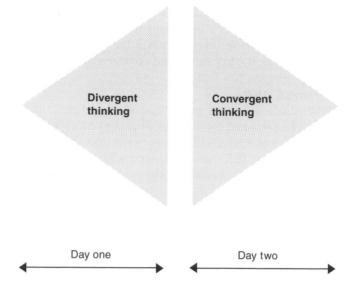

A key to this phase is the mental modelling of the group. It is most important that they not be constrained in the generation of ideas to achieve particular functions and equally important that they withhold judgement until as many ideas as possible have been created. In this way they have as comprehensive a selection as possible from which to choose and mould a more appropriate solution.

This is illustrated in the diagram above. The first phase of the study workshop will see a process of widening of perspectives as the sharing of information builds a better appreciation of the dimension of functions and values that need to be addressed. The second phase starts to bring clarity of direction, selection of solutions and agreement on outcomes for the proposed facility.

Fig. 15.4 Workshop perspectives.

In the process, the group is encouraged to suspend judgement. Group synergy is utilized to create more numerous and more creative options than is possible with the members working individually. During the creativity phase, the questions asked are: how else may the required function be performed or what else will perform the required function; what will the alternatives cost; is the function necessary; and can we eliminate, change, combine, relocate, add, simplify? Creative ideas are generated and recorded.

Phase 4. The judgement phase
During this phase, participants are asked to judge the ideas based on criteria previously identified and to select those that may lead to improved outcomes.

Phase 5. The development phase

Creative ideas judged as having potential during the previous phase are further developed during this phase into potential solutions.

In some cases, the work of this phase will extend beyond the value management workshop and may involve, for example, the development of detailed drawings and cost estimates. Such tasks will be listed in the action plan.

The final task in a value management workshop is to deal with the outcomes of the workshop session. Depending on the length of the workshop and the stage of the project at which it is conducted, these outcomes may take several forms and be expressed as follows.

- Action Plan – this would identify the findings and provide a timeline for ongoing development, evaluation and decision-making. Follow-up meetings for those involved in actioning items in the action plan are recommended to ensure that all value improvement opportunities are fully developed.
- Workshop Findings and Recommendations – with longer studies, where time is available to evaluate options more thoroughly, firm recommendations may be developed. They may be presented to sponsors or management as an implementation plan. They may also be delivered as a formal presentation.
- Formal Presentation   where possible, these recommendations should be formally presented to management and other stakeholders so that issues can be resolved before a final report is prepared. The process of preparing a formal presentation assists in clarifying issues and consolidates the commitments to, and ownership of, the recommendations.

15.2.3 The post-workshop stage

During this stage, the action plan is implemented, a report on the workshop is prepared and circulated and an evaluation is conducted. The responsibility for ensuring that the action plan is implemented is allocated to an appropriate person who is also responsible for incorporating the recommendations into the ongoing project development.

The report provides a clear and comprehensive record of events that led to the development of recommendations. It is sometimes necessary to refer back to events during the workshop for clarification or justification. Finally, a form of evaluation is recommended to ensure that the value management process achieves its objectives in a cost-effective manner.

15.3 The benefits of value management

A value management study generally has some impact on everyone associated with a project, in one way or another. The client is generally most concerned with achieving value for money from the investment. Users are concerned that the project meets their needs as closely and effectively as possible. Designers are keen to meet the expectations of both the client as well as the users and to comply with relevant standards and performance criteria. Project managers are keen to ensure that the project is managed within the constraints of time, quality and budget and contractors are keen to provide services at an adequate profit.

Value management assists in identifying and meeting the needs and interests of all of these groups and, in the process, enhances both the understanding of the project and the communication processes. In a well-managed study, all parties can achieve significant benefit from both the process and the outcomes. Beneficial outcomes are highlighted in Table 15.1.

The impact of a value management study on a project is usually quite significant. Nevertheless, if the study does no more than confirm the project's status and direction, then the stakeholders can be assured that the confirmation is based upon a defensible analysis of function, cost and worth, and on an exploration of possible alternatives.

The potential risks associated with the value management process are minimal and can be effectively managed. They include: improper application of the methodology by an unskilled facilitator or inadequate information leading to incorrect assumptions; inadequate representation of stakeholders in the workshop group; inadequate allocation of time, leading to sub-optimal outcomes; inadequate support by senior management, and late initiation of a study, within the project life cycle, which limits the potential impact.

15.3.1 Initiating a study

A value management study may be initiated by:

- the client (who may be concerned with finding the best solution, developing a clear brief, minimizing risk)
- the project manager (who may be interested in a better brief, involvement of stakeholders in planning process, time savings)
- the design consultant (who may be seeking stakeholder input in the design process, to explore alternative solutions)
- the project sponsor/client (central agency) (who may be concerned with the identification of optimal solutions for delivery)
- the asset manager (who may seek innovative strategies for managing the asset portfolio).

15.3.2 Determining the need for a study

The decision to initiate a value management study will take into account the perceived potential for a value improved outcome, the stage in the project development life cycle, the need to have broad representation and involvement, the benefits that a workshop involving key stakeholders will return, the access and availability of key stakeholders and the cost of the study. The decision is generally agreed in consultation with management.

In broad terms, the factors that underpin the decision to initiate a value management study are the potential for savings from managing the large number of stakeholder interfaces, the complexity of the project or the innovative nature of the project. The goals may be to overcome a difficult problem, to achieve project

Table 15.1 Beneficial outcomes from value management

Outcome	Example
Clarification of stakeholder needs, the separation of needs from wants, a refined definition of user requirements, functional clarification and definition of the project objectives, improved client brief	A proposed training facility to be established within a university campus: initial briefed area was reduced by 30%, bringing it back within its approved budget; this was done without sacrificing any of its required functions in the delivery of training programmes.
Rationalization of outcomes	Proposed establishment of new operational headquarters within a very tight timeframe: several months of detailed planning had already been undertaken by members of the team but they were unable to meet the required deadline for occupation. The VM study's review of the project schedule and activities reduced the project time by four weeks.
Identification of alternative designs, alternative solutions, alternative locations	1. A major hospital redevelopment with a detailed planned staging based on hundreds of hours of work by many people: the VMS found that the schematic design was operationally unworkable for the hospital leading to a revised strategy and changes in scope.
	2. A police facility at pre-tender stage: the original design solution was abandoned after the VMS showed that a vital client requirement had been overlooked and a completely new design was required.
Identification of alternative construction methods or modifications to construction time-lines	A remand facility, under construction, for a Justice Department, was behind schedule and over budget; the VMS review of the unlet trade packages yielded substantial cost and time savings. The participants acknowledged this should have been done much earlier and would have yielded even greater savings.
Enhanced client involvement with project development	1. A major public exhibition facility where the schematic design cost plan was some 30% over budget: the client's active participation in the VMS ensured essential functions were catered for but also allowed immediate endorsement of design changes, which improved both affordability and quality.
	2. A proposed bridge, where the planned length of the span was challenged by local residents: their local knowledge and experience brought information to the process which would otherwise have not emerged until after construction was under way and a much less expensive engineering solution resulted.

Table 15.1 *Continued*

Outcome	Example
Team building leading to improved communication between stakeholders	1. A proposed sporting facility: the operational requirements were tested with active involvement of interstate operators of similar facilities. With better understanding a significantly improved design solution resulted.
	2. A proposed extension to a horticultural facility, where attempts over several years at reaching an agreed design solution had left the stakeholders frustrated and disillusioned: the VMS, including the key stakeholders, clarified everyone's understanding of what was really needed, through a better appreciation of the functions and of alternative methods of satisfying them. The result was a shared solution which proceeded immediately into detailed design development and tender.
Wider ownership of project outcomes and commitment to implementation	1. A major road project requiring a route selection with seemingly contrary sensitive environmental, economic and social issues: the VMS process was able to bring the conflicting interests together, clarify what was important to the various parties and show them how they could best use their perspective to guide the route selection.
	2. A proposed redevelopment of a civic facility was seen as controversial: the VMS process drew out the shared values of the various stakeholders regarding the facility and its future that any solution had to address. More options were generated from which a blended solution emerged, dissipating the controversy and winning support for the final design.
Identification of risk	A proposed dam raising to improve reliability of water supply for a growing community: the VMS identified the risks to the project and with a better understanding of its functions, a clever design and development solution emerged with consequent time and cost savings.

acceleration, to overcome interfacing problems, to develop or improve a brief; to meet statutory requirements, to provide rigorous review or audit, to improve value for money and/or to develop/review design options.

15.3.3 Timing (Figs 15.5 and 15.6)

Value management studies have application at all phases of the asset life cycle, covering planning, procurement, maintenance and management, and disposal. A

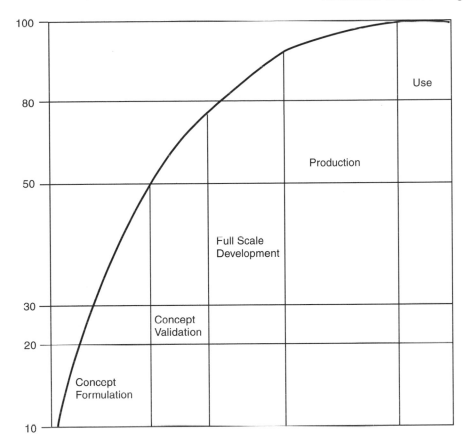

A critical perspective on the timing of a VM study is to appreciate a key principle involving physical assets. This is that which at the time of validating the concept (in this case that a facility is required), sets 80% of the pattern of the life cycle costs of the facility. In other words even before any drawings are developed some 80% of the expected life cycle costs of the asset are being predetermined. This is illustrated in the diagram above which was derived from US research.

The important message that this tells us is that 'getting it right at the start' will give us greater control over the overall cost of this facility. In other words the best time to try and improve value is at the concept stage – not once design work has begun. This is reinforced by Fig.15.6 which shows the relationship between the value of savings that can be achieved with VM and the cost of implementing those savings at various stages of the project.

Fig. 15.5 Cumulative percentage of life cycle costs.

value management study may be initiated at any of the following stages of the project life cycle, and indeed some projects will involve a VMS at strategic planning, concept development and concept evaluation, project definition, project delivery, and review stages.

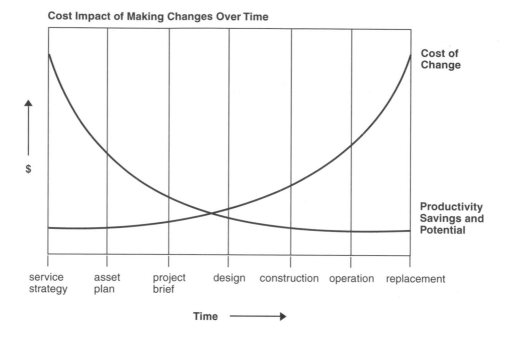

We all know that making changes when in contract is very expensive. When in design the cost and delays associated with redesign work can be enormous for many projects and may mean the difference between a good commercial outcome and a financial disaster. This is not so much because of the cost of the design work itself but rather because of the costs associated with delays to final delivery of the facility, and the effectiveness of relationships throughout the procurement process.

Fig. 15.6 Value management potential.

The greatest potential saving is possible at the earliest stage in the project life cycle, i.e. when the needs are being identified, and minimum commitment of funds has been made. Both the study objectives and the mix of stakeholders in the workshop group are determined by the particular phase at which the study is conducted. It is important that the use of value management is planned well ahead so that the study or studies are conducted in good time so that maximum benefit can be achieved.

15.4 Initiation criteria

There are a number of initiation criteria discussed below. These provide the necessary justification for commissioning a value management study.

Because of the potential for savings through value management, studies may be initiated on projects where it is suspected that there could be savings. These may include options for developing alternative, more cost-effective ways of achieving functions, taking into account the total life-cycle costing.

Projects which contain complex issues may require that decisions be made using input from key stakeholders. The complexity of a project increases with the number of disciplines involved in the design or with the number of stakeholders. There may, for example, be competing interests, or a need to establish priorities amongst a number of contentious items. Value management provides a methodology for addressing such complexities, and for providing a range of potential solutions.

Projects may involve problems or problem situations that are best resolved using a facilitated group process. Such problems may be common, recurring and difficult to resolve without the benefit of stakeholder involvement.

Value management studies can be used to accelerate projects. By bringing together a range of people involved in the project delivery process, a high level of consensus can be achieved in a short space of time. Where, for example, the study is conducted at the project briefing stage, the time can be reduced as a result of stakeholder involvement and the understanding of the project gained during the study. In this case value management can shorten the critical path of the project delivery process.

Value management studies can be used to audit capital works procurement programmes. They can also be utilized to achieve continuous improvement of standard products, or used to check a project rigorously. Value management is used in this way to ensure that value for money is being achieved through the client's capital works programme.

Where such problems exist, value management may be used to find solutions that may be either temporary or permanent. Examples include the need to develop a high degree of understanding between departments jointly involved in a project, or the need to find a bridging solution to a new or existing problem.

Where the general public has a significant level of interest in a project, and issues such as cultural value are involved, value management studies should involve key stakeholders and can be used as a means of addressing the complex issues in a formal manner, and for validating the project.

The cash value alone may be sufficient grounds for initiating a study. It is important, however, that all parties agree to the need for a study. Any attempt at making studies mandatory, based simply on the cost of a project, is likely to encourage resistance to the process and, as a direct result, less than optimum outcomes.

A value management study may be initiated for the purpose of developing the project brief. When conducted at an early stage in a project life cycle, maximum opportunity for value improvement is available. An added advantage is that the client, end user, designer and other key stakeholders are present, participating in a facilitated problem-solving exercise, sharing knowledge and understanding.

Projects which involve new ideas or new ways of doing things may be subjected to a value management study in order to develop and compare options.

15.5 Why use value management?

What sets value management apart from other problem-solving methodologies is that it is both informational, in that it systematically deals with all that is known about the project, including underlying assumptions, givens, perceptions etc., and it is transformational. This means that individual paradigms may be shifted in the process and the creativity may lead to radically different solutions. It is not uncommon for individuals to reverse completely their position on issues as a direct result of their involvement in a value management workshop.

This comes about because the workshop environment involves stakeholders in cooperative problem solving. Enhanced communication and networking are direct benefits of the workshop process which have a significant impact throughout the project.

It can provide a structured mechanism which avoids individual concerns or possible embarrassment over querying aspects of the project or in making suggestions to improve it.

The workshop process 'forces' information disclosure and explanations that might not otherwise emerge by other processes, thus making it available to assist the design team to produce a more appropriate solution.

It is important to note what value management is not. It is not a substitute for cost planning/cost engineering, project management and quality assurance. Nor is it a 'cure-all', a 'witch-hunt' or an excuse for design-by-committee.

For value management to be successful, the important aspects are explicit executive support, the attitudes of participants, the methodology and workshop facilitation, and general management of the process.

15.5.1 The strategic value management study

As the title suggests, this study is conducted at the earliest possible point in the project life cycle, as soon as a perceived need is identified. It involves a broad range of stakeholders, usually representing a high level in the client organization, and includes representation from users, regulating bodies, planners, asset managers and the like. The group is generally large, up to 30 in number, and the workshop is generally of two days' duration. These may be two consecutive days, or other combinations which suit the particular situation. The characteristics of the strategic study include:

- a wide range of stakeholder representation
- a two-day workshop format
- a systems approach focused on the service delivery
- a focus on function
- creative development of delivery options
- assessment of total asset cost which includes life cycle costing
- prioritization of service delivery options
- the use of advanced facilitation methods.

15.5.2 VM studies, buildability and POE

Value management studies conducted at concept and design stages have a different mix of stakeholder involvement, different objectives and different outcomes. They

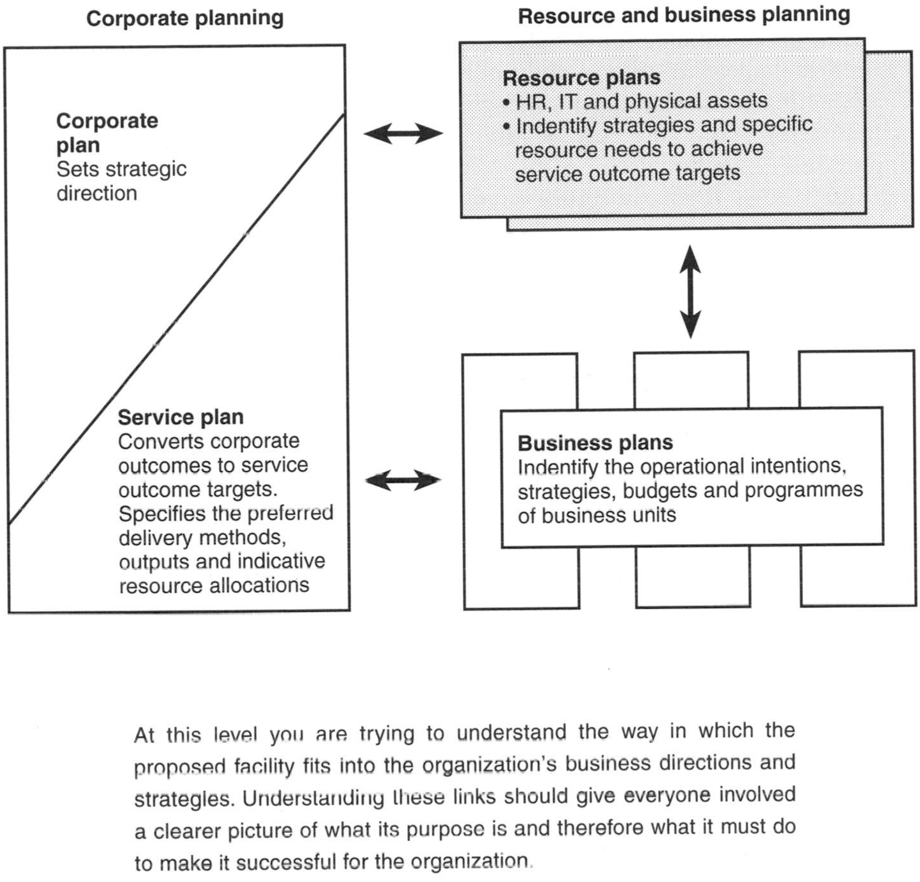
Corporate planning
Resource and business planning
Corporate plan
Sets strategic direction
Service plan
Converts corporate outcomes to service outcome targets. Specifies the preferred delivery methods, outputs and indicative resource allocations
Resource plans
• HR, IT and physical assets
• Indentify strategies and specific resource needs to achieve service outcome targets
Business plans
Indentify the operational intentions, strategies, budgets and programmes of business units
At this level you are trying to understand the way in which the proposed facility fits into the organization's business directions and strategies. Understanding these links should give everyone involved a clearer picture of what its purpose is and therefore what it must do to make it successful for the organization.

Fig. 15.7 Strategic links.

may be two-day, three-day or five-day studies, although not necessarily conducted over consecutive days.

Another form of value management study is conducted immediately after letting the tender but before construction commences. This involves the project team and the contractor(s) in a process of determining the best solution to logistical issues related to the construction.

The value management methodology is most suitable for assessing the issues arising from occupancy and use of the facility. A *post-occupancy evaluation* (POE) can be a powerful tool for assessing success in achieving project goals and client satisfaction.

Each type of study is important to the overall success of the project. In all cases, the standard job plan is followed, representative groups are formed which are appropriate to the type of study, objectives are set and agreed, and the process is professionally facilitated (see Fig. 15.7).

15.6 The cost of a study

The decision to commission a value management study should be based on 'best practice' management criteria and not on cost alone. However, the costs involved in conducting a study must be recognized and include the cost of a consultant facilitator (or team), the cost of attendance for each participant, administrative support and venue hire.

Whilst studies regularly return savings many times the cost of conducting the study, the cost must be substantiated and the need for the study validated. It is also important, however, to value the benefit of the enhanced networking and communication that follows a successful study.

15.7 The study facilitator

The choice of facilitator is fundamental to the success of a study. The decision to engage an in-house facilitator, a consultant facilitator, or a facilitation team will be made taking into account the nature of the study, the complexity of the task, and the size of the stakeholder group. It is important that the facilitator be objective and therefore have no other involvement with the project.

In choosing a consultant facilitator, it should be a requirement that he or she conduct value management studies in accordance with the relevant standards (e.g. Australian and New Zealand Standard AS/NZS 4183:1994). The facilitator should also be asked to provide references from previous clients. Factors to be considered in the selection of a facilitator include experience as a VM facilitator with similar projects and suitable qualifications.

15.8 Performance measures

The two key aspects of a study that require attention in any form of performance assessment are process and outcomes. An assessment of the value management process is made on the basis of how effectively the facilitator managed the process as specified in the job plan, facilitated the group, and liaised with the client throughout the entire value management study.

An assessment of the outcomes will be made by observing that the study objectives were met and there was an enhanced understanding of the project by workshop participants. The participants should have gained a better understanding of the views of others with enhanced networking and communication amongst them. For the client there should be an improvement in value, and savings in cost, where appropriate. Other measures may include an improved client brief, reduced project development timeline and reduced project construction time.

15.9 Conclusions

Value management is a tool that has application at all stages of the asset management life cycle, from the identification of service requirements through to the implementation of the resource and asset plans. It may be used in operational and maintenance planning. It may also be used on projects requiring refurbishment of assets, maintenance of assets or the disposal of assets.

It has application at all stages in a project life cycle from strategic planning, concept development, design brief, design review, through to implementation and operation.

Value management provides the manager with a tool that can:

- ensure a project is cost effective
- resolve a complex problem
- identify a number of options and select a preferred one
- identify the means by which a service may be provided
- review a brief
- identify the means by which a project may be delivered
- identify ways of providing functions at a lower total cost (life cycle cost)
- identify additional functions that improve the outcome of the project
- separate needs from wants and establish priorities
- improve the standard of performance or quality of the project outcomes
- generate commitment to outcomes through structured participation of stakeholders.

At the strategic planning stage, the value management study is most useful in that the widest range of opportunities are available for potential value improvement. By involving a facilitated group in the strategic planning process, there can be input from key stakeholder groups, and ownership and commitment to the outcomes. However, the decision to use a VMS needs to be planned and budgeted for, and a skilled facilitator engaged to assist in managing the process through the three key stages, namely pre-workshop, workshop and post-workshop. To make the most effective use of a VMS requires an understanding of the process. It is hoped that this chapter has provided the reader with sufficient understanding to enable him or her to initiate a VMS, to manage the VMS in conjunction with the consultant facilitator, and to evaluate the outcomes of a VMS.

References and bibliography

Adam, E. (1993) *Value Management Cost Reduction Strategies for the 1990's* (Melbourne: Longman Professional).

Akiyama, K. (1991) *Function Analysis: Systematic Improvement of Quality and Performance* (Cambridge, MA: Productivity Press).

Miles, L.D. (1989) *Techniques in Value Analysis and Engineering* (Value Foundations USA).

Senge, P. (1990) *The Fifth Discipline: The Art and Practice of the Learning Organisation* (Sydney: Random House Australia).

16

Risk management

L.Y. Shen†

Editorial comment

Virtually anything that people do has some degree of risk attached, whether it is something that is inherently dangerous in a physical sense, such as playing a contact sport or driving a fast car, or something that carries a risk of financial loss, from speculation on the stock market to buying a television.

People carry out simple forms of risk analysis in their daily lives when deciding, for instance, on which appliance to buy: a more expensive item might be expected to be of a more dependable quality, or we may decide to take a chance on a cheaper model in the hope of getting similar performance at a lower cost. In such a case the buyer is aware that the cheaper item has a higher risk of early failure but the cost saving is balanced against that risk – we all know someone who has something which they bought cheaply and which has lasted for years, and we hope when deciding to buy the cheaper alternative that we might also be lucky and buy the good one of the batch.

What we are doing when making such decisions is actually risk management – we identify the costs and benefits associated with alternative choices, assess the likelihood of those costs and benefits being realized, and make a balanced decision on the basis of our analysis. When making small purchases or choosing between simple alternative courses of action we may not even be aware that we are carrying out these analyses yet we make these sort of decisions every day, e.g. when choosing between catching a bus or a taxi to reach an important appointment – the bus is cheaper but the taxi will probably be faster and has more flexibility in terms of choice of route, speed and so on. When choosing between actions of greater significance, such as buying a house or a car, some of us may well take a piece of paper and write out a list of the positives and negatives attached to particular choices and try to evaluate, in monetary terms, the effects of those positive and negative factors. We then have to predict, in some way, what we think the likelihood of the range of possible outcomes might be. We also need to consider lost opportunities which may result from making a particular choice, e.g. if we buy a more expensive but more durable and reliable item then we may not have enough money left to buy something else. It is this sort of cost–benefit and probability analysis that is at the heart of risk management.

† The Hong Kong Polytechnic University, Hong Kong

Structured risk management can provide clients with a better basis for project selection and for making choices between alternative approaches to building procurement. In all cases the aim is to give clients information that will assist them in maximizing the value they gain from investment in building while not exposing themselves to unacceptable levels of risk.

Risk management does not eliminate risk, but it does offer decision-makers a range of tools for identifying and assessing risk. Clients can select those tools which are most appropriate to their particular project, with respect to other factors such as their general attitude to risk and prevailing economic conditions.

The following chapter introduces the concepts of risk identification and management, and looks at some basic techniques for a structured and analytical approach to risk management in the construction industry.

16.1 Introduction

By its very nature the construction industry is considered to be subject to more risks than other industries. Getting a project from the initial investment appraisal stage through to completion and into use involves a complex and time-consuming design and construction process. This involves a multitude of people, from different organizations, with different skills and interests; and a great deal of effort is required to coordinate the wide range of activities that are undertaken. A variety of unexpected events may occur during the process of building procurement, and many of them can cause losses to the client or other interested parties. Such events are commonly called *risks*. For example, a downturn in the economy in the future may cause a client to make less than expected profit, or even a loss, from a completed project; such a downturn is a *risk event*. Cost and time overruns in many construction projects are attributable to various risk events, where uncertainties were uncontrollable, or not appropriately accommodated during pre-planning of the project.

The principle of *risk management* is widely used in the construction industry, applied at various stages during the procurement process. It has been shown that proper application of risk management techniques can significantly improve the investment performance of construction projects (Flanagan and Norman, 1993). The phases of a project during which risk management can be applied have been identified as

- initial appraisal
- outline or sketch design
- detailed design
- contract strategy
- construction.

In particular, it is widely accepted that it is during the initial appraisal phase that risk management is most valuable as a great deal of flexibility in design and planning remains that allows consideration of ways in which various risks might be avoided or controlled (Thompson and Perry, 1992). It is also at this stage, however, that there is the greatest degree of uncertainty about the future, yet the client must make decisions about such fundamental concerns as the investment budget, the size and quality of the project, financing strategies and so forth. Traditional estimates of expected profits, calculated by applying a projected percentage return on investment, have been shown to have serious limitations as a means of providing information on which clients can base investment decisions (Shen, 1993). It has been demonstrated, however, that risk management techniques can be an effective tool that clients can use to assist them in making allowance for future uncertainties. Consequently, the client can be more confident in his or her decision-making as (s)he has information which identifies possible uncertainties and their likely impact on a potential project (GCIS, 1994).

The proper application of risk management methods can also improve the effectiveness of other project management techniques to a large extent. For example, risk analysis can improve the accuracy of project discounted cash flow (DCF) analysis by assessing properly and systematically the future uncertainties and risks.

The client will make his or her decision to invest in a project on the basis of achieving the greatest *value* for his or her money; and it is during the initial project appraisal stage that the client will try to establish project parameters that will ensure that (s)he realizes this basic aim. A client may have a maximum sum to invest and wants to determine the best investment option, or may require a specific building and want to know the likely construction cost; in any case clients want the most reliable information possible as to the likely value and/or profit which they can expect to gain if they proceed with the project they are evaluating. At the very least they want to be certain that the potential returns from the project will balance the risks. If a project proposal includes a *risk profile* for various investment options this will be of great benefit to the client in his appraisal of the proposal, giving information such as the likely outcomes if all possible risk events occur, i.e. everything goes wrong. Obviously a client can make a more sensible decision about which option to pursue when (s)he can compare the risk profiles for each option. For example, the various options might be the construction of an office building, or the construction of a housing estate, or some other market investment opportunity (such as simply keeping the money in a bank), or it may be a combination of several options.

The analysis of risks during the initial project appraisal stage enables the client to make two important decisions.

- *Whether to invest in a project, or reject it* – this is clearly the most important decision as it determines whether the project proceeds at all. This decision will be based on the constructive analysis and evaluation of the potential risks: technical, economic, financial, political and legal.
- *What the project objectives will be* – the client must decide his or her objectives in terms of budget, time scale, function, return on investment, quality standard, etc. These decisions will be based on the client's investment expectations, and his/her evaluation of various alternative investment options where such evaluation depends on projections of performance, cost and schedule with in-depth risk analysis carried out on those projections.

This chapter will examine risk management as it applies to initial project appraisal, and will describe the various approaches and techniques of risk identification and risk analysis.

16.2 What are the risks?

The concept of risk is related to the activities that flow from decisions made by the client, where the outcomes of those activities may differ from expectations. These differences are the result of uncertainties that are inherent in the information on which the client bases his or her decision-making. This information includes historical data, predictions about the future, and the decision-maker's subjective judgement; and therefore, by its nature, it displays varying degrees of uncertainty. In a broad sense, risk, as it applies to building, can be defined as the possible occurrence of an uncertain event or outcome, which, should it occur, will cause significant variations or consequences such as extra cost or delayed completion.

The basic uncertainties which concern clients on construction projects are:

- will it be finished on time (*time risk*)?
- what will it finally cost (*cost risk*)?
- will it perform as it was intended to (*quality risk*)?

Thus, the typical risks in a construction project include:

- cost overruns – failure to keep final cost within the cost budget
- time overruns – failure to keep within the time stipulated for design, construction, and handover
- poor quality – failure to meet the required technical standards in relation to quality, functionality, safety, and environment control (internal and external).

These typical risks indicate the consequences that may be caused by uncertainties; they are commonly referred to as *risk effects*. These effects are, in turn, the result of possible occurrences called *risk causes* or *risk factors*. As listed by Flanagan and Norman (1993) those risk factors which may affect construction projects include:

- failure to obtain approvals from relevant authorities (government agencies, statutory bodies etc.) within the time allowed in the project programme
- unforeseen adverse ground conditions (rock, sand, sub-surface water)
- inclement weather resulting in extensions of time
- industrial action (strikes, work to rule)
- unexpected price rises (labour, materials)
- failure to let to a tenant on completion
- accidents on site resulting in injury or death, causing delays and/or extra cost
- latent defects due to poor workmanship, or inadequate supervision
- *force majeure* (flood, earthquake, armed conflict)
- late production of design information leading to claims by the contractor for loss or expense (idle plant/personnel, delayed completion)
- labour, material and/or equipment shortages
- disputes between project parties causing extra cost and/or project delays.

The chances of any of these risk factors affecting a project, and the significance of their impact on the project, will vary between projects depending on project location, duration, total cost, resource limitations, complexity, project type, type and number of participants in the project team and so on. Generally, these risks, if they eventuate, will result in financial loss to the client and often to other team members. Clients' advisers at the initial project appraisal stage are expected to be able to identify all possible risk causes, to analyse their implications for the project, and to develop a risk management strategy to assist their clients in their evaluation of project proposals.

16.3 The nature of risk

As risk refers to a lack of predictability about outcomes in a management decision-making situation it is directly related to the concept of chance or probability. An identified risk displays three elements or attributes (Shen, 1990):

- *A range of possible outcomes* – a risk factor will result in a range of possible outcomes or consequences; typically there are three outcomes considered, i.e. the *optimistic* outcome, the *pessimistic* outcome, and the *most likely* outcome. All the possible outcomes may be in discrete or continuous distribution; however, only one possible outcome in the range will actually happen.
- *Individual consequences* – the consequences of each possible outcome can be assessed; for example, the worst possible case (the pessimistic outcome) may result in extra cost to the client, while an assessment of the best possible case (the optimistic outcome) may predict that the client will make a substantial profit.
- *Probability* – the probability of the occurrence of each outcome can be assessed and allocated. For example, the probabilities for the occurrence of the three outcomes mentioned above may be 0.3, 0.5 and 0.2 respectively. In any case the sum of the probabilities for all possible outcomes must equal 1.

Obviously the decision-maker's subjective judgement will have a significant effect on the assessment of the nature of risks; in general, those risks with lower probability of occurrence will have greater impact on a project, while those with higher probability of occurrence will have smaller impact.

16.4 The advantages of identifying risks

Risk management does not remove the risks attached to construction projects; however, the degree to which a client may have confidence in his or her investment decisions can be significantly increased if, during project appraisal, answers are available to such questions as:

- What are the major risks that may affect project objectives?
- What are the possible outcomes and consequences should any of these risks occur?

- What is the likelihood of the project objectives being attained, given that there are attendant risks?
- What methods can be used to control or accommodate these major risks?

The earlier that the client knows the answers to these questions the more confident (s)he will be about his or her chosen investment option, and a proper understanding of the risks allows informed decision-making based on more realistic scenarios. Additionally, the client will be able to develop an appropriate mechanism for responding to risks, and include suitable risk allowances in project proposals.

Risk analysis not only assists clients in decision-making but also provides other parties involved in the project, such as the contractor, with an appropriate framework for managing and responding to risk. It allows construction managers to identify not only the risk allocated to him or her in the contract, but also those risks inherent in the nature of the construction work. A better understanding of the forward risk situation can improve decision-making for all project participants.

16.5 Risk identification

Risk identification is a diagnostic process in which all the potential risks that could affect a construction project are identified and investigated, thus enabling the client to understand the potential risk sources at an early stage in the project. Such understanding at the project proposal stage will help clients concentrate on strategies for the control and allocation of risk, strategies such as choice of procurement system or the amendment of the terms of contract.

Risks can be broadly grouped into the following categories (Shen, 1990):

- *Business risk* – associated with capital expenditure, possible income, operating expenses and property value. Business risk indicates the probability that the expected level of investment return will not be achieved.
- *Pure risk* – also referred to as static risk, non-market risk, or unsystematic risk. Pure risk indicates a potential outcome that has no potential gain. It is related to physical and technical causes and subsequent losses occur at random, and is beyond the control of the decision-maker. For example, an accident on site is a pure risk, that would bring loss and damage if it happens.
- *Speculative risk* – sometimes referred to as dynamic risk, market risk or systematic risk. It involves the possibility of either gain or loss should an uncertain event occur. This kind of risk is related to changes in general project conditions, such as changes in the economic environment, politics or technology. Changes in these factors could cause variations in project development cost, operating cost, or the value of built property, thus changing the rate of investment return.
- *Financial risk* – relates to the loss of financial capital. Financial risk increases whenever the amount of debt or related charges increases. If the client sponsors the project from retained earnings, the financial exposure to risk may threaten the existence of his or her business if the project fails.

From a management viewpoint, risks are sometimes classified as *controllable* and *uncontrollable* risks. Controllable risks are those risks that a decision-maker accepts voluntarily, and where the risk outcome is, at least in part, within his or her direct control. For example, a decision-maker might voluntarily accept the risks associated with the application of new technologies whose real technical capabilities are uncertain, and such risks may be termed performance risks. These risks may be tolerated if additional benefits such as prestige or favourable financial outcomes are deemed likely to occur. By exploiting available expertise and through careful planning the decision-maker may be able to control the eventual outcome, or reduce the negative consequences.

By contrast, uncontrollable risks are those that the decision-maker cannot influence. Uncontrollable risks usually emanate from the external environment, or the political, social or economic spheres. These risks may be associated with such factors as weather conditions, material prices, or taxation changes. While these risks are clearly beyond the influence of the decision-maker, (s)he can, however, take precautionary measures if (s)he identifies the risk properly.

When trying to identify the risks attached to a specific project, greater detail is required. For example, typical risks that may be identified in a construction project may include (Flanagan and Norman, 1993):

- *Political risks* – changes in political policies, changes in government, changes in international relations, environmental protection, public safety regulations.
- *Economic risks* – inflation, changes in interest rates, changes in exchange rates, other changes in financial markets.
- *Technical risks* – design errors or variations, latent site conditions, shortages of material/labour/plant, uncertainties due to use of new technologies, bad weather, uncertainties in other geographical conditions, *force majeure*.
- *External relations risk* – poor communication with other project participants (e.g. contractors, suppliers), poor communications with professionals (e.g. engineers, surveyors and other consultants), poor relations with relevant government departments, poor relations with the general public.
- *Management risks* – inappropriate management structures, confused management policies, poor contractual relations, negative attitudes of participants, poor communication between project team members, poor quality of managerial personnel.
- *Design risks* – uncertainty in application of new technologies, incomplete or inadequate specifications, sub-optimal buildability.
- *Environmental risks* – ecological damage, pollution, inadequate waste treatment.
- *Legal risks* – liability for the acts of others, changes or variations in relevant government laws and regulations, contractual disputes.
- *Operational risks* – fluctuations in market demand for service (for example, an unexpected decline in demand in using hotels developed by the client), unexpected rise in maintenance costs that varies from that estimated at the pre-design feasibility study stage, variations in functions from client's expectations.

Previous works have suggested many examples of risks involved in a construction project. A comprehensive list of risks in construction projects prepared by Perry and Hayes (1985) is given in the appendix to this chapter.

A number of techniques have been developed for risk identification. The most common method involves compiling a list of risks for a particular project based on records of past projects (historical data). Particular attention must be given to the unique characteristics of the project that is being evaluated when data from earlier projects are used. Other common methods for risk identification include *brainstorming, tree diagrams* and *influence diagrams.*

16.5.1 Brainstorming

This technique is used to produce imaginative potential solutions which would be less likely to emerge in formally structured discussions. This technique is applied by a team who put forward all their ideas about potential risks, with all ideas accepted and written down; participants are restricted by their knowledge domain, and the flow of ideas is 'lubricated' by humour, exaggeration, imagery, and any other tools of creativity. No evaluation of the ideas is made at the initial stage, and the group members do not criticize one another's ideas. The aim of this technique is to list a large number of potential risks across a wide range.

The major advantages of brainstorming are that groups can generate more ideas and information than individuals can, and that group involvement can increase commitment, motivation and satisfaction. There are barriers, however, to the successful application of this technique: individual participants fear looking foolish, there may be problems arising from having people of varying seniority working together, e.g. juniors may be unwilling to put forward unusual ideas for fear of damaging their prospects, while senior personnel may be equally reluctant to make wild suggestions that could damage their image or credibility.

16.5.2 Tree diagrams

Tree diagrams are used to trace the causes, origins and consequences of risks. This involves multi-level analysis, and at the baseline level all risks will be identified against each risk effect. Tree diagrams are a graphical means of bringing together the risk information needed to make project decisions. They show all future possible outcomes, and each outcome is given a probability value indicating its likelihood of occurrence. An example of a tree diagram is shown in Fig. 16.1.

16.5.3 Influence diagrams

Influence diagrams relate risk effects and risk causes explicitly. Risk effects are marked on a horizontal axis and risk causes are identified against each risk effect. Risk causes can be categorized as main causes, sub-causes, sub-sub-causes and so on, depending on the level of detail in the available information. An example of an influence diagram is shown in Fig. 16.2 (Shen, 1990).

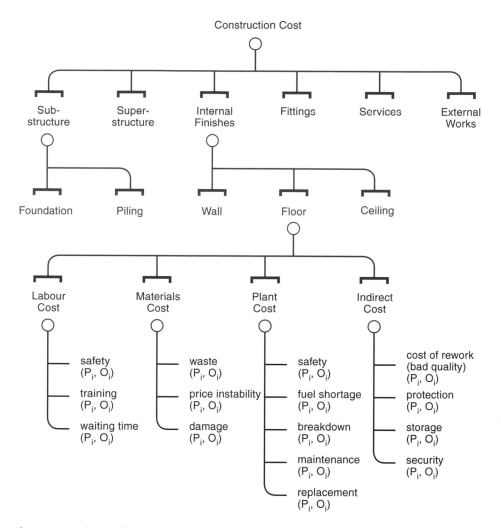

Fig. 16.1 Tree diagram for risk identification.

16.6 Risk analysis

Risk analysis is used to evaluate risks quantitatively, and to ascertain the importance of each risk to the project, based on an assessment of the probability of its occurrence and the possible consequences of its occurrence. Risk analysis assesses both the effects of individual risks, and the combined consequences of all the risks on the project objectives. The major purpose of risk analysis is to provide a project risk profile that the client can use to look ahead to possible future events and see the probability of those events occurring. The client can then decide whether or not to invest in the project, or to adopt specific strategies for dealing

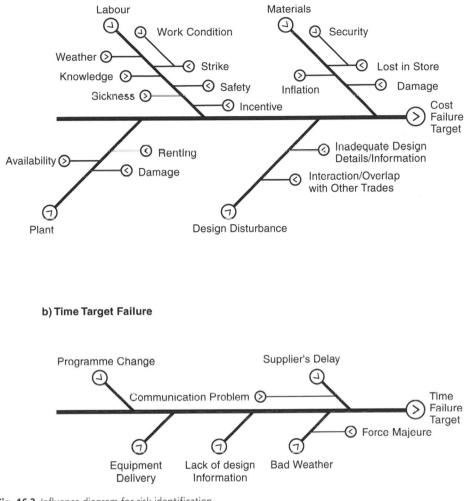

Fig. 16.2 Influence diagram for risk identification.

with the major risks. The techniques for quantifying risk must be based on a proper understanding of concepts such as probability and probability distributions. Many techniques have been developed to assist risk analysis; those commonly used are discussed below.

16.7 Risk measurement

When risk analysis is carried out it is expected that it will provide answers to questions such as 'which option or project is more risky?', or 'what is the magnitude of the risk?'. There is a need to have a *risk measure* that incorporates

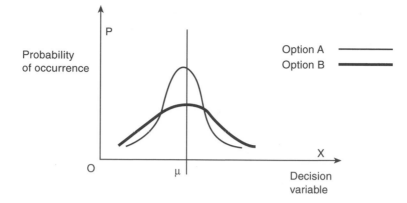

Fig. 16.3 Probability distribution and risk measurement.

risk natures, i.e. the number of possible outcomes, the consequence or value of each outcome, and the probability of the occurrence of each outcome. The *expected value* from a risk probability distribution is a simple measure: it averages all risk outcomes and indicates the central tendency of the distribution of the possible outcomes. The measure of expected value, however, fails to take into account the type of probability distributions to which different risks will be attached. The value of the standard deviation has been suggested as a more effective risk measure: the standard deviation describes the dispersion of all possible outcomes around the expected value. A smaller standard deviation value indicates a smaller variation between all possible outcomes, and graphically it indicates a tighter probability distribution. On the contrary, a higher risk is indicated by a widening spread of the outcome distribution, which gives a larger standard deviation value. For example, in Fig. 16.3, there are two risk options A and B, and option B is more risky as it gives a larger standard deviation value.

16.7.1 Sensitivity analysis

Sensitivity analysis is used to assess the effects of individual risks on the project objectives. This technique examines variations in one risk factor while holding all other risk factors constant in order to find out the degree to which that one particular risk factor will impact on the project. This type of analysis may be performed using quantitative models, or by asking 'what-if' questions. The importance of this analysis lies in the identification of those risks that will have the greatest impacts on the project, allowing decision-makers to focus their risk management strategies on those risks that are of greatest significance (Flanagan *et al.*, 1988). The inherent disadvantage of sensitivity analysis is that each risk is considered independently; therefore, interrelationships between individual risks are ignored, and assessment of the combined effects of a number of risks is necessarily limited.

16.7.2 Probability analysis

Another important part of risk analysis is the assessment of the probability of the occurrence of each possible risk outcome, and thus to provide a probability distribution across all possible outcomes. Defining a probability distribution for risk outcomes is not only a matter of applying mathematical probability techniques, but is also a creative process requiring experience and subjective judgement. The probability assessed for any risk may be classified as *objective probability* or *subjective probability*. Objective probabilities are calculated from statistical data or repeated observations, whereas subjective probabilities are based on assumptions and judgements made by decision-makers. Based on their experience, decision-makers can assign probabilities to possible outcomes, e.g. when past experience suggests that a risk event is likely to occur, a probability value close to 1 may be appropriate. Subjective probabilities are commonly applied in the construction industry as few building projects have repetitive processes or identical activities, although there will always be similarities that allow experienced decision-makers to use their knowledge of past projects.

There are a number of standard probability distributions that can be applied to various risks. Typical standard probability distributions include uniform distributions, triangular distributions, normal distributions and beta distributions. Examples of these distributions are given in Figs 16.4(a)–(d) (Shen, 1990). The most important consideration is that the chosen distribution type represents the nature of the risk as precisely as possible.

16.7.3 Simulation technique

The main purpose of risk analysis is the assessment of how a combination of individual risks can affect a project in total. In order to achieve this, individual probability distributions that describe individual risks are combined. The most widely accepted tool for doing this is a special numerical method called the *simulation technique*. This is a sampling method performed by running an analysis a number of times using a risk analysis model containing values taken randomly each time from individual risk distributions. By using this sampling approach the analysis model serves only as an experimental medium, and model usage of this type distinguishes the simulation technique from optimization, in which the model is treated analytically rather than experimentally, and the values of variables are fixed for calculation. By using the simulation technique the experiment can be repeated many times, with each run producing a result, and thus a large number of results can be obtained. Obviously, it is essential to have the assistance of computer power in adopting such an iterative approach. The results from the iterative experiments will form a distribution indicating the combined impact of all risks on the project. Figure 16.5 shows the distribution results from a project cost simulation analysis that includes 500 iterative simulations. Figure 16.6 shows the cumulative distribution of this analysis (Shen, 1990).

The simulation technique enables the systematic evaluation of the consequences of alternative investment strategies within different risk environments before decisions are made. The great benefit of the simulation technique is that the results generated provide a range of possible combined consequences of risks. The technique provides the client with a forecast of the likely range of outcomes for the

a) Uniform Probability Distribution

$$p(x) = \begin{cases} \dfrac{1}{b-a} & (a \le x \le b) \\ \\ 0 & \text{others} \end{cases}$$

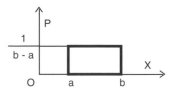

b) Triangular Probability Distribution

$$p(x) = \begin{cases} \dfrac{2(x-a)}{(b-a)(c-a)} & (a \le x \le b) \\ \\ \dfrac{2(c-x)}{(c-b)(c-a)} & (b < x \le c) \end{cases}$$

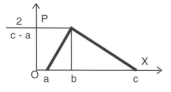

c) Normal Probability Distribution

$$p(x) = \dfrac{1}{\sqrt{2\pi}\sigma} \exp\left[-\dfrac{(x-\mu)^2}{20^{-2}}\right] \quad (-\infty < x < +\infty)$$

(i) Normal Distribution with different μ

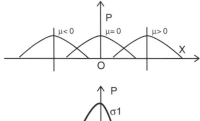

(i) Normal Distribution with different σ

$\sigma1 < \sigma2 < \sigma3$

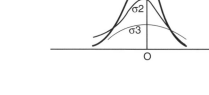

d) Beta Probability Distribution

$$p(x) = \dfrac{1}{B(p,q)} \dfrac{(x-a)^{p-1}(b-x)^{q-1}}{(b-a)^{p+q-1}}$$

$(a <= x <= b;\ p, q > 0)$

Where: a = minimum value
b = maximum value
p, q = parameters of the distribution
B(p,q) = beta function

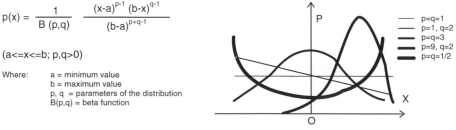

- p=q=1
- p=1, q=2
- p=q=3
- p=9, q=2
- p=q=1/2

Fig. 16.4 Risk probability distributions.

project rather than the traditional single forecast, and it enables assessment of the combined effects from all identified project risks. Effective decision-making then depends on the client's ability to use this information appropriately.

Number of simulation = 500

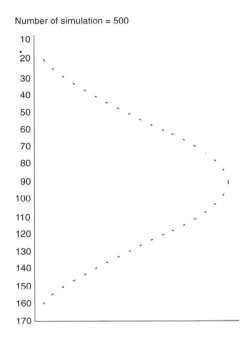

Fig. 16.5 Example of distribution results from simulation.

Number of simulation = 500

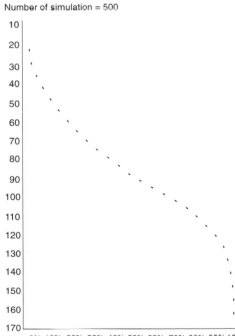

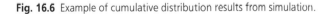

Fig. 16.6 Example of cumulative distribution results from simulation.

16.8 Risk management

Risk management is typically defined as a sequential system consisting of risk identification, risk analysis, and risk response (Lifson and Shaifer, 1982). Based on the results of the risk identification and analysis stages, the client's risk response determines whether the level of each risk is acceptable or not, and what action can be taken to account for the perceived risks. The most common actions in risk management include avoiding risks, reducing risks, transferring risks to other parties, and retaining them but minimizing their effects if they occur.

During the earlier stages of a project the client may take preventative action to reduce, avoid or transfer risks. Rejecting a proposal is an obvious way of avoiding risk; however, if the client wishes to proceed with a project, then risks should be reduced wherever possible. This will normally be done through a variety of actions including detailed design review, further geographical and/or geotechnical investigation, more detailed study of the project environment (legal, financial, economic etc.), the use of alternative contractual arrangements and strategies, closer coordination of the project team, or the application of different technologies.

Clients can also consider transferring risk to other parties through contractual methods such as the adoption of a lump sum or project management contract. This approach requires some care, however, as allocation of risk to another party may result in substantial loss to the client if the other party does not have the financial, technical or managerial capacity to successfully absorb the additional risk, and therefore withdraws from the project due to insolvency or lack of resources (Shen, 1997).

16.9 Risk management and risk attitude

The discussion in this chapter has stressed that risk identification and analysis are important tools that clients can use to improve the effectiveness of their decision-making. Regardless of the procedures adopted, different decision-makers will choose different options that are subject to different risk environments. This occurs because different people have different perceptions of the value of alternative risk options, and these differing perceptions are related to the *risk attitude* of the various individuals.

Generally, different people have different attitudes towards a particular risk environment: some are risk-takers by nature, willing to take additional risks on the expectation of higher returns, i.e. they are prepared to pay a possible premium in exchange for the possibility of gaining a greater return on their investment. Other people, however, are risk-averse, and they are willing to sacrifice the possibility of a higher return, even if the attached risk is assessed as being relatively small, and they may even pay a premium to avoid risk and have it carried by someone else. Some people are risk-neutral, that is they are neither risk-averse nor do they welcome risk, rather they tend to base their decision-making on expected values obtained by statistical analysis.

Utility theory can be used to quantify a decision-maker's attitudes towards risks, and thereby provides a basis for comparing the risk attitudes of different individuals, or for comparing an individual's risk attitudes under different risk situations. In applying utility theory, utility value is used to measure an individual's evaluation, or implicit value, or preference, for each risk option, where the implicit value reflects the individual's attitude towards that risk option. For example, a risk-taker will normally place a larger utility value on an option that has the expectation of higher return but which carries greater risk. To illustrate the use of utility theory, consider a simple example in which a decision has to be made between two strategies, A and B:

Strategy A has a certain (probability 100%) payoff of $150 000.

Strategy B is more risky. There is a 50% probability of a payoff of $300 000 and a 50% probability that the payoff will be zero.

We first can calculate the expected returns from the two strategies. For strategy A, the expected value μ_A is a certain outcome of $150 000. For strategy B, this is done using standard statistical techniques, by which the expected payoff of strategy B is μ_B:

$$\mu_B = 0.5 \times \$300\,000 + 0.5 \times 0 = \$150\,000$$

It is important to note that none of the outcomes from strategy B actually gives a payoff of $150 000. What this expected value (μ_B) tells us is that, if strategy B were to be repeated many times, say, 10 or 100 times, then on average the decision-maker would achieve a payoff of $150 000.

Now the two strategies can be presented to decision-makers who have different risk attitudes. It would appear that a risk-averse decision-maker would adopt strategy A, with 100% certainty. His/her utility value (or implicit value) for strategy A is larger than that for strategy B. By contrast, a risk-taking decision-maker would most likely place a larger utility value on strategy B, as (s)he will place a higher value on the potential payoff of $300 000 even given the 50% probability of getting nothing. The risk-neutral decision-maker will be indifferent between the two strategies.

The calculation of utility value is described in a number of standard works (Lifson and Shaifer, 1982; Shen, 1990). This sort of analysis of risk attitude enables decision-makers to quantify what appears to be a totally subjective choice, based purely on human behaviour. An individual's personal risk attitude is an important factor when that individual exercises personal judgement on the risk analysis information used for decision-making. Utility theory suggests that the decision-maker's choice will be the option which promises the highest expected utility value.

In summary, the application of risk management is the process that integrates systematic risk analysis with a subjective perception of the risk environment. A decision-maker's subjective risk perception gives some degree of realism to the process, and may increase his or her confidence and commitment to the control of risk.

16.10 Conclusions

The benefit of managing risks during initial project appraisal is to increase the confidence and the possibility of success for project clients to receive value for money from their investment decisions. The importance of applying risk management techniques during initial project appraisal is that they allow flexibility in consideration of alternatives in design and planning whilst the greatest degree of uncertainty exists. Although risk management does not remove the risks attached to construction projects, it provides assistance to clients in decision-making in an appropriate framework for understanding and responding to risks.

This chapter provides a basic understanding of risk management principles and their application to initial project appraisal. The discussion emphasizes a systematic risk management approach consisting of risk identification, risk analysis, and risk response. Some common approaches and techniques of risk identification and risk analysis have been described. On the other hand, the application of these techniques involves high levels of judgement but differs from the usual method of assessing risks, which is largely based on past experience. Thus, the effectiveness of applying risk management techniques will depend on proper integration of systematic risk analysis and an individual's personal risk perception.

References and bibliography

Flanagan, R. and Norman, G. (1993) *Risk Management and Construction* (Blackwell Scientific Publications).

Flanagan, R., Shen, L.Y. and Moor, K. (1988) Life cycle costing and sensitivity analysis. In: *Proceedings of British–Israeli Conference on Building Economics*. The Building Research Station, Haifa, pp. 100–16.

GCIS (The Government Centre for Information Systems) (1994) *Management of Project Risk* (Norwich, UK: CCTA).

Lifson, M.W. and Shaifer, E.F. (1982) *Decision and Risk Analysis for Construction Management* (John Wiley & Sons).

Perry, J.G. and Hayes, R.W. (1985) Construction projects – know the risks. *Chartered Mechanical Engineer*, **32** (2), 42–5.

Shen, L.Y. (1990) *Application of Risk Management to the Chinese Construction Industry*. University of Reading, UK: PhD Thesis.

Shen, L.Y. (1993) *Simulation in Construction Estimation* (*Working Paper*). Department of Building & Real Estate, Hong Kong Polytechnic University, ISBN 962-526-008-0.

Shen, L.Y. (1997) Project risk management in Hong Kong. *International Journal of Project Management*, **15** (2), 101–7.

Thompson, P. and Perry, J. (1992) *Engineering Construction Risks: a Guide to Project Risk Analysis and Assessment Implications for Project Clients and Project Managers* (Telford).

Appendix. Risks in construction projects

An example of a comprehensive detailed checklist of risks in construction projects prepared by Perry and Hayes (1985).

1. Physical
Force Majeure (Acts of God) – earthquake, flood, landslide, etc.
Pestilence
Disease

2. Construction
Delay in possession of site
Productivity of equipment – possible failure
Availability of equipment – spare parts, fuel
Inappropriate equipment
Weather
Quality, availability and productivity of labour
Capability of professional staff: BANKRUPTCY
- competence
- unreasonableness
- partiality

Industrial relations
Labour – sickness and absenteeism
Suitability, availability and supply of materials
Supply of manufactured items   OF   SPECIALIST MATERIALS.
Quality, availability and productivity of sub-contractors
New technology or methods – application and feasibility
Safety – accidents
Extent of change
Failure to construct to programme and specification
Poor workmanship
Ground conditions:
- inadequate site investigation
- inadequate information in documents
- unforeseen problems

Mistakes
Coordination of all construction contractors
Liaison with public services
Irregularity of work load
Theft   OF   MATERIALS.
Errors or omissions in bills of quantities
Insufficient time to prepare bid tenders
Communication
Delay in information from designers
Poor design and shop drawings
Site access
Damage during transportation or storage
Damage during construction due to:
- negligence of any party
- vandalism
- accident

3. Design
Incomplete design scope
Availability of information
Innovative application
New technology
Level of detail required and accuracy
Appropriateness of specification
Likelihood of change
Interaction of design with method of construction
Non-standard details
Non-standardization of suppliers
Quality control exercised – inspection and approval
Temporary design – quality, responsibility and supervision

4. Political
Changes in law
War, revolution, civil disorder
Constraints on the availability of labour
Customs and export restrictions and procedures
Requirement to use local labour or management
Requirement to joint venture with local organizations
Inconsistency of regulations within country or organization
Embargo
Requirement for permits and the procedures for their approval

5. Financial
Availability of funds of client
Cash flow of client – particularly effect of delay
Loss due to default of contractor, sub-contractor, supplier or client
Cashflow problems for contractors due to:
- slow payment by clients of completed works
- dispute

Adequate payment for variations
Failure of low bidder to enter construction contract
Inflation
Exchange rate fluctuation
Repatriation of funds
Local and national taxes
Insufficient insurance
Bid and construction bond unfairly called

6. Legal – Contractual
Direct liability
Liability to others
Local laws and codes

Conditions of contract:
- liquidated damages
- maintenance
- hold-harmless clauses

7. Environmental
Ecological damage
Pollution
Waste treatment
Preserving historical finds
Local environment regulations

PART 3

Implementation and management

17

Procurement law

John Twyford†

Editorial comment

An understanding of the basic nature of contracts is essential to anyone who intends to be involved in the construction industry, as virtually all work in that industry is carried out under some form of contract. Such contracts may or may not be in writing: when a householder agrees to pay a tradesman a sum of money for the performance of some routine maintenance, e.g. the replacement of a fence, there is a contract formed when that agreement is made even though the contract may only be verbal and may involve only a relatively minor sum of money. In coming to that agreement, however, certain obligations are accepted by both parties; all too often, though, one or more of those obligations is not met by one of the parties and a dispute ensues.

Disputes that arise as a result of breaches, real or imagined, by one or other of the parties to a building contract can have a significant effect on the final cost of the works. Clients who intend to build must look closely at the type of contract which they enter into, particularly with regard to the allocation of risk between the parties. This does not mean that it is merely a matter of drawing up a contract which assigns all risk to the building contractor as this may simply result in the contractor becoming insolvent in the event of major risks occurring and the contractor being unable to absorb the losses which result. In the long run this situation will mean that the client is faced with the costly business of finding another contractor to complete the works.

While this outcome may not have a great impact on the quality of the finished product, and the building may still be a valuable asset, well suited to its intended purpose and so on, the financial loss which the client suffers will obviously reduce the value he has gained from his investment.

The precise nature of any building contract will obviously depend to some extent on the type of procurement method chosen, e.g. design and build, lump sum, BOOT. The underlying principles, however, relating to performance, negligence or breach of contract remain the same.

It is clear that any client who is interested in getting maximum value for money must examine the contractual situation with great care and ensure that the contract which they sign provides them with the best opportunity of achieving the time, cost and quality outcomes they anticipate.

† University of Technology, Sydney, Australia

In many parts of the world contract law is firmly based on the common law of England. The following chapter provides an overview of contract law as it applies to building works, and describes those basic foundations upon which that law is based.

17.1 Introduction

All buildings and civil engineering projects are procured by means of a contract. This is so even where an existing building is purchased. The procurement regimes include hard money contracts for the construction of what a consultant has designed, turnkey projects or procurement by project management, and each requires a different contractual arrangement. This chapter seeks to give an overview of the basic legal principles that relate to such contracts. The following topics are considered: the tendering process, standard form contracts, the position of architects and engineers, the need for parties to cooperate to realize the contractual aim, and finally damages and negligence in the execution of the contract works.

Promises are the building blocks of contracts. The law of contract identifies those promises that the State requires the promisor to keep. It has been said (Arendt, 1958) that promising confers 'the capacity of disposing of the future as though it were the present.' All sophisticated political systems need to accommodate promise making, as without it economic activity would be reduced to barter and exchange. The law of contract is derived from decided cases. Most of the rules dealing with formation, interpretation, breach and damages are found by consulting the appropriate authorities. The exception is Malaysia where the law has been codified[1] and in other jurisdictions the law has been modified by statute.[2]

In common law countries, a contract may be entered with a minimum of formality. Essentially, all that is required is an exchange of promises in a commercial setting. In addition, the transaction thereby embodied needs sufficient precision to allow enforcement should one of the parties default. Contracts need not be reduced to writing, although in a transaction of any complexity or where the obligations or performance might be questioned, common sense dictates that the parties commit their intentions to writing.[3] Construction industry contracts have certain characteristics that require the law to develop in particular ways to support the needs of the parties. Those characteristics include: a long period of time between the formation of the contract and the completion of the project, the potential for changed circumstances and the fact that the party having the work executed (referred to as the principal in Australia and in this chapter although, in other jurisdictions, the title employer is frequently used) acts through a third party (architect or engineer).

A contract comes into existence at the time when one of the negotiating parties unequivocally accepts the offer of the other. Only those terms that are contained in the offer become part of the contract thus formed. The final offer will of course include terms that have been contributed by both of the parties during the negotiation.[4] A contract is made at the instant when the offer is accepted.[5] The rules of offer and acceptance in the construction industry have been formalized into a structured tendering process. It is appropriate to examine tendering in greater detail as the law has recently undergone some refinement.

17.2 The tender process

It is well established that a principal, in calling tenders for the execution of work, is requesting offers to execute that work. This is so even though the principal may (and almost certainly will) have dictated the form of the offer. Any of the offers is capable of becoming a contract by acceptance. It follows that the principal is not bound to accept the lowest or indeed any of the tenders.[6]

Generally, an unsuccessful tenderer is not entitled to recover the cost of tendering; 'he undertakes this work as a gamble, and its cost is part of the overhead cost of his business which he hopes to meet out of the profits of such contracts as are made as a result of tenders which prove to be successful'.[7] There are exceptions to this rule. Where a successful tenderer has, at considerable expense, carried out additional work outside the original tendering process at the request of the principal there will arise an obligation to pay for this work on the basis of unjust enrichment.[8] A second exception arises where the principal has imposed conditions in the invitation to tender. In these situations the courts have been prepared to hold that a collateral contract has come into existence upon the contractor accepting the invitation to tender and lodging a conforming tender. The principal will be obliged under such a contract to open all of the tenders and consider each fairly.[9] A tenderer who submitted the lowest price that was overlooked in favour of a seemingly more unattractive bid can recover as damages, not only the cost of tendering, but also the loss of profit on the project.[10] A principal who has stated that it will use identified criteria to evaluate the tenders is liable in damages if it fails to do so. The damages could include the cost of tendering and loss of profits.[11]

A principal who supplies inaccurate information to tenderers as part of an invitation to tender thereby causing loss is potentially liable under the law of negligence. This is increasingly the case as the law of negligence develops to encompass statements causing economic loss. The courts have not been supportive of attempts by principals to restrict their potential liability by the use of disclaimers or exemption clauses.[12] In Australia, where the information supplied can be characterized as 'misleading or deceptive' there exists a potential liability under section 52 of the *Trade Practices Act 1974* (Australia) and the various state counterparts.

Agreements between the tenderers, which have the effect of 'substantially lessening competition', attract a heavy monetary penalty and expose the participants to civil action.[13] Where the agreement has the effect of 'fixing, controlling or maintaining' the price, the lessening of competition need not be proved.[14]

17.3 Standard form contracts

This discussion concerns standard forms of construction industry contracts. At the outset the point is made that the standard documents here under consideration form part only of the transaction between the principal and the contractor. In addition, there will be annexed to the document a set of drawings, a specification and perhaps a bill of quantities. It is the sum of these documents that form the contractual obligation. Sometimes standard forms of contract are referred to as conditions of contract.[15] The drawings and specification will delineate the technical obligation and the bill of quantities is an economic tool to assist in estimating and the administration of the contract.

A standard form of contract fulfils several important functions which include:

- A statement of the basic obligations of the parties: the contractor's main obligation being to execute the work and the principal's to pay for the work.
- The parties may require that terms that would otherwise be implied by law be 'negatived'.
- A term overcoming the rule of law that construction contracts are entire contracts and the contractor's right to payment does not arise until all of the work is executed. This is done by making provision for progress payments.
- The provision of administrative procedures necessary to achieve the contractual aim.
- Allocation of risk inherent in the transaction.
- Provision for resolution of disputes.

Some standard contracts result from negotiations between professional bodies representing principals and contractors. Documents produced this way usually attempt a balance between the interests of both parties although there is a tendency for the result of the negotiation to reflect the prevailing state of the market. Other documents are produced unilaterally by government instrumentalities or the professional bodies. These documents favour the interests of the party who commissioned the document, although the courts have been sympathetic to those forced to sign them.

Standard form contracts have been produced in all jurisdictions and are frequently identified by letters designating the source and numerals identifying the particular edition. The following samples were selected from a wide range of documents available. JCT 1980 (prepared by the Joint Contracts Tribunal, United Kingdom), AS4000 (prepared by the Standards Association, Australia), SIA88 (prepared by Singapore Institute of Architects), PAM69 (prepared by Pertubuhan Akitek Malaysia) and SCC1 (prepared by New Zealand Institute of Architects).

It is said that standard form contracts contain the distilled wisdom of the organizations that have participated in the preparation of a particular contract and therefore should not be altered lightly. It is doubtful if this is strictly correct and standard forms are more likely to include compromises on questions like risk allocation. These matters should be carefully examined before a contract is entered and care taken to ensure that the end product represents the parties' agreement on these matters. Almost without question, the result will reflect the bargaining power of the parties. Attempts to conceal the risk allocation within a document should be avoided as the end result will lead to litigation. Equally, incautious attempts to amend one clause of a standard contract without regard to the other clauses can lead to difficulties. In one spectacular example the parties agreed to delete the variation clause from the contract because the principal had only limited government funding and could not afford variations. The parties found themselves in severe difficulty when, with the building almost complete, the authorities refused to allow the building to be used for its intended purpose because fire-rated doors were omitted from the original design.[16]

17.3.1 The role of architects and engineers

Traditionally, contracts for building and engineering works have interposed between the principal and contractor an independent third party who is usually an architect or engineer (in this work referred to as the superintendent). The stated role of the superintendent is to administer the contract. It is a dual role, with the superintendent required to act as an agent of the principal in giving instructions and as an assessor, valuer and certifier[17] where called on to do so by the contract. The latter functions have the potential to create a conflict of interest for the superintendent in as much as he or she is appointed and paid by the principal but must keep the principal at arm's length when exercising a discretion in dealing with matters such as variations or extensions of time. For many years it was thought that the superintendent enjoyed the immunity from claims for negligence of an arbitrator when exercising this function. The matter has now been settled by the House of Lords and an architect held to be negligent in certifying payments to a failing contractor in excess of the value of the work executed.[18] A certificate issued by a superintendent should represent the honest unfettered opinion of the superintendent on the issue in question.

Some care needs to be taken in drafting clauses imposing this role on the superintendent (in both the head contract and the contract for employment of the superintendent). It must be made clear that the superintendent is administering the contract rather than supervising the execution of the works. If the latter course is followed the superintendent may become subject to unintended responsibilities in relation to the works. In one instance an architect was held negligent in respect of an injury to a workman.[19]

The extent and effect of a superintendent's power to issue instructions and certificates will depend upon the contract. A dissatisfied party may go behind the decision of a superintendent where the certificate was given or withheld as a result of fraud or collusion with the other party. The power to issue a certificate is determined by the provisions of the contract. A variation cannot be ordered unless the power to do so is expressly reserved in the contract. The common law adds the further qualification that the new work must be of a character and extent contemplated by the contract. That is, work that is not beyond the scope of the contract or expressly excluded from it. The question of scope would be determined according to 'objective assessment by an independent bystander'.[20] It will depend upon the drafting of the arbitration clause as to whether or not an arbitrator will have the power to review the decisions of the superintendent. Presently there is doubt as to whether or not a court would have the same power to review a superintendent's decision given a clause in the contract authorizing the issue to be arbitrated and the fact that the parties elected instead to bring their dispute to court.[21]

17.3.2 Obligation of the parties to cooperate

This matter (and incidentally, the difficult position of the superintendent) is well expressed by the authors of one of the standard text books (Dorter and Sharkey, 1990) in this field:

> [T]he contract administrator's role is an invidious and almost impossible one. Apart from duties to both principal and contractor, he has a duty to achieve the contractual aim. Although the principal is supposed to be cooperating in that achievement, in practice they are very soon evidencing their competing commercial concerns. Yet he is required to hold the balance between those two contenders.

The courts are prepared to imply a term into contracts that both the principal and the contractor promise each other that they will do all that is necessary to achieve the contractual aim.[22] Principals have been found to be in breach of this requirement on a variety of occasions including where a principal:

- required a superintendent who was also the principal's employee to exercise a discretion in its favour[23]
- failed to give reasonable consideration to whether or not the contractor had shown cause why the principal should or should not take over the works or cancel the contract because of the contractor's alleged breach[24]
- failed to carry out preliminary work which was necessary before the contractor could commence work.[25]

This is an area of the law where there has been considerable judicial activity and it may be expected that the courts will further refine the law to protect a contractor where the principal has, through neglect or as a conscious policy, made the contractor's performance more onerous. Care should be taken in drafting clauses that confer rights to determine a contractor's employment and, equally, care should be taken in administering those clauses.

17.3.3 Claims and damages for breach of contract

The basis of assessment of damages by the courts for breach of contract is

> [S]uch [loss] as may fairly and reasonably be considered as either arising naturally, that is, according to the ordinary course of things, from such breach of contract itself or such as may reasonably be supposed to have been in the contemplation of both parties . . . Now if the special circumstances under which the contract was made [and accordingly the potential loss greater] were communicated by the plaintiff to the defendant . . . the damages from the breach of such contract are recoverable.[26]

Although simply stated, the formula is more complex in its application to the construction industry on the basis that the obligations imposed on the parties are complex and failure to carry out an obligation can have a multiplier effect.

In terms of the rule stated above, a principal who has suffered as a result of a contractor failing to execute the work in accordance with the technical requirements of the contract is entitled to have the work brought to the stage where it does comply. If necessary, the rectification work could be carried out by another contractor at the expense of the defaulting contractor. To this sum might be added a reasonable sum to compensate the principal for other losses associated with the failure to comply with the contract. This could include damages for delay. A serious issue arises where the cost of rectification exceeds the diminished value of the building due to the defective work. Normally it would be expected that the principal would be required to accept the lesser sum on the basis of a general duty on all plaintiffs to mitigate their losses. This is not so and a principal can expect to get what it bargained for even if at greater expense to the contractor. It was held that a plaintiff was entitled to have a house constructed with defective brickwork demolished and reconstructed at a greater cost than the diminution in value of the house. The court however, added that in requiring rectification, the work must be necessary to conform to the contract and rectification must be a reasonable course to adopt.[27] The House of Lords subsequently decided that it was unreasonable to require a contractor to demolish a swimming pool that was slightly less than the specified depth.[28] It is less likely that a principal will be awarded the cost of rectification if it proposes to sell the building and thereby gain a windfall profit.[29]

How the actions of one party to a construction contract have affected the timely performance or financial result of the other is another cause of disputation in the industry. If the principal has delayed the contractor it will frequently be alleged that the contractor is entitled to claim 'loss and expense' as a result of the principal's breach of contract. The breach will be said to comprise some obstructive delay on the part of the principal (or superintendent) or a failure to respond to an obligation under the contract in a timely manner. Similar claims can arise from delays caused to the contractor where the risk has been allocated against the principal. The basis for calculation of these claims is frequently entrepreneurial and the following examples are drawn from the same dispute.

- A subcontractor had discontinued its operations in South Australia and the builder had to employ another subcontractor at an increased price.
- A subcontractor encountered liquidity problems and to avoid liquidation of that subcontractor the builder negotiated a higher price, which was still lower than the price that would have been payable to another if the first subcontractor had not had help with his liquidity problems but had gone into liquidation.
- The builder was liable to a subcontractor for delay costs arising from the execution of the subcontract works at a later date.
- The builder had to replace, at a higher price, a subcontractor who went into receivership and refused to complete its subcontract.
- And finally, the builder paid out more (both with his own labour and with a substitute subcontractor) to make up for a subcontractor who declined to complete his subcontract because of the delay and increased costs.[30]

It is desirable to avoid these disputes and care is needed in drawing both the contract clauses at the outset and in administration of the contract. The problem is not solved by an attempt to force one party to take unreasonable risks. It is suggested that the appropriate risk allocation model is that proposed by the Dublin construction lawyer Max Abrahamson. He suggests that the risk ought to be accepted by the party best able to control it.[31] By way of example, in a lump sum contracting situation, the principal would assume the risk for the performance of the superintendent and the contractor that for labour relations.

An analogous but nevertheless related basis of claim is for acceleration costs. This can arise where a contractor makes a legitimate claim for an extension of time within the requirements of the contract that is rejected or not answered by the superintendent. The contractor is obliged to continue work during the period of uncertainty regarding the extension of time. Here the failure of the superintendent to deal with the application in a timely fashion may be a breach of contract on the part of the principal. The damages suffered by the contractor would be the cost of accelerating the progress of the works to complete within the originally stipulated time in order to avoid the need to pay liquidated damages. In order to succeed in such a claim it would need to be shown:

- a delay which entitled the contractor to an extension of time
- a specific claim for an extension of time within the prescribed period
- a failure or refusal to grant the extension
- an express order to keep on the programme or evidence that this was required
- evidence of actual acceleration.

Global claims

The increasing complexity of construction projects has led to difficulties in preparation and presentation of claims. This has encouraged the legal advisers to contractors to prepare claims on a more general basis without differentiating between the effect particular causes of delay have on the overall claim. These are the so-called 'global' or 'rolled up' claims which are, understandably, of great concern to developers and financiers. Even so, the courts have accorded limited recognition to this form of litigation and, accordingly, a short discussion is appropriate. There need to be three conditions present before the court will recognize a global claim. These conditions are:

- The loss and expense attributable to each head of claim cannot in reality be disentangled.
- There is a complex interaction between the consequences and the events.
- The inability to disentangle the consequences of these events is not the result of a delay on the part of the contractor in making the claim.[32]

These claims have been recognized in England, Hong Kong,[33] Australia[34] but not in Malaysia.[35]

Liquidated damages

Almost without exception, the parties to a construction contract will agree on the *quantum* of damages payable in the event of the contractor failing to finish within the agreed construction time. Such an arrangement has potential benefits for both parties in that the contractor knows the extent of its liability in the event of a default and the principal has access to payment of the liquidated damages without the need for legal proceedings. The courts have always held that liquidated damages clauses needed scrutiny as the law would not permit parties to privately penalize each other. Accordingly, liquidated clauses were construed as penalty clauses unless the *quantum* of damages met prescribed criteria. In particular, the damages could not exceed what was a genuine pre-estimate of the likely loss viewed at the time the contract was entered.[36] A decision of the English Court of Appeal[37] has given currency to a proposition that where a contractor is incurring liquidated damages for running over time the ordering of a variation (or any other act on the part of the principal further delaying the contractor) will deprive the principal of the benefit of liquidated damages. It is said that the original time provisions no longer apply and time is said to be 'at large' with the contractor now required to complete within a reasonable time. As a corollary the *quantum* of the liquidated damages no longer applies except to serve as a cap on the ordinary damages that might become payable at the expiration of the 'reasonable' period of time. The doctrine has been described as the 'prevention principle' and is said to apply unless there is a provision to the contrary in the contract. The doctrine has been supported by several commentators[38] and criticized by others.[39] A decision of the Supreme Court of Victoria suggests that the principle may not be of universal operation in Australia.[40] Contractors should not assume an immunity from liquidated damages merely because a principal has caused a delay.

17.3.4 Negligence

Between 1972[41] and 1983[42] the English courts opened the question of whether or not persons involved in the construction process owed a duty of care to a party who suffered pure economic loss as result of a concealed defect in a building. If the duty existed then there was a potential liability in negligence. The liability extended to the consultants who designed the building, contractors and subcontractors who executed the building work and the local authority that approved and inspected the work. The courts held that such a duty of care existed independently of any contractual obligations provided that the plaintiff and defendant were in a relationship of proximity. To date, no satisfactory definition of the concept of 'proximity' has been evolved.[43] This view of the law was widely taken up in other parts of the Commonwealth, although abandoned by the House of Lords which reversed the earlier authority. In England, there is now no liability in negligence for economic loss arising out of defects in a building.[44]

In Australia, although there is still some doubt about the liability of an approving authority,[45] it is clear that a contractor is liable in respect of economic loss to a third party arising out of the construction of a dwelling.[46] In New Zealand, the courts have held that a duty of care exists and, accordingly, the potential for parties to be found liable in negligence; however, the duty is based more on the reliance by the plaintiff on the authority's inspector than proximity.[47] The Malaysian courts have followed the English authorities.[48] In Singapore, the courts accept that the duty of care exists and recently held a developer liable to a condominium management corporation in negligence for damage suffered through defective construction of ceilings in a car park.[49]

17.4 Conclusions

What has been said is intended to give the reader an understanding of the rules of contract law so far as those rules relate to building and engineering contracts. The most significant feature of the rules is the opportunity afforded a dissatisfied contractor to make claims to recoup losses on a project. Although the claims will be characterized (in the main) as damages for real or imagined defaults on the part of the principal or those who advise the principal, it seems clear that many of the problems have their genesis in the fact that the contract price was too low in the first place. This is not surprising as contractors must compete for work in a fierce market. Accordingly, transactions that are structured in a way that the contractor has a reasonable opportunity of making a profit have the best chance of reaching a conclusion without litigation. If the contractor does not make a profit there is a whole industry waiting to assist in making claims.

Endnotes

1. *Contracts Act 1950*. There the decided cases interpret to code.
2. See *Trade Practices Act 1974* (Australia) which proscribes misleading and deceptive conduct generally and, in the contractual setting, supplements the law relating to misrepresentation.
3. There are exceptions where building contracts are required to be in writing; see *Home Building Act 1989* section 7 (New South Wales). This requirement is essentially consumer protection legislation.
4. *Butler Machine Tool Co Ltd v. Ex-Cell-O Corp (England) Ltd* [1979] 1 All ER 965.
5. There have been attempts to modify this rule. Lord Denning MR suggested that if parties regarded themselves bound at the conclusion of a negotiation the formal acceptance of an offer should not be essential, *Gibson v. Manchester City Council* [1978] 1 WLR 520. The decision was reversed on appeal to the House of Lords. A similar principle appeared to have support from McHugh J. of the NSW Court of Appeal in *Integrated Computer Services Pty Ltd v. Digital Equipment Corporation (Australia) Pty Ltd* (1988) 5 BPR 97, 326.
6. *Spencer v. Harding* (1870) LR 5 CP 561 and see *Hudson's Building and Engineering Contracts*, p. 216.
7. Per Barry J., *William Lacey (Hounslow) Ltd v. Davis* [1957] 1 WLR 932 at p. 934.
8. *Sabemo v. North Sydney Municipal Council* [1977] 2 NSWLR 880. The same principle was given a wider conceptual basis by Deane J. in *Pavey & Matthews v. Paul* (1989) 162 CLR 221.
9. *Blackpool & Fylde Aero Club Ltd v. Blackpool Borough Council* [1990] 3 All ER 25 where the Council accidentally omitted to examine the plaintiff's tender at all and was held liable for damages.
10. *Pratt Contractors Ltd v. Palmerston North City Council* [1995] 1 NZLR 465.
11. *Hughes Aircraft System International v. Airservices Australia* (1997) 146 ALR 1.

12. *Morrison–Knudson International Co Inc v. Commonwealth* (1972) 46 ALJR 265.
13. Section 45 *Trade Practices Act 1974* (Australia), section 27 *Commerce Act 1986* (New Zealand).
14. Section 45A *Trade Practices Act 1974* (Australia).
15. As a matter of law this is not correct, as all of the terms are not conditions, i.e. those essential terms, the breach of which would give the other party the right to treat the contract as being at an end.
16. *Update Constructions Pty Ltd v. Rozelle Child Care Centre Ltd* (1990) 20 NSWLR 251.
17. This language is taken from the JCC document where it is submitted that, as a matter of legal drafting, this relationship is best dealt with.
18. *Sutcliffe v. Thackrah* [1974] AC 727.
19. *Florida Hotels Pty Ltd v. Mayo* [1965] 113 CLR 588.
20. *Wegan Constructions Pty Ltd v. Wodonga Sewage Authority* [1978] VR 67.
21. In England the answer seems no, *Northern Region Health Authority v. Derek Crouch Construction Co Ltd.* [1984] 1 QB 644 and perhaps yes in NSW, KBH *Construction Pty Ltd. v. PSD Development Corp Pty Ltd* (1990) 21 NSWLR 348.
22. *Mackay v. Dick* [1881] 6 App Cas 251.
23. *Perini v. Commonwealth of Australia* [1969] 2 NSWLR 530.
24. *Hughes Bros Pty Ltd v. Trustees of the Roman Catholic Church of the Archdiocese of Sydney* (1993) 10 BCLR 355.
25. *Teknikal dan Kejurkuteraan Pte Ltd v. Resources Development Corp* [1994] 3 SLR 743.
26. Per *Alderson B. Hadley v. Baxendale* (1854) 9 Exch 341 at p. 354.
27. *Bellgrove v. Eldridge* (1954) 90 CLR 613.
28. *Ruxley Electronics & Construction Ltd v. Forsyth* (1995) 3 All ER 268.
29. *Carosella v. Ginos and Gilbert Pty Ltd* (1981) 27 SASR 515.
30. Per Bray C.J., *Taylor Woodrow International Ltd v. The Minister of Health* (1978) 19 SASR 1 at p. 14.
31. These principles have been taken up in Australia by a Joint Working Party comprising representatives of Government Agencies, major clients and the industry.
32. On this subject see the article by Hon David Byrne, 'Global Claims Maze or Way Forward?' (1997) 15 ACLR at p. 113.
33. *Wharf Properties Ltd v. Eric Cumine Associates* (1992) 52 BCLR 1.
34. *John Holland Construction and Engineering Pty Ltd v. Kvaerner R. J. Brown Pty Ltd* (1997) 13 BCL 262.
35. *Syarikat Tan Kim Beng and Rakan-Rakan v. Pulai Sdn Bhd* [1992] 1 MLJ 42.
36. *Dunlop Pneumatic Tyre Co Ltd v. New Garage & Motor Co* [1915] AC 59.
37. *Peake Constructions (Liverpool) Ltd v. McKinney Foundations* (1970) 1 BLR 111.
38. Dorter and Sharkey (1990) at paras 9.80 and Duncan Wallace (1979) at p. 653.
39. Perhaps the most spirited criticism came from Davenport (1995) at p. 81.
40. *Aurel Forras Pty Ltd v. Graham Karp Developments Pty Ltd* [1975] VR 202.

41. *Dutton v. Bognor Regis Urban District Council* [1972] 1 QB 373.
42. *Junior Books Ltd v. Veitchi Co Ltd* [1983] 1 AC 520.
43. The most comprehensive attempt to date has been that of Deane J. (as he then was) in *Sutherland Shire Council v. Heyman* (1985) 59 ALJR 564 at p. 595; however, this might be contrasted with the dissenting judgment of Brennan J. (as he then was) in *Bryan v. Maloney* (1995) 128 ALR 163.
44. *Murphy v. Brentwood District Council* [1990] 2 All ER 908.
45. *Sutherland Shire Council v. Heyman* (1985) 157 CLR 424.
46. *Bryan v. Maloney* (1995) 128 ALR 163.
47. *Invercargill County Council v. Hamlin* [1994] 3 NZLR 513.
48. *Karajaan Malaysia v. Cheah Foong Chiew* [1993] 2 MLJ 439.
49. *RSP Architects Planners and Engineers v. Ocean Front Pte Ltd & Anor* [1996] 1 SLR 113.

References and bibliography

Arendt, H. (1958) *The Human Condition* (University of Chicago Press).
Cremean, D.J. (1995) *Brooking on Building Contracts.* Third edition (Butterworths).
Davenport, P. (1995) *Construction Claims* (Sydney: The Federation Press).
Dorter, J.B. and Sharkey, J.J. (1990) *Building and Construction Contracts in Australia: Law and Practice.* Second edition (The Law Book Company).
Duncan Wallace, I.N. (1979) *Hudson's Building and Engineering Contracts.* Tenth edition (Sweet & Maxwell).
Robinson, N.M., Lavers, A.P., Tan, K.H. and Chan, R. (1996) *Construction Law in Singapore and Malaysia.* Second edition (Butterworths Asia).

18

Information and information management

John Mitchell†
Roger Miller††

Editorial comment

The great Gothic cathedrals of Europe were constructed with little documentation over periods ranging from 20 to several hundred years. In spite of their structural complexity and intricate detailing, information regarding their construction was largely passed by word of mouth, with the designer/builder – usually the master mason – directing the works in person. Physical models were made, some sketches were done on parchment or paper but the pace of construction was such that there was plenty of time for information to be conveyed verbally to those who needed it.

Construction at the end of the 20th century, however, bears little resemblance to that leisurely process of the Middle Ages. Information in physical and electronic form is generated in great abundance and must be quickly and efficiently conveyed to those who need it. The management of information in the construction industry in the 1990s involves the handling of vast quantities of information, and the dissemination of that information to a great many interested parties: consultants, clients, contractors – in fact, all those concerned with the inception, design, construction, commissioning and ongoing operation and maintenance of the buildings that are the final products of the construction process.

As buildings and the activities and services which they house have become more and more complex, so the range and number of participants in the procurement process has increased. Similarly the quantity and diversity of building-related information has expanded almost beyond measure. For example, the Great Synagogue, in Sydney, Australia, was built in the late 19th century from only a handful of architectural drawings; by comparison, for a large city development project on a site adjoining the Great Synagogue, completed roughly 100 years later, and valued at A$600 million, nearly 57 000 drawings were issued in less than 18 months. In the same period, over 10 000 items of correspondence were generated. This resulted in an average of 167 documents being handled each day. Each and every transaction

† Woods Bagot Architects, Australia and South-East Asia
†† University of New South Wales, Australia

involving these items had to be recorded, and the actions which flowed from them monitored and reported.

Time constraints, economic pressures, changing procurement systems – all these factors, and many more, have contributed to an increase in the demand for timely and efficient collection, dissemination and storage of accurate and detailed information. Poor communication leads to disputes, abortive work, redesign, on-site clashes, and many other costly problems. It is a sound strategy to put in place, at the outset, a clearly defined information management framework. This should establish the pattern of information flow though the life of the project, and identify the manner in which information will be managed – preferably through a single point of responsibility, whether the architect, the project manager or a separate entity who is designated as the information manager for the project.

The benefits to clients are undisputed: better information flow saves time and money, disputes are avoided, a better product is obtained in a shorter time, and the client's aim of value for money is achieved. The following chapter examines the nature of information flows in the industry and, in particular, looks at emerging electronic methods of information management.

18.1 Project information strategy at the briefing stage?

Every building project is unique. Not only are the site, layout, materials and construction methods unique, but a different set of design professionals, contractors, subcontractors and material suppliers typically come together for each new project. In this sense, buildings are 'hand made', yet are extremely complex, have long lives, and are expensive both to build and to operate. These factors create unique problems in the management of project information.

This book is about the very early stages of projects.

Why is it relevant to understand information management in such detail at the early stages? The reason is that, if we are to make good early decisions about information management strategies, it is essential that we understand the nature of information, its flow and usage throughout the entire project and its long-term importance in the life of the project. One must understand the opportunities that are available, and the pitfalls that must be avoided, if we are to lay the foundations for an effective information strategy.

The technology to achieve excellence in project information is available and affordable. However, one needs to understand the information technology concepts that are to be used throughout the entire project to make decisions on information management. This chapter will firstly present these concepts and discuss their management. It will then review those tools that are available to assist in adding value specifically at the pre-design and briefing stages.

18.2 The nature of project information

The nature of the building process necessitates the generation and collection of a wide range of information at each phase of the project cycle: feasibility; sketch

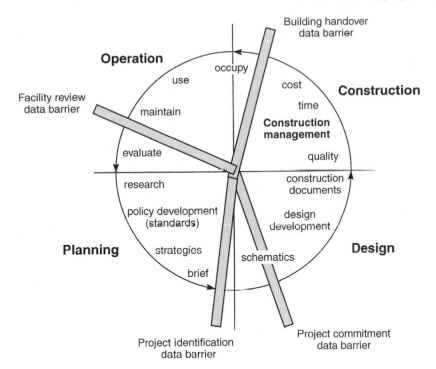

Fig. 18.1 Facility life cycle perspective.

design; design development; engineering; cost planning; detailed design; construction planning and estimating; component manufacture; construction; commissioning; and facilities management. Information passes from one phase to the next. We will call this the *longitudinal axis* of information flow.

Of course, these phases overlap but they can nevertheless be differentiated. The process is illustrated in Fig. 18.1: information is passed forwards through each phase of the project cycle. This figure also shows the location of four significant barriers to information transfer:

- at Project Identification, when the project brief is first handed to the designers
- at Project Commitment, when it is decided to proceed to design development
- at Building Handover, when the client takes over the project from designers and builders
- at Facility Review, when facility managers review the project with a view to demolition, major refurbishment or significant change-of-use.

At any particular phase in the project, information in a variety of forms is passed between participants on a daily or even an hourly basis. For example, detailed architectural design, engineering design, cost analysis and preliminary work on-site will normally all be proceeding simultaneously. Information will constantly be flowing between managers and designers within and between each of these activities. We might term this the *lateral axis* of information flow in the project.

Initially, the information is vague and imprecise. It may only describe a concept. As it is passed from one stage to the next, the information is developed, transformed and added to, and usually becomes more detailed and exact. In a sense, the information may be seen to evolve in the same way that a statue is chiselled out of a block of stone: large rough cuts to begin with, becoming finer and more precise, leading eventually to the fine polishing stage. The difference is that, for project information, the process is neither predefined nor linear as the process moves towards the final objective.

During the entire design and construction process, there are many changes to the design and documentation of the project. Some of these are unavoidable, but many occur because of errors or misunderstandings. Sometimes a considerable amount of backtracking is required. Both design and construction work must be redone. This is inefficient in terms of cost and time; it can also be a serious demotivator. Imagine the attitude of tradesmen who must tear down and redo the work they have just done because someone gave them an obsolete drawing by mistake. Will they be as motivated to do a good job next time? And what if there is a further change after that? The *accuracy, currency* and *timeliness* of information are vital for efficiency and quality in construction.

Another complication is that information at any specific phase is presented in a variety of document formats: sketches; layout drawings; detail drawings; schedules; written briefs and specifications; reports and spreadsheets; requests for information; tenders and invoices; notices and responses; codes and work method statements, operating manuals . . . the list is endless. Our strategy must take into account this variety of formats, and allow a variety of time scales for the transmittal and turn-around of various kinds of information.

Information is generated and used in different work places and work organizations, by people with widely different agendas, backgrounds, education levels and objectives. Increasingly, for large projects these work places are in different countries; even the work team for a single profession may be in different companies. For example: within the interior design function, an international specialist in hotel kitchens may be briefly involved; for the curtain walls, a firm in one country may build and test the prototype of the wall while a different company actually manufactures the final product, all for a project that is perhaps in a third country. Our strategy must also take into account this diversity of participants and work places.

18.2.1 The transfer of project information

People and machines use information for a variety of purposes: it may only be analysed and approved, but typically a participant adds to or enhances the information ('adds value' to it), or uses it to create, manufacture, construct or operate some component of the building. This means that, after being processed, the information is normally passed back to the originator or passed on to one or more other participants. Some of the more common flow paths are illustrated in Fig. 18.2.

For example, the structural engineers will receive information about the layout and function of the building from the architect. This will normally be in the form

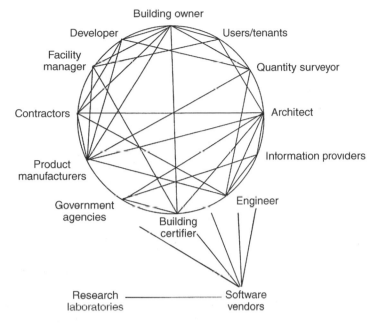

Fig. 18.2 Building project information flows.

of a CAD (computer-aided design) database or a set of drawings, and schedule of spaces. At the predesign stage, this information will be quite general and approximate. Later, it will be more refined and developed. In either case, the engineers will use the information they receive as input for structural calculations, using their specialist in-house computing resources and their particular professional expertise. They will then feed back to the architect information on materials for the structure, sizes of structural components, and comments on how the structure has been, or will be, impacted by various design decisions that have been made or are contemplated. The level of detail the engineers provide will be appropriate to the particular project stage and to the level of precision in that information they have been given.

Most information is transferred in paper form. Even documents that are generated using computers are commonly committed to paper by printer or plotter, and sent or faxed to other parties. The transfer of information in electronic form between most participants is still very limited and is illustrated in Fig. 18.3. When electronic means are used, the information is often converted to less structured information formats such as text or drawing files since these are the common information exchange standards between different computer software systems.

18.2.2 Differing views of project information

Different participants will 'see' the same information in different ways. Traditional architectural drawings include elevations and floor plans. Also, notional cuts through the building (sections) are made at various key locations to show the

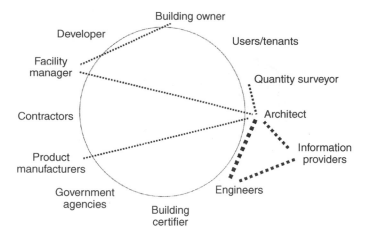

Fig. 18.3 Current electronic data exchange.

internal appearance of the building. On these drawings the architect will 'see' the data that describe a wall, for instance, as a divider of space. The way the wall achieves this division, the doors and windows in the wall, its appearance and texture, its fire, sound and heat insulating properties in relation to the spaces on each side, are all of special interest to the architect.

The construction manager, however, has a completely different view of the same data: what is documented and specified, as (s)he sees it, is a two-storey pre-cast wall with two window assemblies and a door assembly in it. Each assembly implies work items that have material and labour components. These will be inputs to his/her costing and resource planning systems. For example, the construction manager may be interested in recording that three workers are needed to install the assemblies; that they will need a quantity of sealant and gasket around the perimeter; that they will need fixing bolts and lugs that are not shown on the drawings; they will need various tools and scaffolding; they will need a crane of a certain type and capacity to lift the assemblies safely into position; and they need some temporary propping while they fix the assemblies to the structure. The construction manager looks at it from a 'how do I build it' point of view and 'how do I manage the on-site processes'. That is a totally different view from the architect. This view is illustrated in Fig. 18.4.

The owner of the building has a different point of view once again: (s)he may ask questions like: 'How is each window assembly painted? What is the life expectancy of the gaskets that will be used? What is the thermal rating of the glass? Is that door on a secure external wall? Is that door on a firewall and therefore requires annual certification to prove its continuing utility?' (S)he will want a view of the data that give answers to these types of questions.

18.2.3 Life cycle uses of project information

The facilities managers will require a different information view once again. During the life of a building, spaces may be subdivided in various ways and used for a variety of functions. Space may be leased or costed to particular tenants

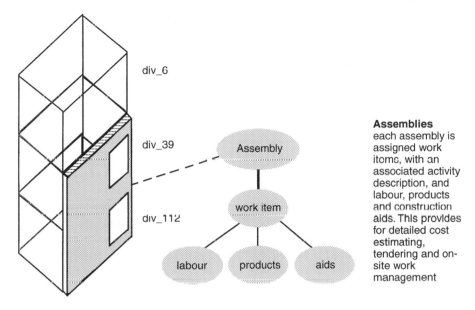

div_6

div_39

div_112

Assembly

work item

labour   products   aids

Assemblies
each assembly is
assigned work
items, with an
associated activity
description, and
labour, products
and construction
aids. This provides
for detailed cost
estimating,
tendering and on-
site work
management

Fig. 18.4 Construction manager's view.

or departments. Usage of the space must therefore be planned and recorded. Furnishings and fittings must be tracked and listed on asset registers.

The extent of maintenance expenditure is often not fully appreciated. Facility managers routinely let maintenance contracts worth millions of dollars. For instance, one government department in a state in Australia has recently let a contract to maintain the buildings for just six of their schools. The contract sum is 20 million dollars over five years. This department has over 1600 schools to maintain so one can get some insight into the magnitude of their maintenance task. They need information that enables them to better manage this maintenance work and to feed back maintenance information to designers. This requires a different view of the information. Facility managers want to know what was built, the specific properties of the materials used, where each major component was obtained and what maintenance will be required on it.

Owners and facilities managers have a 'life-long' interest in some project information compared with other project participants. For their purposes, the information must be readily transferred from one computer generation to the next several times over, yet still be accessible to, and manipulated by, the current computer hardware and software systems when required.

It will not always be known at the design stage what information will be required 20 or 30 years into a project's life. Who could have foreseen 30 years ago that building owners today would be vitally interested in knowing the location of every piece of asbestos in their buildings? Who would have thought we would be converting quite modern city office and warehouse buildings to residential uses?

How can we match the 'information cycle' with the 'life cycle' of a project? As mentioned earlier, there are currently four major blocks to the transfer of information in a project:

- at project identification, when the idea of the project and the project requirements are first handed to the designers
- at project commitment, when the client decides to proceed with design development for the project
- at hand-over of the completed building, when the owner or users occupy the building and assume responsibility for its operation
- at facility review, when facility managers review the building in detail to decide if and how it should be continued.

Typically there are only 'paper transfers' of information at each of these points, despite the fact that during construction (for example) all the design documents will have been prepared digitally.

At the briefing stage, usually little structured information is prepared; cost plans, facility requirements, regulatory or development constraints are not normally comprehensively collated, nor are they stored digitally. At the commissioning stage, the different nature of facility management data, the lack of clear asset management standards and the absence of experienced users all hinder the exploitation of the vast amounts of 'hidden' digital data.

18.3 Modern design information systems

The CAD systems now available are based on different concepts and have vastly improved capabilities compared with earlier systems. Firstly, they store and manipulate objects in a true three-dimensional geometric environment. This means that once positioned in the design, an object can be viewed from any angle, that clashes with other objects can be detected, and if required, instructions can be generated for machines to position it.

Secondly, CAD systems can now store other information about these objects in addition to size, shape and position. Depending on the object, we may wish to store properties such as: strength, weight, colour, supplier, heat transmission, fire rating, sound coefficient, required fixings, cost profile, when fixed, who fixed, and maintenance profile. These properties are called up, stored and manipulated by designers and builders. They may be introduced, expanded or enhanced at different phases of the project, by the appropriate participant.

Thirdly, many CAD systems can now 'understand' the relationships that an object may have to other objects. These relationships may be direct and explicit: for example, windows and doors 'know' that they belong to particular walls. Or they may be implied: a gutter implies a down pipe; a light fitting implies a power source and a switch.

18.3.1 The practical implications of modern CAD systems

What do the capabilities of modern CAD systems mean in design practice? Say the architect wishes to place a wall containing an opening window in the design of a building. Non-object-oriented CAD systems require you to draw two parallel lines on the plan and close the rectangle to indicate the window. You then must

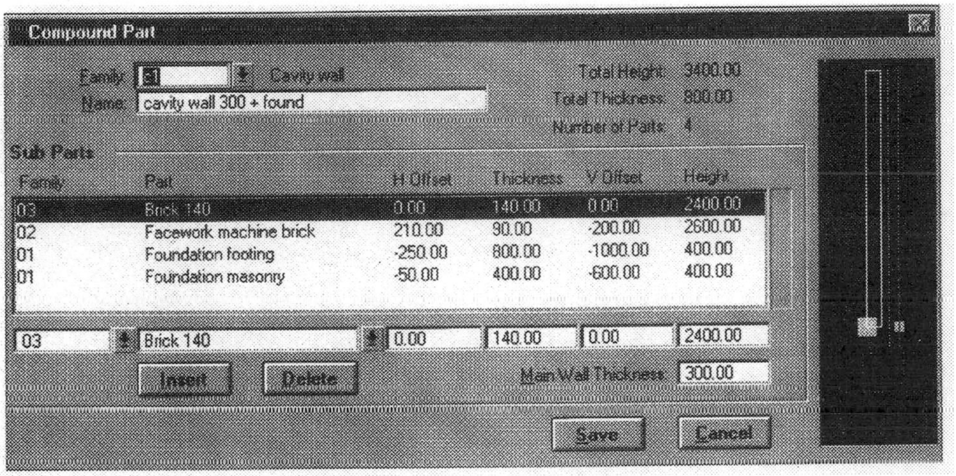

Fig. 18.5 Wall object.

insert and trim an arc to show how the window opens, etc. Next, you draw the wall and window into the appropriate elevations and sections. Finally, the window has to be entered on the appropriate schedule with the correct size, fire rating, weather properties, window hardware, and so on. This is both tedious and error prone.

In an object-oriented system, there will be a family of wall objects for which properties have been previously described. For example, in a given situation, the architect may want the wall to consist of a brick leaf on the outside, an air gap of 15 mm, a stud wall, insulation and plasterboard lining on the inside. An example of an object editor used to build up or customize a wall object is illustrated in Fig. 18.5.

The architect selects the appropriate wall object (let's call it 'wall') and it is then just a matter of indicating where 'wall' is located. Immediately there is a more powerful entry in the project information. The computer knows a lot about 'wall' and will show it automatically on the elevations and sections, as appropriate. It will terminate it appropriately with other wall objects at either end, and with the selected floor and ceiling system objects.

When the architect calls down and positions a window object in 'wall', the relationship between the two is generated. Such a window object would also be called down from a library or built up from other elements (see Fig. 18.6). If we subsequently move the wall, the window will also move (of course, you may say, isn't this what one would expect from an intelligent helper?). The window also appears on other views, as appropriate, and also on the window schedule. Further, the heat flow and sound transmission properties of wall and window are known to the computer and can be called down into thermal or sound analysis software later if required. Instead of just drawing, the architect is *designing*.

Architects, engineers and other designers are not consumed by the arcane demands of two-dimensional (2D) graphics. Drawing stairs, for example, used to require considerable skill and effort. Multiple 2D views had to be manipulated simultaneously on the drawing; many human and building code requirements had

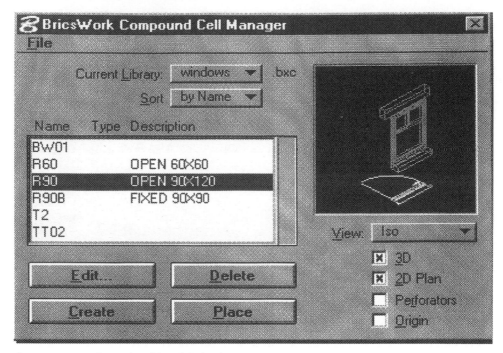

Fig. 18.6 Parametric opening objects (windows).

to be respected during this process. Now, designers can call down intelligent stair objects that 'know' when they come to a landing, that are customized to use treads and handrails that conform to the standard that is appropriate for the particular building. Such an object will 'know' certain things about stairs already; it will retrieve other information directly from the project database; it will have an interface to prompt the designer for additional parametric information before generating the stair (see Fig. 18.7).

Architects no longer have to go through the process of working in a 2D plan here and a 2D elevation there. The architect works and visualizes in three dimensions (3D) and design intent is apparent at all times, since the architect is working with real things, not 2D representations. Architects are now actually manipulating, editing, and designing objects with a purpose: to design a building that suits a functional requirement. These objects are intelligent. They know what they can do and what they are not allowed to do! They have material properties, they have dimensions, and they have manufacturers' product codes.

When the time comes for the quantity surveyor to give advice on cost, the whole design database can be made available to him/her and he/she will use specialized software to extract those properties of the components that are important to cost. Later the construction manager may add further information related to who is to supply and install the wall and the window, who will paint them and with what, and when this will happen on the construction schedule. All this information is recorded for future management of the facility.

BricsWork Stair Maker
Family: 04
Part: Reinforced concrete
Type: Closed
Level: 1
Symbology: Active
By AA: 90.0°
View: Iso
Clockwise
Step Width: 29
Rise: 6
Tread: 8
Stair Height: 84
Stair Length: 109
Stair Width: 61
Vertical Angle: 34.9919°
Number of Steps: 15
Start Steps: 3
Space Between: 3

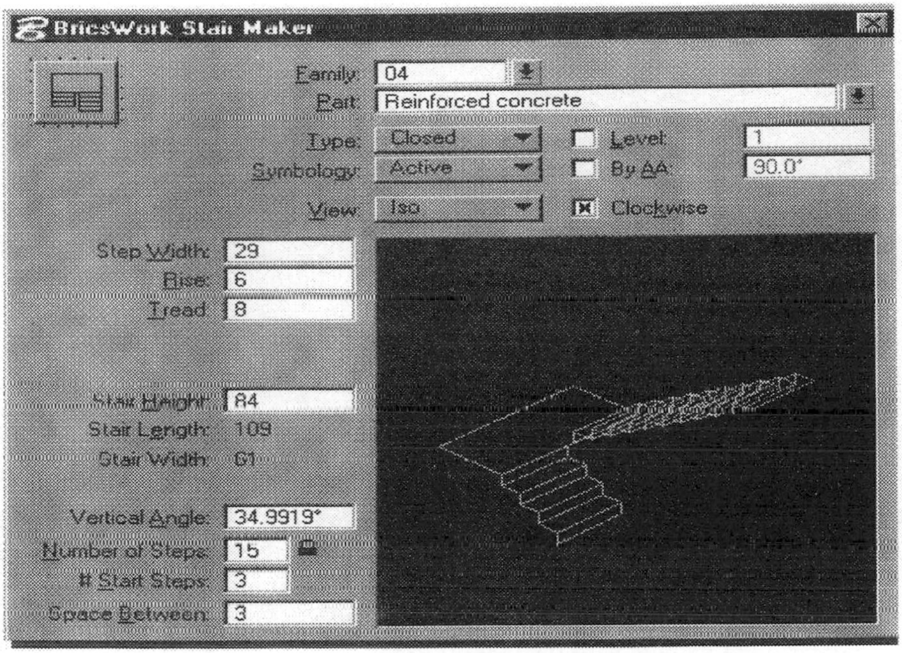

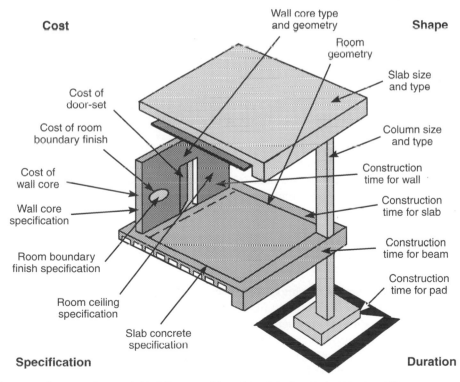

Fig. 18.7 Parametric stair object.

Cost
Wall core type and geometry
Shape
Room geometry
Slab size and type
Cost of door-set
Cost of room boundary finish
Column size and type
Cost of wall core
Construction time for wall
Wall core specification
Construction time for slab
Construction time for beam
Room boundary finish specification
Construction time for pad
Room ceiling specification
Slab concrete specification
Specification
Duration

Fig. 18.8 Information for a typical building assembly – CONCUR Project, Taylor Woodrow UK.

Some of the information items that may be built up for a typical building are illustrated in Fig. 18.8, which comes from a project undertaken by a large UK contractor. Various professionals working independently but to a common end put information shown on this figure into the database. A complex relational structure between objects in the database will result, since all the information on the project will become interrelated. One model of this structure is shown in Fig. 18.9.

Although this structure may appear daunting, it must be remembered that it is the computer that does all the work in establishing and maintaining the links within the database. Each object 'knows' when it is introduced into the design data, that it should establish links with other objects and pieces of information. It will imply these linkages and relationships automatically, or require the appropriate project participant to establish them explicitly at the relevant time.

18.3.2 The ideal of the virtual project

What is required is a 'virtual project' that represents the real project in information terms. Each participant would have an interface with the virtual project (a 'view' of the information) which is consistent with and appropriate to their role in the creation and operation of the real project (see Fig. 18.10). The virtual project would have a parallel life with the real project.

Just as Internet users care little about the technical intricacies associated with how information physically flows from sender to receiver, project participants would not need to be concerned with the details of how or where the virtual project is physically stored, or how the internal relationships between objects in the data are maintained. Provided each participant can access accurate information, in useful forms and in timely fashion for their purposes, the benefits will follow. Of course this goal is still some distance away, but it is closer now than ever before.

18.3.3 Information management and project value

All of these factors make the generation and management of information for building projects difficult, yet absolutely critical for success. Information should be regarded as the property of the project, not the individual who created it. It is a shared resource. There are tremendous opportunities for improved quality and process efficiency if:

- information is created in a way that makes its reuse easy and routine
- information can be transferred completely and accurately between parties
- information can be tailored to the needs of particular participants.

The chances of errors, misunderstandings, and rework will be minimized. Even errors in project scope, that is building 'the wrong project', will be diminished because clients can play a more informed and constructive role if information is provided for their purposes in forms that are appropriate to them. For example, photo-like generated images with correct lighting, shades and colours, are much more useful and meaningful to clients than elevations and artist's perspectives. The comment later that 'I didn't realize it would be like that . . .' can therefore be avoided.

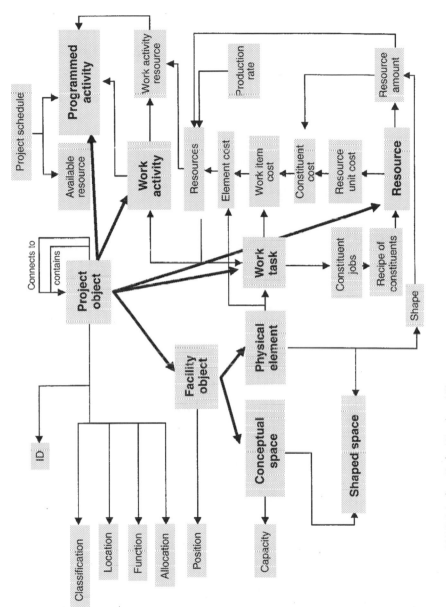

Fig. 18.9 Information structure – CONCUR Project, Taylor Woodrow UK.

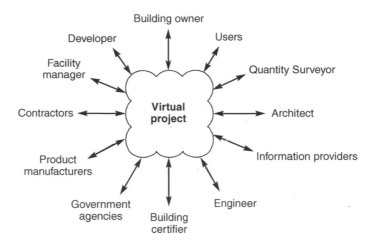

Fig. 18.10 Building project interoperability.

18.4 Existing standards for information interchange

As discussed earlier in this chapter, the same information in a project is viewed and used by the various participants in different ways. It is also represented internally in computer software and hardware systems, which have been developed by the computing industry for those participants, in vastly different formats and structures.

This statement holds true for differing applications: for example, structural design versus architectural design; costing versus scheduling; space layout design versus space management. It is also true for software within a single application from different vendors; for example: AutoCAD^{TM} versus ArchiCAD^{TM} for architectural design; PrimaVera^{TM} versus Microsoft Project^{TM} for scheduling. Usually, software packages cannot understand or manipulate information in each other's internal formats. In fact, these internal formats are usually not published and may change from one release of the software to the next.

The situation is somewhat different when it comes to formats that are used to store information externally on disk or for transmission over the Internet. In some cases, especially where there is a dominant software vendor, certain external storage formats have become default industry format standards. Vendors have incorporated software modules into their packages that can translate from this default standard into their internal format. For example, many packages that deal with text (word processing, presentation, web page authoring) can recognize and load the basic text from documents written in MS Word^{TM} or WordPerfect^{TM} formats. However, often more advanced formats in these documents such as footnotes, headers and footers, and even emboldening in the text may be lost in the transfer process.

18.4.1 Electronic transfer of CAD information

In the case of CAD information, there are two proprietary file formats that have been widely adopted for transferring information from one computer to another.

These are the Drawing Exchange Format (DXF) and Drawing format files (DWG). When information is transferred between CAD systems in these formats, much information may be lost. When one tries to transfer information from the CAD system into or back from other application areas (for example, into or back from costing or scheduling packages) even greater difficulties result. This is because these low-level default standards were intended to transfer graphic representations only; they were never intended for exchanging information between object-oriented systems or transferring highly structured project information. Nevertheless, they are widely used for CAD information exchange in the absence of better standards.

18.4.2 Electronic business transactions

Another area in construction where there are emerging data transfer standards relates to 'business transactions', and is generally called Electronic Data Interchange. EDI is already used widely for business transactions in other industries such as warehousing and distribution, and in some areas of manufacturing. Standards for EDI include EDIFACT (sponsored by the UN), ANSI X12 (USA), Tradacom (UK), and CI-net (Japan).

The use of EDI in construction has been largely a follow-on from mature commercial applications in associated manufacturing industries. It is mainly used for materials management, especially for tender, order, delivery, invoicing and payments related to materials and components. The process is illustrated in Fig. 18.11. EDI presents few problems in principle but the greatest benefits will accrue if it can be standardized and universally adopted, so that nearly all project transactions of this type are handled by EDI. Transactions could then be electronically cross-referenced to other project information, information reused, and experience stored and retrieved.

There are relatively few business transactions in most projects (compared with banking or retailing, for example). The real benefits of EDI in construction do not lie in the saving of transaction costs, but in other factors. EDI demands well-structured administrative processes and must be integrated with good supporting information technology. It helps the administrative processes to better serve the production processes of the project. Its use cuts time and results in optimum order quantities. Perhaps most important of all, EDI supports 'just-in-time' relationships with suppliers that are so important for efficient construction.

18.4.3 The Internet and the Web

The World Wide Web and the Internet are based on a number of standards, but these standards do not relate to file formats or data structures. However, both are significant in the electronic transfer of project information because they are:

- inexpensive to acquire and operate
- very fast, compared with traditional transfer methods
- universally available, both locally and internationally

Project functions

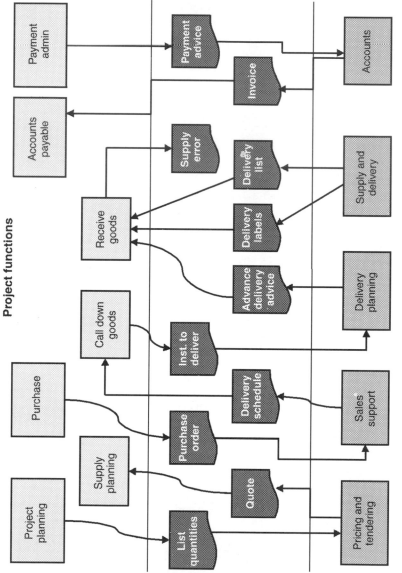

Supplier functions

Fig. 18.11 EDI document flow.

- easy to use and reasonably reliable
- well integrated with generic personal computer software
- make information transfer largely hardware and software independent.

Some projects have web sites that provide services such as progress information, contact lists, document registers, project news pages, and invitations to tender. It is possible to use web sites to transmit copies of the documents themselves, both for tender and for construction, although this presents some non-technical problems related to legal requirements and confidentiality. Web sites are also used for public relations and for marketing of the project, and may contain continuously updated digital images of the project for this purpose. However, transferring information does not make it immediately useful to another application. Significantly higher levels of standardization are required. The applications must be 'interoperable'. This concept will be discussed later in the chapter.

18.4.4 The importance of using electronic mail

Electronic mail (e-mail) is the beginning of the electronic commerce learning process. E-mail teaches processes for sending and receiving digital information, the most fundamental of tasks. E-mail leads participants to understand how documents are created and to think more carefully about their structure and format. E-mail users must be able to distinguish between application files, graphic files, word processing files, and other typical file types.

As this understanding is developed, the ease of electronic communication becomes apparent and awareness of the benefits increases. Electronic authoring of all work has more logic when that work can be exchanged, shared, built upon rather than continually recreated. This can have a profound impact on project work as information transfer is fast, and in a format that can be used directly.

The use of e-mail leads participants to understand that they no longer run separate islands of business activity and process, the situation that often prevails today. For example, the position of the 'backroom' specifications expert should not persist: specification information is only a particular view of the common design database. Specification data are always being implicitly edited by many participants and should be readily available at any time during the project phases. In a 'hard copy world', this division of expertise has become entrenched; in a digital environment it is by no means necessary. We have the opportunity to increase efficiency by working with information stored centrally and shared.

E-mail, using the Internet, is becoming widely used for text messages, especially between designers. It is common to attach quite large files containing CAD or other data to these messages and to send these internationally between participants.

18.5 Current structures in CAD databases

Traditionally, the principal output from a CAD system is a *plot* (an output drawing originally produced on a pen plotter, hence the name). Separation of the work of various members of the project team is through the use of *layers* in the

CAD data. Typically, a layer is a view of the building showing a subset of components that a particular designer is concerned with (e.g. structural elements, air-conditioning layout, landscaping). Layers enable designers or contractors to isolate their aspect of the project from the others so as to reduce complexity and allow work to progress independently. There are several useful standards that have been developed for layering CAD information, but these have been disappointingly ignored. Instead, *ad hoc* conventions are arrived at by consultants or agreed for a particular project.

The layer is usually determined when a component is entered into the database: a code is entered or generated to identify the component's layer. Designers can more easily differentiate their part of the project from all the others by displaying only their layer while they are working on it. However, to resolve design conflicts, designers must display or plot drawings that concurrently show several layers. For example, structure and air-conditioning layers may be displayed to ensure that there are no geometric clashes between them and that services penetrations in the structure are properly sized and located.

Although layering greatly assists designers and contractors, some problems remain:

- Layering is basically concerned with views of graphic information (plans); it is limited; it does little in itself to enable other uses of the CAD information.
- Layering is concerned with organizing the drawings, not the information; it does not help organize non-graphic information associated with the components.
- Resolving conflicts within the design is still a tedious manual procedure that must be redone whenever there is a major change, and is therefore error prone.
- When the component is entered, it is determined which layer(s) it is to appear on; this can predetermine how the information can be easily used.
- It is difficult to generate views of the information that show only some aspects of several designers' output (for example, a 'formwork view' of the data showing information only about penetrations and blockouts from several CAD layers).
- Layering does little to assist quantity surveyors or contractors with their tasks. For instance, one cannot automatically highlight all the components of a particular type, or an entire work package.

18.6 The product model approach to project databases

Product models are conceptual structures that specify what kind of information is used to describe complex assemblies and how such information should be structured. They were first used in the manufacturing industry for products such as automobiles and offshore oilrigs.

An information model defines the project data, where the domain of knowledge comprises built facilities throughout all phases in the project life cycle. The information model defines objects, attributes and relationships of things such as activities, spaces, building components, service systems and so on.

Standardized product models of buildings have proved useful in data structuring. The methodology used in building models is object-centred and based on advanced concepts from object-oriented database systems, knowledge-based systems and programming language theory. The central concepts are objects, attributes of objects and relationships between objects.

A building product model should fulfil a number of criteria in order to provide the core of the computer integrated construction environment of the future. It should be:

- *Comprehensive.* The model should be all-embracing. It should be capable of containing all types of data and information.
- *Cumulative.* The model should cover the information created and needed during all stages in design, construction, operation and maintenance of the building.
- *Non-redundant.* The model should not contain redundant data; information should be stored only once, to avoid data inconsistency and waste.
- *Output independent.* The structure and information content of documents should be determined independently of the information structure of the model itself.
- *Hardware and software independent.* The standard should specify only what information is to be included in the model. It should not specify how this information is physically stored in files and records.

The goal of the product data model definition is to standardize the terminology and define the necessary parameters of all kinds of project information. In this way, the model will allow unambiguous transfer of information between applications software systems and therefore between project participants.

18.6.1 Standard for the exchange of product data

A leading international standardization project for product models in building is the Standard for the Exchange of Product Data (STEP). The STEP development originated in Europe and has resulted in a significant amount of construction-related product modelling. Particular subsets of the overall standard have been undertaken by research organizations in various countries as projects. The global aim of these projects is the creation of an Integrated Building Design (IBD) environment to cover all aspects of building.

STEP is based on the proposition that a key feature in computer-integrated construction is the use of neutral standards for the structuring of data describing a building. Such international or national standards may, in the future, replace current building classification systems as the primary method for structuring information in advanced CAD systems, knowledge-based systems and databases.

An integrated database of the type envisaged by STEP is a structured collection of graphics and data, based on known entities in the building model and their relationships, for example a 'room' or 'door'. Schedules of finishes or specifications can be automatically generated from the entities defined in the database. This underpins a new level of functionality that will result in improved quality through integration and sharing of information between facility project participants.

STEP is a data modelling and exchange standard. It does not require object orientation, of the kind discussed earlier, in the software itself. However, object orientation is the natural flow-on effect of its use: it would be difficult to imagine a software package that is STEP compliant in data exchange with other systems and is not object oriented itself. It therefore fits well with the other modern approaches to project information.

18.6.2 Computer Integrated Construction

The concepts and standards discussed up to this point are leading towards a computer integrated built development, often called Computer Integrated Construction (CIC). Full CIC is still a 'concept for the future', characterized by both the use of computing for all kinds of applications and by the integration of these applications by data transfer networks and transfer standards.

Integration in this concept can primarily be understood to mean efficient information sharing and data exchange using information technology as the enabling technology. Data are shared over communication networks (Internet, WANs, LANs) using centralized and distributed databases.

The requirements of Computer Integrated Construction include:

- widespread computer literacy (e-mail, digital documents, enabling hardware)
- industry-wide standards for information interchange
- communications infrastructure (internal and external networks)
- discipline specific application software
- work practice changes ('re-engineering' construction).

18.7 Data sharing and exchange

Whilst CAD of one type or another is ubiquitous in the design professions, the sharing of information is rare and hopelessly under-exploited. After many years and several attempts to set data exchange protocols, DWG and DXF remain the only standards that are widely used for CAD data. As discussed above, these are de facto standards and have serious limitations. Even if there are specific user agreements about their use on a project, a satisfactory project information environment is not assured.

Information failures are responsible for many quality problems and add to the cost of building and construction projects. Although the Architecture, Engineering and Construction Industry (AEC) has made substantial investment in CAD and information technology, effective use of the capital is hampered by lack of compatibility between proprietary systems and databases.

Typical problems in AEC include:

- lack of understanding between project participants
- lack of coordination between disciplines and life cycle stages
- discrepancies between graphical and textual processes
- lack of consideration of building processes.

The AEC industry, even in industrialized countries, has so far failed to assert its requirement for an 'integrated solution' comprising common definitions of objects and functions, supported by software applications written for dedicated use in the sector. One of the reasons for this is that the industry has not been sufficiently forceful in its demand on software vendors for integration. Another is the fragmented nature of the industry: there is little incentive for one participant to structure information to improve the efficiency of another participant; there is no dominant player who can force standards on the others.

18.7.1 What is interoperability?

Efficient information transfer using modern object technology would enable secure, consistent and accurate access and sharing of common data between project participants, so that:

- Information is created once and is then available over the life of a project. The same information can be used in different project phases.
- The same information can support different discipline processes, and be used by applications from a number of vendors.
- The focus of the design and documentation process moves to incremental refinement of project information by successive project groups.

This capability is called *interoperability* because it allows data access operations across different software applications and network computers. Interoperability will support a natural integration of application processes over the whole life cycle, reducing costly duplication of data entry and manipulation, and allowing designers to spend more time in design optimization and client consultations. It is estimated that this kind of technology could bring cost savings of up to 30% in AEC project development (Latham, 1994).

18.7.2 Industry Foundation Classes

As discussed previously, the time is at hand when it is possible for design teams to develop a design model on a computer for an assembly of standard objects. An object will incorporate physical characteristics such as weight, colour, fire resistance, electrical or thermal properties, as well as representations of how it appears in space in various design and management situations. The definitions of these common objects constitute Industry Foundation Classes (IFC) as a common language for the AEC industry around the world. IFCs provide a common interpretation of real world objects so software applications can be developed in a consistent and interoperable manner.

18.7.3 International Alliance for Interoperability (IAI)

The International Alliance for Interoperability was founded in 1994 with the objective of achieving and promoting interoperability by developing IFCs. At the time of writing, there are ten IAI regional chapters established in North America, Europe, Japan, Korea, Singapore and Australasia. Around the world, in 1998, there were more than 650 firms in the IAI movement, each producing and using IFC standard object definitions.

Software company and research centre members include Autodesk, Bentley, CSTB, Fujita Business Systems, Intergraph, Lawrence Berkeley Laboratory, Nemefscheck, Primavera and Timberline. These companies already participate in the development of software that is used internationally for a variety of tasks over the entire project cycle. Their software products are tested and certified for interoperability to a stated version of the IAI Standard. Purchasers of certified software are assured of its capability to import and export information to other similarly certified software products.

IFC development emphasizes incremental delivery of interoperable information infrastructure through its IFC object definitions rather than the conduct of lengthy research. The idea is to deliver some useful definitions and products as early as possible and then to build on these in later versions.

Currently, the IAI operates as follows:

- Through collaboration, coordination, and division of labour, IFC development work is apportioned between participating countries.
- Each participating country forms a regional IAI chapter which draws on expertise and supporting funding for the work that is its focus.
- Each chapter has access to the outputs from similar groups around the world which it can incorporate in its own outputs.
- Scalability issues such as technical integration and conformance testing are jointly developed through the coordination of the international peak body IAI.

18.8 Construction and manufacturing technologies

A final influence is the growing trend by construction companies towards the adoption of manufacturing technology. Numerically controlled machines, driven by digital instructions stored as files, are used to fabricate components for structural steel, aluminium curtain walls and timber joinery. Basically, information technology is used to make these machines more intelligent and more flexible. This approach is also used in the assembly of components in some instances.

A major stimulus in Europe (identified in the Latham Report, for instance) has been the recognition of the benefits of the application of industrialized high technology to construction. This trend is to move from craft-based on-site work to off-site factory-based manufacture of integrated assemblies. These are then shipped and delivered to site, and installed in one-off operations. The benefits of this approach are higher quality and more efficient production in the more controlled factory environment, reduced wastage and pollution, improved opportunities for skills development, better use of information technology and faster on-site assembly.

On-site construction equipment itself is also becoming more intelligent. Advanced models of excavation equipment, cranes and hoists, concrete placing booms, and floor finishers, for example, use information technology in their control systems to help operators improve machine performance. However, except where there are extreme hazards to construction workers, the use of fully automatic robotic devices has been generally restricted to prototype projects, particularly in Japan. Nevertheless, there can be little doubt that the use of this technology, even on-site, will increase in the years ahead.

These influences further support the need for whole-of-life concurrent engineering systems that can drive a project from a centralized model, wherever the workplace. The instructions for operating specific machines, either in the factory or on-site, will be generated by specialized software processors that extract the required product information directly from the model. Once again, the standards and technologies of the types previously discussed are important for the economics of these technologies. Their use will mean that data will not need to be taken-off or recreated to generate the instructions for the machines. A time-intensive and expensive activity will be abolished; a possible source of error eliminated.

18.9 Need for a clear information management strategy

Information management is too important to be allowed to 'just happen' in a project. The client should specify, as part of the initial brief, and subsequently in various contractual arrangements throughout the project, not only what information is to be transferred back at the completion of the project, but also how information is to be shared and transferred within the project.

A good information strategy will add value and reduce cost at every stage of the project by:

- allowing participants to build new information on to that of other parties
- enabling the rapid and economic copying and distribution of information
- eliminating errors in transcription and minimizing information re-work
- encouraging better use of information technology in all project phases
- improving the quality of information, and so the quality of the actual project.

Specifically, the information strategy for the project should require that all participants:

- use specific industry standards to transfer information at a high level (such as STEP or IFCs, rather than DXF or DWG formats)
- structure project information logically and completely so it is useful, even for unforeseen applications (i.e. ensure interoperability)
- use electronic storage and cataloguing so information can be easily found, reused and enriched, rather than recreated
- ensure common ownership of non-confidential project information through enforceable contractual agreements between participants
- require electronic transfer of information via such technologies as e-mail and project web sites
- adopt appropriate attitudes to information quality, security and confidentiality.

The strategy should set out specific information deliverables to be interchanged:

- between consultants
- to and between consultants and contractors
- from contractors to clients and facility managers.

There should also be enforceable warranties on the correctness and completeness of information exchanged on the project so that others can reasonably rely on it. If information is important, then should not the quality assurance regimes for the project be extended to cover it?

This may require the use of specific quality regimes and technologies. For example, an as-built drawing of below-ground services may be more of a liability than an asset if it is incorrect; perhaps the contractor should be required to use Global Positioning System (GPS) technology to record the position of piping before it is covered.

It is equally important not to constrain participants unnecessarily in such a way that they are prevented from making best use of their investments in knowledge and computing technology. It is important that participants are left free to choose the software and storage standards they use internally. For example, the strategy should not require the use of a particular CAD or scheduling package, only that it be compliant with the chosen interface standards.

18.10 Information technology during briefing

In principle, the briefing process should provide:

- an early, rigorous testing of the idea behind the project
- clear documentation of the owner's expectations and requirements
- a procurement strategy for the project.

The resulting brief should define the client's functional needs, arrive at a cost plan, establish project feasibility, and set out the client's detailed technical requirements, including a clear statement of the services required in the project.

The principal benefits of a properly prepared brief are that:

- a clear statement of user requirements will be available to all participants, rather than designers having to guess or infer those requirements
- the spatial requirements of the project will be known before detailed design begins, and design proposals can then be measured against them
- the basis for a reliable cost plan will be laid, although this plan must obviously be followed up at later stages in the project if it is to be effective
- building services and support systems will be identified early in the design process, so they can be incorporated in early design decisions.

An added benefit is that the briefing requirement presents an opportunity for the organization to examine precisely the need for the facility: studying and questioning internal business and operational processes challenges the working of the organization. This may result in the reworking of processes so that it becomes apparent that an entirely different type of facility is required, or perhaps no new facility at all!

Briefing is a key activity for quality assurance because it determines the scope of the project. A project that is delivered on time and on budget, with good quality materials and workmanship, is of little value to the owner if it is the wrong size or does not satisfy the owner's basic functional requirements. Errors in scope are common, and are difficult and expensive to fix later in the project life cycle.

Briefing data can be expressed and presented in three ways:

- *spatial data*: the physical areas and characteristics of spaces in the building
- *organizational data:* the functional purposes to which the building will be put
- *process data:* the decision-making and transformations that are made.

The briefing procedure may be *systematic* (i.e. following a well-defined logical path from inception to conclusion), or it may be *ad hoc* with many twists, turns and backtracks in the process as stake-holders and designers learn from their dialogue. Each of these briefing scenarios will now be discussed more fully, with reference to the role of Information Technology in the process.

18.10.1 In systematic briefing procedures

Formal systematic briefing procedures have been established in certain classes of projects such as hospitals, schools and police stations. While these projects can be very complex, they are well understood and follow standard forms. As a result of this, and because the clients for these projects are highly experienced, the project brief can often be expressed in a systematic manner. A 'functional briefing hierarchy' can be established to arrive at the brief on a 'top down' basis.

For example, the brief for a hospital may begin with decisions about fundamental parameters, such as which medical specialities the hospital is to support and how many patients in various categories are to be serviced in each speciality. From this 'highest level' information, the requirements for facilities at the next level can be decided; for example, how many operating theatres and of what capability, or what size dispensary should be provided in the hospital. At the third level in the hierarchy, a particular type of operating theatre will imply spaces for particular activities – for example, a space for clinical hand washing and gowning, and a space for patient recovery. These are termed *activity spaces*.

Activity spaces can be defined in detail from an industry reference library which will contain specific information on space, layout, equipment, services and relationship to other spaces. Even the relative location and reference height above the floor level for equipment may be defined, as well as lighting, isolation and acoustic requirements. An activity space from a reference library is illustrated in Fig. 18.12.

Specialized information technology can be used to great advantage in this briefing scenario. The brief will include schedules of activity spaces in digital form. Detailed cost and operational analysis can be performed on the proposed facility before it has been laid out spatially by the designers. The operation of the facility may even be simulated before it is 'designed' in an architectural sense.

During detailed design, the task of calling down and positioning the required activity spaces within the building form can be done systematically. Associated

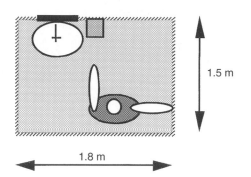

Activity spaces
an activity space is the finest detail level in the functional briefing hierarchy. It provides for some activity to be carried out, e.g.:
activity: clinical hand washing and gowning
equipment: surgeon's basin with elbow action taps, soap dispenser, mirror and gowning bay
area: 2.7 m²

Fig. 18.12 Industry reference libraries.

equipment will be automatically included as required. The architect can work directly from the digital brief and will tailor the standard library activity space for the particular instance. Other non-briefed (yet essential) spaces will be added. Software can assist with this schematic planning (see Fig. 18.13).

Other required characteristics of the space (for example, required lighting or acoustic isolation) will be automatically included in the design database and will be available later for other designers (for example, the lighting engineer or acoustic consultant).

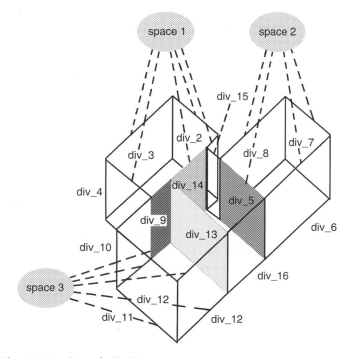

Schematic planning
the 'wireline' design develops by realizing the geometry of each space; abutting spaces (rooms) inherit more complex requirements, and require extra space such as circulation to complete a practical design. Spaces 1 to 3 are briefed, the 'corridor' has to be added explicitly

Fig. 18.13 Intelligent facility data.

18.10.2 In *ad hoc* briefing procedures

Other projects are unique, or at least not routine, except perhaps in broad terms. In those cases, the briefing procedure and structure must be drawn up neatly from first principles. Generally this is done by understanding and documenting the processes of the client organization.

A well-formed brief will include comprehensive requirements about:

- services systems, including environmental and energy standards to be followed
- materials and finishes performance
- security, safety and access for various classes of users
- capital and operating costs, including life cycle analysis.

The information is often developed in stages. It is refined and interpreted as the facility proposal is fleshed out. These data are the basis for evaluation during design, and should later become the basis for the commissioning and operational phases of the project. Well-developed and accessible requirements data can significantly inform the design and construction teams.

Apart from physical information about the proposed facility, briefing can tap into other sources of information: climatic, geographic, regulatory environment; the products and processes of the organization. In a complex organization, briefing will involve breaking down communication barriers between departments and functional areas, and between hierarchies in the organization. It is often a matter of uncovering 'what the organization knows about itself'! Communication is crucial within the client organization, within the design team, and between the two.

Information technology can play an important role in this briefing scenario by:

- transforming and tabulating information in various ways, using spreadsheets, databases and flow diagrams
- presenting information in easily understood formats and using terminology that is particular to the various participants
- making information accessible to a wider community within the organization and the project team via web sites and other electronic means
- creating business process simulations and virtual reality animations to expose the brief's assumptions and requirements.

Information technology also plays a role in controlling and disciplining the briefing process. It can be used to provide a logical structure for briefing, and to require users to think through and explicitly document their operational processes. In this sense, it can be seen as an instrument for 'liberating' the organization's information about itself, information that is locked away in obscure operating manuals, laptop computers, databases and even only in people's heads.

18.11 Conclusions

The effective use of information technology will increase the value of a project in terms of cost, quality and scope. This is because accurate and timely information, in a form that can be easily shared by the project participants, is essential for efficiency. Good information helps everyone to do the job better.

Interoperability between software and the effective use of networks make sharing of information fast, cheap and error-free. It also allows analyses and simulations that would otherwise be precluded by cost or time constraints. This results in projects that more closely fit the clients' needs. It also allows project participants greater insight into proposed construction and operational processes.

The project will benefit from an effective information strategy not only during the design phase, but throughout the entire life cycle: construction, commissioning and the life-long operation of the facility all have information requirements, although the information will need to be presented in differing formats and views in these phases. While it is important to use the best tools available for tasks specific to the preliminary phase, the establishment of an effective information strategy for the life of the project is the most important consideration in this phase.

References and bibliography

Björk, B.-C. (1995) *Requirements and Information Structures for Building Product Models.* VTT, Technical Research Centre of Finland, Espoo, Finland.

Crawford, J., Newton, P., Wilson, B. and Wylie, R. (1996) Collaboration on design & construct projects using high-speed networks: from vision to reality. In: *Proceedings of InCIT96 Conference*, Sydney, Australia, April.

de Valence, G. (1996) Construction networks and project intranets. *B.E.S.T. Magazine*, Sydney, Australia, March.

Duyshart, B.H. (1997) *The Digital Document* (Oxford: Butterworth-Heinemann).

Latham, M. (1994) *Constructing the Team, Final Report of the Government Industry Review of Procurement and Contractual Arrangements in the U.K. Construction Industry,* (London, UK: HMSO).

Maher, M.L., Simoff, S.J. and Mitchell, J. (1997) Formalising building requirements using an activity/space model. *Automation in Construction, 6*, 77–95.

Negroponte, N. (1995) *Being Digital* (Sydney: Hodder & Stoughton).

STEP: *ISO 10303 Industrial Automation Systems and Integration – Product Data Representation and Exchange.* ISO TC184/SC4.

19

Integrated design

Rick Best†

Editorial comment

It is axiomatic that better building design leads to better buildings; and similarly that those better buildings will be more valuable to their owners. There is ample room for debate, however, as to what exactly constitutes 'better' design, and how better design may be realized.

Ultimately the success or otherwise of a building design will be decided according to a complex mix of judgements offered by a range of interested parties. These will include building owners, occupants, tenants, critics, various authorities, and even passers-by. These people will appraise the building from many viewpoints: some will base their opinion of the building's success on purely visual considerations, others may judge it according to functionality, occupant comfort or by the amount of energy it uses. The building owner may judge success simply on the basis of return on investment.

Building design is something of a balancing act, part technical skill, part art, involving many compromises and trade-offs. Various parties to the process must cooperate and coordinate their inputs to the design so that the final design fulfils all the requirements of the client's brief, obeys all legal constraints, and is both functional and pleasing to the eye. The Australian architect and author Robin Boyd suggested that architecture is the only art which always starts with a puzzle – the puzzle being the unique set of problems set by the client and a host of other factors including site, legal constraints, function, budget and so on. Within the limits imposed by this 'puzzle' the designer is then faced with using 'the sternly practical business of providing bodily shelter as a medium for artistic expression' (Boyd, 1965).

Foremost among the constraints which act on building designers is that of cost – with few exceptions buildings are designed to meet a budget, and that budget obviously places very definite restrictions on the nature of the design. The linear design process that has typically been used in recent years has certain drawbacks which militate against the adoption of optimal solutions. The process is fragmented, the participants are numerous, and often in conflict, communication is limited, and the final results too often reflect the shortcomings of this approach. Most importantly, this method tends to hamper innovation and promote the implementation of

† University of Technology, Sydney, Australia

conventional solutions rather than encourage a more fundamental analysis of the design problem, which can then lead designers to solutions that better serve the needs of the client.

An integrated team approach to solving the complexities of design can produce superior buildings, i.e. buildings that represent better value for their owners, are more efficient in operation, are more economical to run, and provide a better indoor environment for those who occupy them. The following chapter explores the concept of integrated design and illustrates the benefits that such an approach can produce.

Reference
Boyd, R. (1965) *The Puzzle of Architecture* (Carlton, Australia: Melbourne University Press).

19.1 Introduction

It has been suggested (Peck, 1993) that 'design is the process that determines the *value* of a completed project'; however, in the traditional approach to building design and procurement it is at the earliest stages of that linear process that the life cycle cost is established, often by only one or two members of a team that will eventually number 30 or more. In fact it is most commonly the lead consultant, whether architect or project manager, who, in consultation with the client, makes early decisions about siting, envelope materials and building form which severely restrict other members of the team, whose input is not considered until the nature of the building has already been determined.

If it is reasonable to equate 'value' in building with 'better buildings', as discussed in Chapter 2, and 'better buildings' are those which achieve greater success in terms of satisfying clients' expectations, then it must be recognized that such success depends heavily on the decisions made in the early stages of the design process.

19.2 Integrated design

In simpler times a single consultant, usually the architect, could successfully combine the roles of designer, engineer, project manager, construction superintendent and cost manager.

In the 1880s, a US army engineer, General Montgomery C. Meigs, both designed and supervised the construction of the Pension Building in Washington, DC.[1] His aim was to create a healthy environment for the clerks of the Pension Bureau and to this end he paid particular attention to lighting and ventilation. In this 3–4 storey building, 120 m × 60 m (400′ × 200′) on plan, the purely passive ventilation system (Fig. 19.1) could produce, under optimum conditions, 30 air changes per hour in the central atrium (Lyons, 1993). After just one year of occupation the number of sick days taken by the bureau's 1500 employees was reduced by 8622 compared with the previous year (National Building Museum,

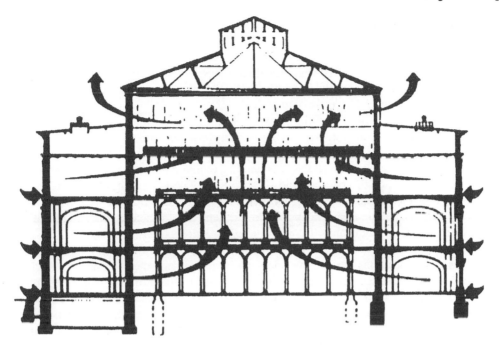

Fig. 19.1 Pension Building – ventilation (Lyons, 1993).

1996). Using current costs this represents a saving, based solely on reduced absenteeism, of approximately 2.6% p.a. (see Table 19.1)   other savings which would accrue in a modern building, such as avoided costs for purchase, maintenance and operation of mechanical plant and artificial lighting, would add significantly to those savings.

What the Pension Building demonstrates in a simple manner is the basic nature of 'integrated design', with the building envelope and the ventilation system forming an integrated system. The result was a building that was fundamentally 'better' than others of its time; better because it better served the requirements of the client by providing the space necessary to house the activities required by the client but by doing it in such a way that the productivity and well-being of the occupants were improved.

Table 19.1 Pension Building, Washington DC

Total employees 1500
Assume average salary of $35 000 p.a.
Total wages $52.2M p.a.
Average working days per employee 220 p.a.
Average cost per day $160 per employee (excluding on-costs)
Number of sick days 8622 p.a.
Savings 8622 × $160 = $1.38M, i.e. 2.6% p.a. reduction in costs

19.3 The design process

Since Meigs' time, building procurement has become far more involved: buildings have become much more complex in themselves, particularly in their engineering services, and there are many more players, not only in the design team, but in the overall project team. There are also an increasing number of external pressures that must be accommodated, including environmental regulations, public perception of development proposals and so on. It is clearly not feasible to expect that a single person can address all the aspects of the design and development process, and the traditional linear approach to the problem has evolved in response to the increasing complexity of both the process and the product. The fragmentation of the design process has led to common scenarios such as that outlined by Lovins and Browning (1992) where 'the architect draws a building then throws the drawings over the wall to the mechanical engineer with the injunction, "Here, cool this".'

This process sees many consultants working in isolation, from inadequate briefs, with many variables that have powerful effects on their design – such as orientation, the decision to seal and aircondition the building, or selection of envelope design or form – already having been set before their work even begins. Any attempts to change such decisions later in the process are generally perceived to be too costly (although that may not be true if the impact of the decisions is considered in terms of the life cost of the building) and less than optimum choices made early in the process tend to remain in place and become harder and harder to alter as the process continues.

With the increased complexity of the process, and the largely unsatisfactory results of the standard design process, new approaches are needed. The establishment of a multi-disciplinary team at the earliest possible stage of the design and development process can significantly improve the buildings that result. When all facets of building design are addressed in an integrated or holistic way then the solutions to individual questions can be optimized as part of the whole design rather than being made in isolation and without reference to the impacts of such decisions on the rest of the design. As Berry (1995) puts it, we can 'move from the culture of problem solving to that of problem elimination', i.e. by agreeing on answers to fundamental questions at the outset, potential problems can be avoided completely. In this way problems are not simply generated by one part of the design team and then left to others to solve.

The decision to pursue integrated design does not automatically lead to the elimination of mechanical plant, or the creation of an environmentally benign building and thus 'better buildings'; what it does do, however, is to give the opportunity for a wider appraisal of the client's needs, and to allow the design team to approach the design problem from more directions.

19.3.1 The linear process

With regard to the design of energy efficient buildings, Eley *et al.* (1995) assert that '. . . the traditional process of planning, designing and constructing new buildings is significantly flawed'. While energy efficiency is only one aspect of building design the same flaws which militate against arriving at low energy designs also contribute to the adoption of less than optimum solutions to a wide range of design problems.

The traditional process has been based on a linear approach with a lead consultant coordinating the overall design process with a range of other specialist consultants providing expert input at various stages of the process.

Before the emergence of the discrete engineering disciplines which we are familiar with today, architects, such as Montgomery Meigs, were responsible for most aspects of building design, from footings to furniture. While modern building designs are the product of many consultants it is still the architect who is largely responsible for major decisions relating to siting, internal planning, exterior appearance, and selection of structural and envelope materials. While the supervisory role has been largely assumed by the project manager, much of the design process is still driven by the architect.

19.3.2 The flawed process

The flaws in the traditional approach to which Eley refers are numerous:

- Functional isolation of consultants – different disciplines use different vocabularies and measure success in different terms. 'Developers speak dollars per square foot; financiers, risk and return; bankers, spread . . . mechanical engineers, square feet per ton and kilowatts per ton; and so forth unto Babel' (Lovins and Browning, 1992).
- Little utilization of the 'collective energy' of the team – individual consultants working in isolation cannot draw on the momentum that can be generated by group working, and there is little opportunity for achieving agreed decisions on optimum solutions to problems that have complex interrelationships between factors that are generally seen as being each within the purview of different disciplines.
- Many basic parameters of the design are set by a few players without regard for the effects that those decisions have on other consultants who will have input to the design at a later stage of the process – input particularly from services engineers and contractors in the earliest stages of design can have great influence on the design outcome in terms of form, efficiency, and buildability yet these people are seldom involved in the process until after the most far-reaching decisions, e.g. orientation, and envelope material selection, have been made,[2] '. . . unless the building services engineer is involved from day one the chances of making any significant contribution to the overall design process are practically zero' (Berry, 1995, p. 29).
- Little perception of the shared ownership of the design – although the architect's role may have been diminished in recent times it is still the architect who enjoys the praise, or shoulders the blame, when the final product is evaluated. The fragmented nature of the process provides scant incentive for the other consultants to feel any sense of ownership or responsibility for the overall design.

. . . in the future, architects and engineers [should] present themselves as members of the design team with shared responsibility for the total building design. (BD&C, 1996, p. 53)

- Lack of communication between consultants – as consultants from different disciplines work on their particular part of the design they may not fully appreciate the effect their decisions will have on other aspects of the design. There is often a pre-occupation with intricate or exotic solutions or devices that actually have less impact on building performance than fundamentals such as building orientation (Vaughan and Jones, 1994). Isolated activities have also tended to be computerized by different software companies with the result that programs used by different team members are often unable to communicate directly with each other. Information exchange is then slowed, and the likelihood of corruption of data, or use of superseded information, is increased.
- Lack of client and user/occupier involvement in the design process – many clients, whether individuals or organizations, are largely ignorant of the building procurement process. This leads them to seek the advice of professionals in the construction industry who will take on the responsibility for initiating and completing the procurement of a facility. A possible consequence of this, however, is a reluctance on the part of clients to involve themselves in the planning and design process, yet it is the needs of the client, and of the eventual occupants of a facility, which are of paramount importance if the final product is to be a success.
- Inappropriate fee structures – consultants' fees have usually been calculated as a percentage of the value of that which is constructed or installed. This has little relation to the time spent by consultants on preparing their designs, and actually rewards consultants who overdesign and specify building components or plant which cost the client more '. . . traditional scales and budget breakdowns [do] not encourage integrated work' (Batty *et al.*, 1996, p. 118).
- Narrow or obsolete approaches to cost planning and management – traditional cost planning approaches were developed in response to the procurement process as it already existed and this has resulted in cost planning procedures now commonly used which begin with costing a sketch design. These sketch designs have usually progressed to a point where decisions regarding factors such as siting, form and envelope have already been made. This results in cost planning being replaced by cost estimating and cost cutting with '. . . an improper emphasis on wringing cost out of each building component separately rather than out of the building as a whole' (von Weizsacker *et al.*, 1997, p. 23).
- Use of 'rules of thumb' in design based on conservative and/or conventional thinking – this is also a consequence of the conventional method of determining fees based on cost of work installed, and the hiring of design professionals based on competitive fee proposals. As these fee structures do nothing to motivate consultants to look for innovative or fine-tuned designs, and often make it uneconomic for them to try, so designs are based on past experience and/or 'standard' practice.

● Little appreciation or application of life-cost approach to design costing – '. . . interaction between the architect and other consultants is mostly such that little thought is given to . . . life cycle costing' (Prasad, 1994, p. 154). This is as much the fault of clients as it is of designers as all too often they take a very short-term view of building costs, with initial capital cost generally being the focus of the cost planning process.

19.4 The benefits of integrated design

If approached in a positive, cooperative manner a holistic integrated design process can greatly reduce the problems associated with current building design practice. Improved communication and closer collaboration among a greater number of consultants from the very beginning of the process promotes the creation of a shared vision for the project.

Commonality of purpose and a sense of joint ownership of the total product will produce a more robust and unified design solution, with a range of measurable benefits – in short, better, more valuable buildings (Fig. 19.2).

A broader and deeper understanding of the requirements and expectations of clients and occupants by the design team results from the identification and evaluation of design problems by a wider range of participants. The brief can be

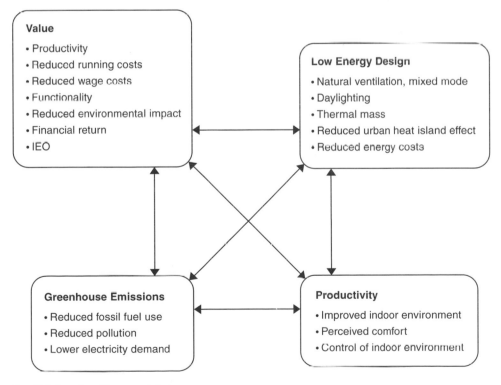

Fig. 19.2 Benefits of integrated design.

developed with multiple inputs from the whole range of consultants and other interested participants, e.g. the facilities manager who will be responsible for the building after it is occupied, before any ill-considered decisions or preconceived ideas are set in place. Revisiting and changing such decisions and ideas can become too costly once the design has progressed to the detailed design stage.

Closer cooperation between consultants, particularly if supported by computer-based information exchange and a shared electronic project database, improves quality control during design. This results in fewer discrepancies in documentation, fewer clashes on site, and, particularly if the construction manager is included in the design process, improved buildability and reduced construction time and cost.

An open approach to problem-solving with more participants' input encourages innovation in design and allows input from a wider range of sources on questions of new materials, components, and techniques. Lenard (1997, p. 174) suggests that '. . . idea generation in construction relies on teamwork and a multi-disciplinary approach.'

Cost planning and management is based on a broader view of the project. The costing of alternative systems or components in the building can be related to their contribution to the complete design rather than be viewed in isolation and treated as discrete mechanisms that can be interchanged or eliminated without consequences and which extend beyond their primary function. It is vital that the team fully understands the interrelationships that exist between fabric and services for instance, and how cost-cutting measures applied late in the design process without proper appreciation of all the functions that a component fulfils, may have catastrophic effects on building performance.

There is wide agreement that complete cost analysis should consider all the functions and implications of individual components. Lovins (1995) speaks of the 'magical economics [which] come from single expenditures with multiple benefits', while Berry (1995) says that '. . . the economic truth [is] that if a device performs more than one function then its future is assured.'

Lovins' 'magical economics' can only become reality if everyone involved is aware of the *totality* of the design; it is no good if a component's primary function is seen as non-essential and can therefore be deleted when such elimination renders other systems less functional or inoperative. Vaughan and Jones (1994, p. 40) point out that a '. . . design is a holistic entity, succeeding or failing on the presence and interaction of all its parts – indeed it can fail totally if one part is omitted or implemented incorrectly.'

An appreciation of the designers' intent is equally important for those who actually construct the building, as some systems and details may be unconventional and appear superfluous or unimportant to the uninitiated. Such a lack of understanding may result in the built product not being exactly as designed with consequent sub-optimal performance.

Lovins illustrates his 'magical economics' using the example of superwindows and explains the capital savings and operational savings which flow from the selection of what appears to be a more expensive window option. Increased daylighting and natural, rather than mechanical, ventilation techniques not only reduce running costs but provide other benefits: a better workplace environment, happier, healthier employees, reduced absenteeism, and improved productivity (see

Chapter 20). These strategies can generally only succeed when the building's services and fabric work closely together (Evans, 1992) and for this to occur design must proceed in a more integrated way.

Further benefits accrue from this approach:

- Smaller HVAC plant as heating and cooling loads are reduced.
- Reduced space requirements for ducting and plant means reduced floor to floor height (and, therefore, reduced wall areas), possibly additional floors in areas where building height restrictions apply, and increased usable floor area.
- Money saved in construction costs can be used to offset possible higher costs of more environmentally benign materials or components, e.g. carpets with lower outgassing of toxic substances, or photovoltaic arrays for electricity production.
- Inherently more efficient ratios such as circulation to gross floor area, perimeter to floor area, and wall area to floor area, which translate directly to monetary savings.
- An enhanced public image with possible rental premiums.

19.5 Implementing integrated design

Integration is much more than mere coordination – it is concerned with combining a number of disparate components into an organic whole. The benefits that can be realized from truly integrated working have been acknowledged for some time, yet it has not been generally implemented. Even when there is agreement, in principle, to adopt such an approach it often does not occur as '. . . putting a multi-disciplinary team together does not necessarily lead to multi-disciplinary working' (Batty *et al.*, 1996, p. 119).

19.5.1 Assembling the team

It is obvious that there must be a clear commitment to the concept of multi-disciplinary working on the part of the initiators of the project, and of all the other participants. While the design process must actively encourage all team members to contribute ideas and expertise it is essential that someone takes the responsibility for ensuring that the overall design concept is clearly understood by everyone involved throughout not only the design phase but during construction.

The team leader may come from any discipline; Kahl *et al.* (1993, p. 161) feel that architects '. . . can assume the integral function that produces a concept in the holistic sense from interdisciplinary co-operation', but the individual's particular field of expertise is less important than their capacity for facilitating successful collaboration.

Any member of the team should be free to comment on any aspect of the design; successful collaboration requires a non-adversarial or non-disciplinary approach, based on mutual respect, with traditional inter-disciplinary jealousies and rivalries abandoned at the outset.

19.5.2 Starting out

'. . . for integration to succeed there must be a common starting point' (Berry, 1995, p. 29).

The aim of changing the way in which buildings are designed is to produce better buildings. Part of the purpose of integrated design is to encourage the application of new ideas and techniques. This requires that the design team approach each project with a collectively open mind, and traditional or conventional means and methods should be put to one side before planning begins. Buildings such as the IRS complex in Nottingham (Anon, 1994 Webb, 1994) and the Queen's Building at De Montfort University (Bunn, 1993; Littlefair, 1996) have emerged because the designers looked for ways to satisfy their clients' generalized requirements, such things as the nomination of natural ventilation or reduced environmental impact, rather than simply reusing and adapting conventional solutions.

19.5.3 Communication and information management

Effective information management is of great importance throughout the design and construction of any building, but it assumes even greater significance in the context of an integrated design team. The setting up of information management systems and communication channels is explored in detail in Chapter 18.

19.5.4 Fee structures

As outlined earlier, fees based on the cost of installed works provide design professionals with what Lovins (1995, p. 81) describes as '. . . a perverse incentive that rewards inefficiency.'

Under conventional fee agreements, designers who do spend the necessary extra time to design systems that are actually less costly but more efficient often do not expect to earn any profits from their work, but rather treat such projects as research, or as promotional exercises, or are involved because they believe that there are strong social reasons attached to reducing the environmental impact of the buildings they design.

Performance contracting provides an alternative to this generally unsatisfactory situation. It provides a clear economic incentive to designers and contractors to produce buildings that meet, or better, targets for building performance, usually stated in terms such as energy usage or occupant comfort levels.

Performance contracts make provision for bonus payments for bettering targets but also include penalties for failure to satisfy stated requirements. This approach is being trialled in the USA (Peterson and Eley, 1996; BD&C, 1996) with the aim of delivering a suitable system for compensating architects and engineers for the extra time they spend to make buildings more efficient. One of the stated goals is to 'promote better integration between building designers, builders, owners, and operating and maintenance staff' (Eley *et al.*, 1995, p. 32).

19.5.5 Client and occupant involvement in the design process

The client or their representative, project manager or facilities manager, should have a definite presence on the design team for the following reasons:

- It assists the team in fully understanding the client's needs and preferences, and promotes a thorough investigation of the design problem.
- It enables the client fully to appreciate the differences in process and outcomes which are the result of the integrated design process. The facilities management director of the Longman Group, speaking of the group's recently completed headquarters in Harlow (UK) explained:

    The design process was approached in a holistic way. This was frustrating at first, because the client could not see what the building would look like until the architects had determined what it would need to 'do'. (Standley, 1995, p. 50)

- It increases the client's awareness of how their building should function – occupants of buildings such as those of the IRS at Nottingham need to be trained in the operation of the systems which make the building work, e.g. when to open and close windows, or how to manipulate blinds and lighting. Without adequate occupant training the building may appear to fail, so building owners must be aware of their role in the successful operation of their buildings.

19.5.6 Selection of procurement systems

Successful integrated design requires a procurement framework which supports that approach; the traditional design/tender/construct/handover procedure is hardly satisfactory.

Whatever the chosen system it must support integrated working, with some notion of partnering whereby all of those who will contribute to the design, construction, commissioning and operation of the facility can be involved.

The system should include adequate provision for testing of alternative solutions throughout the design phase, and particularly so if passive systems such as daylighting are proposed. It is also important that all participants have a clear idea of the design philosophy, as all too often there is a lack of appreciation by some team members of the reasoning behind vital design decisions, with the result that due care is not exercised when building parts are assembled or installed leading to unsatisfactory performance. It has been suggested that '. . . contractors are tending to provide the feature without thinking about the function' (Bunn, 1997, p. 20).

An example of this would be inadequate supervision of window installation where the detailing has been precisely designed to avoid cold bridging but poor construction makes the detail non-functional. In such cases the designer has taken great care to achieve the desired result but a lack of understanding on site has caused that part of the design to fail.

Any system should facilitate input from builders as well as designers, and encourage innovation; however, for this to provide maximum benefit to the client it should occur from the earliest stages of procurement.

19.6 Case study

In 1987 the ING Bank (formerly the NMB) of the Netherlands moved into its newly completed headquarters buildings in Amsterdam. The new buildings were unique in many ways and their success in a number of areas has provided a fine example of the ways in which multi-disciplinary design teams working towards an integrated design solution from the commencement of a project can produce a wide range of benefits for building owners and occupiers.

The bank's brief was simple and open-ended: it was to provide a facility that would:

- integrate art, natural materials, sunlight, green plants, energy conservation, low noise levels and water
- be functional, efficient and flexible
- be human in scale
- have low running costs.

The buildings were designed by a team consisting of architects, construction engineer, landscape architect, energy expert (a physicist) and artists, as well as the bank's own project manager. All team members were involved from the project's inception, and all were free to comment on any aspect of the design. The results are exemplary: energy consumption is less than one-tenth of that of the bank's previous headquarters with attendant savings in running costs of around US$2.4 million per annum, absenteeism was reduced by around 25%, and the bank's corporate image was enhanced to the extent that it moved from fourth to second ranking amongst Dutch banks. Careful integration of the building envelope, services and extensive daylighting has led to significant reductions in the greenhouse emissions associated with the operation of the building (Holdsworth, 1989; Romm and Browning, 1994, 1995).

Based on reduced running costs alone the payback period for the extra money spent on design and construction was only four months. The value of the buildings to the bank has been greatly increased because of the design approach adopted; employees voluntarily work longer hours, productivity is improved, business has increased, and the buildings will continue to serve the bank's needs for many years to come.

19.7 Conclusions

Building Services (1995, p. 29) has suggested that 'integrated design is not a marriage between architecture and engineering, but a reunion after a long period of separation.' It requires that designers view buildings as 'dynamic, inter-dependent systems' (Levin, 1996) and that many conventional design practices be revised.

The benefits which flow from this approach are many, and not the least of these is that the owner achieves more value from investment in new buildings. This enhanced 'value for money' results from the procurement of buildings that satisfy client and occupant needs more fully, which cost less, often initially as well as over time, are constructed more quickly, and produce higher returns on investment.

Integrated design should be embraced as a fundamental component of any strategy for maximizing value in the construction of new facilities.

Endnotes

1. It now houses the National Building Museum.
2. Newton and Hedges (1996) point out that the relatively small cost of design, compared with construction costs, has led to much greater emphasis being placed on improved construction performance. They go on to cite various studies (Glavan and Tucker, 1991; BEDC, 1987) which have shown that design-related problems are the source of the majority of construction problems.

References and bibliography

Anon (1994) Energy efficiency and commercial building. *Constructional Review*, February, 20–5.

Batty, W., Milford, I., Page, B. and Powell, G. (1996) Interdisciplinary working during the strategic phase of building design. In *Proceedings of CIBSE/ASHRAE Joint National Conference*, pp. 112–19.

BEDC (1987) *Achieving Quality on Building Sites* (London: Building and Economic Development Committee).

Berry, J. (1995) Integrated design. *Building Services*, November, 29–31.

BD&C (1996) Energy strategies that make – and save – cents. *Building Design & Construction*, February, 52–4.

Bunn, R. (1993) Learning curve. *Building Services*, October, 20–5.

Bunn, R. (1997) Book review. *Building Services*, July, p. 20.

Eley, C., Peterson, A. and Wentworth, S. (1995) New building energy performance contracting. In *Proceedings, Third International New Construction Program for Demand Side Management Conference*, Boston, 26 29 March.

Evans, B. (1992) Passive solar offices, integrated design. *Architects' Journal*, 6 May, 41–5.

Glavan, J. and Tucker, R. (1991) Forecasting design-related problems, a case study. *Journal of Construction Engineering and Management*, ASCE, **117**, 47–65.

Holdsworth, B. (1989) Organic services. *Building Services*, March, 20.

Kahl, J., Kerstens, J.-P. and Lehmann, J. (1993) Low energy concepts for office buildings. In *Proceedings of the International Symposium on Energy Efficient Buildings*, Leinfelden-Echterdingen. Edited by H. Erhern, J. Reiss and M. Szerman (Stuttgart: IRB Verlag).

Lenard, D. (1997) *Innovation and Industrial Culture in the Australian Construction Industry*. University of Newcastle, Australia: PhD Thesis.

Levin, H. (1996) Ten basic concepts for architects and other building designers. *Environmental Building News*, http://www.ebuild.com/Greenbuilding/Halpaper.html.

Littlefair, P. (1996) Daylighting under the microscope. *Building Services*, April, 45–6.

Lovins, A. (1995) The super-efficient passive building frontier. *ASHRAE Journal*, June, 79–81.

Lovins, A. and Browning, W. (1992) Green architecture: vaulting the barriers. *Architectural Record*, December.

Lyons, L.B. (1993) *A Handbook to the Pension Building*, 2nd edition (Washington DC: National Building Museum).

National Building Museum (1996) Museum display visited by author, Washington DC.

Newton, A. and Hedges, I. (1996) The improved planning and management of multi-disciplinary building design. In *Proceedings of CIBSE/ASHRAE Joint National Conference*, pp. 120–30.

Peck, M. (1993) Role of design professionals. *Building Owner & Manager*, February, 54–7.

Peterson, A. and Eley, C. (1996) New building performance contracting: lessons learned and new ideas. In: *Proceedings for the ACEEE 1996 Summer Study on Energy Efficiency in Buildings*, **5**, pp. 199–208 (Washington DC: American Council for Energy Efficient Economy).

Prasad, D. (1994) Energy efficiency and the non-residential building sector. *Global Warming and the Built Environment*, Samuels, R. and Prasad, D. (eds.) (London: E & FN Spon).

RMI (1998) *Green Development: Integrating Ecology and Real Estate* (Rocky Mountain Institute, John Wiley & Sons).

Romm, J. (1994) *Lean and Clean Management*, cited in RMI (1998) *Green Development: Integrating Ecology and Real Estate* (Rocky Mountain Institute, John Wiley & Sons) 43.

Romm, J. and Browning, W. (1994) *Greening the Building and the Bottom Line* (Snowmass, Colorado: Rocky Mountain Institute).

Romm, J. and Browning, W. (1995) Energy efficient design. *The Construction Specifier*, June, 41–51.

Standley, M. (1995) A breath of fresh air in building design. *BOMA Magazine*, August, 50.

Vaughan, N. and Jones, P. (1994) Making the most of passive solar design. *Building Services*, November, 39.

von Weizsacker, E., Lovins, A. and Lovins, L.H. (1997) *Factor 4, Doubling Wealth – Halving Resource Use* (UK: Earthscan).

Webb, R. (1994) Offices that breathe naturally. *New Scientist*, 11 June, 38–41.

Occupancy cost analysis

Peter Smith†

Editorial comment

The main point of the following chapter is that the cost of designing and constructing buildings, while apparently high in the first instance, is overshadowed in reality by the ongoing costs of maintaining not only the building itself but, more importantly, of maintaining the functions that the building accommodates. The greatest of these ongoing costs are, in fact, the costs associated with maintaining the people who occupy the building (e.g. salaries, sick leave). Consequently a building which, through its physical nature reduces employee costs, will provide a greater return to those who finance its construction.

Research cited in this chapter has shown that various aspects of the indoor environment produced by buildings have measurable effects on the physical and mental well-being of the occupants, and on their productivity. Factors such as individual comfort controls at workstations, elimination of materials such as carpets and manufactured timber products which emit hazardous gases after installation, and the provision of appropriate mixes of daylighting and artificial lighting all contribute to the mental and physical health of building occupants, and all have been shown to have an impact on their ability to work quickly and efficiently.

Unfortunately, to date there has been little serious attention paid to incorporating these ideas into general building design. There is, however, some evidence now emerging that suggests that clients, particularly if they are planning to occupy the buildings that they procure, are becoming increasingly aware of the implications of a design that is sensitive to the needs of those who will ultimately use their buildings.

It follows that a building which promotes higher productivity and reduces unproductive time related to sick building syndrome, employee illness, and rework due to inaccuracies in completed work, should be of greater value to its owner than a 'standard' building. Clearly the same emphasis that is placed on the physical layout of buildings, with the aim of providing spaces that will allow optimum performance of the activities housed in those spaces, should be placed on planning with the aim of minimizing occupancy costs. Only in this way will the client achieve the greatest value from their investment.

† University of Technology, Sydney, Australia

20.1 Introduction

The total life cost of a building is the total cost of creating and maintaining the building over a specified period of time. This time period may relate to the useful life of a building or may relate to the period over which the proprietor has a financial interest in the building. In most cases, operational costs dwarf the capital costs involved in initial procurement. This is now widely acknowledged and life (operational) cost considerations have received increasing attention during the design stage. However, a largely overlooked fact is that, for many building types, the greatest life cost element is *occupancy costs*. This single element can greatly exceed not only initial capital costs but all other operational costs combined, yet typically receives scant, if any, attention during the design stage. For these building types, effective *occupancy cost analysis* during the design stage has the potential to have the single most significant effect on achieving maximum 'value for money' for proprietors and building users.

As terminology varies widely, occupancy costs, for the purposes herein, are defined as the operational costs associated with the functional use of a building. Hence, they are classified as only one, albeit important, element of building operational costs. Typical examples include the costs of management, staffing, operational supplies, plant and equipment, manufacturing and other costs associated with functional use. Maintenance, repair, energy and other operational costs are considered as distinct from occupancy costs.

20.2 The dominant life cost element

The dominance of occupancy costs is illustrated by the following examples. They demonstrate that, for particular building types, the greatest cost lies in functional use particularly with respect to salaries. This suggests that this area should be the most important design, life cost and value consideration, yet functional-use analysis, an integral part of good design practice, has traditionally not incorporated rigorous cost and/or value analysis.

A detailed life cost study of an office building in Sydney over a 50-year life revealed the following life costs. Of the 67% relating to functional use, the largest component was salaries (Fig. 20.1).

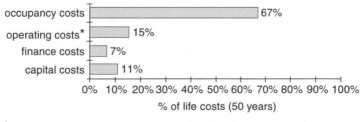

Fig. 20.1 Occupancy costs – CBD Office Building (Hatzantonis, 1992, p. 25).

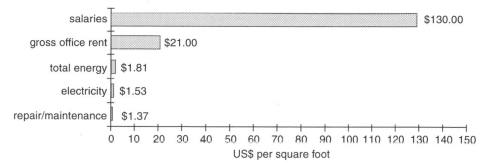

Fig. 20.2 Average annual office expenditure (Romm and Browning, 1994, p. 30).

These results are compatible with a study carried out in Australia by the NSW Department of Housing and Construction (1980, p. 22):

When compared with the total salaries bill for the occupants of the building, the initial capital cost and the running costs appear almost insignificant. For example, over the life of a typical office building for 50 years, studies have shown that 92% of the total cost is for the employees housed in the building, 6% is for running costs and a mere 2% for the capital cost.

BOMA (1991) carried out a national survey of large office buildings in the United States in 1990 and found that the cost of employees, an integral component of occupancy costs, pales other operational costs into insignificance. In terms of average annual expenditure expressed in US$ per square foot, operating costs were identified as: repair and maintenance $1.37, electricity $1.53, total energy $1.81, gross office rent $21.00 and office workers' salaries and associated costs $130.00. Staff salaries cost approximately 72 times the cost of energy (Fig. 20.2).

Service (1998) undertook a life cost analysis for a typical hospital, a specialist facility. Salaries and associated worker/occupational costs were the dominant factor. Figure 20.3 provides a detailed breakdown of these costs.

Worker cost categories were broken down into employee wages and salaries, employee benefits, professional medical services and contracted medical services. Together they accounted for 71% of life costs. Associated work-related occupancy costs (medical supplies, drugs/pharmaceuticals, food and other medical expenses) accounted for a further 17%. The actual building capital cost accounted for only 6% of costs and the building operational costs also only 6%.

Contrary to mainstream thought, the design of buildings can have an enormous effect on occupancy costs and worker costs in particular. As outlined later, building design can greatly influence user productivity and performance and this can often have the most substantial impact on reducing total life costs. The design of a building determines its value, its life costs and its level of operational performance. Despite this importance, many clients attempt to reduce project costs by reducing design and documentation fees to minimum levels. This short-sighted approach will normally end up costing the client and/or end users much more. The reality is that fees for full and expert design development are

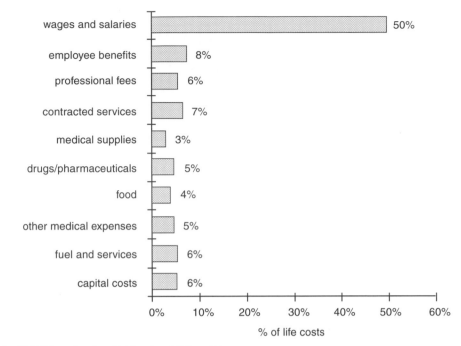

Fig. 20.3 Life costs – hospital (Service, 1998, p. 10).

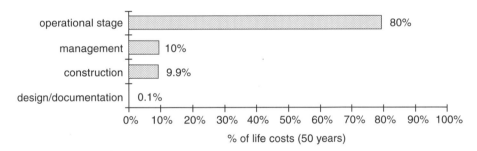

Fig. 20.4 Design fees as a percentage of project life costs (Peck, 1993, p. 56).

insignificant compared with other life costs and, more importantly, the savings and benefits that can result from productivity/performance enhancing design can pay for these fees hundreds, if not thousands, of times over during the building's life.

Peck (1993) compared typical design and documentation costs with other project life costs (Fig. 20.4).

They accounted for only 1% of project life costs and the fact that building design can drastically reduce costs during the operational stage pales these actual fees into further insignificance. More time and attention to the design process and appropriate design fees are essential for optimizing value and reducing life costs.

20.3 Applicability of occupancy costs

Occupancy cost analysis is ideally suited for buildings procured for business or work-related purposes. In these circumstances, 'buildings' would be better viewed as 'dynamic business facilities' where the focus is on the functional needs and performance of the actual users. The quality of user/worker performance has traditionally been attributed to individual, managerial and organizational factors but increasing evidence has been gathered to show that building design can have a profound effect on user/worker performance. Green *et al.* (1986) contend that the design of buildings with highly specialized functions can greatly influence staffing costs. A hospital, for example, has very intricate and specialized design requirements where factors such as supervision and travel distances are important. If these can be optimized for each functional area then it is reasonable to assume that across an entire hospital the total time savings could result in a reduction in staff numbers. This may result in substantial savings in direct staff and administration costs. Even if no staff cuts are made, savings are still available through the minimization of unproductive time. However, the greatest potential for reducing life costs lies in enhancing worker productivity through improving the working environment within a building.

In these circumstances, the building is actually a vital business tool, not simply a dormant product. This is now being recognized by businesses and governments worldwide; managers and organizations are increasingly incorporating the quality of the work environment created by a building's design as a key element in improving work/business performance and increasing profits. This requires designers to place more emphasis on the effect of their designs on the actual performance of the end users. Building designers have a key role to play in the future business performance of the end-users and this is where the greatest value to the proprietor may well lie.

However, many buildings are not procured for the purposes of owner-occupation, where occupancy cost analysis is most beneficial. The greatest value for developers intent on a quick sale will usually lie in maximizing profit through maximizing the difference between total capital cost and selling price. Developers in this situation are far less concerned with operational costs and the end-user. The proprietor who retains ownership of the building upon completion with the intention of leasing will clearly be more concerned with operational costs particularly with respect to energy, maintenance and repair costs. However, these proprietors are not primarily concerned with the end-user, the tenant or lessee's operations. At the design stage, they may not even be able to identify readily the type of activities that will be undertaken by future users.

Nevertheless, building occupiers are becoming more astute and aware of the effect of building design on business/worker performance. More attention to occupancy cost analysis may enable these proprietors to use the resultant benefits as a marketing tool to increase the selling price or lease price, thereby increasing the value of the building not only to the immediate proprietor but also to future owners and users.

Owing to the typical significance of salaries, occupancy costs may not be an important consideration for buildings whose occupants are not in income-earning positions (e.g. residences, community centres). Strictly financial occupancy analyses are suited to buildings procured for government, office, retail, manufacturing, services and the like. However, the process of occupancy cost analysis during the design stage can be used to influence other intangible benefits that are difficult to express in monetary terms but nevertheless lead to better value for money. Building designs that produce optimum environments for end-user performance can, for example, enhance student learning in educational facilities or improve the recuperative rates of patients in a hospital. Additionally, productivity gains are not always measured in terms of work output and quality. Better building design can lead to improved business performance through increased sales and improved corporate image or reputation.

20.4 The influence of facility design on user performance

The accurate measurement of occupancy costs and user-performance changes remains at an embryonic stage. Although there is now increased awareness of the effect that building design has on user performance little progress has been made thus far in terms of accurate economic analyses. Most proprietors will generally seek economic justification before commissioning the use of performance-enhancing design features. Developments in such economic measures will be discussed later in the chapter.

At this point, research findings and case-study examples are largely used to demonstrate the benefits and influence proprietors. The following findings and cases studies provide good examples of how appropriately targeted building design can improve user performance. The case studies include both retrofits (refurbishments) and new buildings. Retrofits probably provide the best means of measuring productivity changes as the 'before and after' analyses are more readily comparable than is the case with new buildings. Interestingly, the majority of these works and design changes were carried out solely to reduce energy, maintenance and repair costs. Subsequent improvements in user performance and/or productivity were an unexpected bonus.

20.5 Improving user performance – new building design

Lockheed Missile and Space Company's commercial office building in California, completed in 1983, provides a good example of how good design can reduce costs and improve business performance. Housing 2700 engineers and support people over a total floor area of 55 000 m^2, a primary design objective was to reduce energy usage by at least 50%. Total annual energy costs were halved with daylighting alone reducing lighting costs by 75%. The energy-efficient improvements were estimated to have added US$2 million to the US$50 million construction cost. The energy savings were worth US$500 000 per year representing a pay-back period of only four years. However, this was overshadowed by improvements in worker productivity. The average absenteeism rate dropped by 15% which, in itself, paid for the extra energy-efficient construction costs in less than one year (Romm and Browning,

1994). The architect, Lee Windheim, cited in Thayer (1995, p. 28), states that 'a mere 2% of the total cost of Building 157 over its useful life will be for the initial design and construction. Another 6% will be spent on maintenance costs, including energy, and employee salaries will account for the remaining 92%. This is why the contribution of the building to employee productivity is the most important design consideration. Small gains in productivity can make huge differences in corporate profits'.

The significance of productivity gains in increasingly competitive markets is further highlighted by the claim by Lockheed management that the improved productivity gave them the competitive edge that helped them win a US$1.5 billion contract. The profit made from this contract more than paid for the entire construction cost of their building!

The West Bend Mutual Insurance Company's new $15\,000\,m^2$ office building in Wisconsin, completed in 1991, also has a number of energy-saving design features that have had a significant impact on operating costs and worker productivity. These features include an energy efficient lighting system (task lighting and occupancy sensors), larger window areas, shell insulation, 'environmentally responsive workstations' with individual temperature and airflow control and a more efficient HVAC (heating, ventilation and air-conditioning) system. Whilst annual electricity costs were reduced by 40% the real benefits were found in improved worker performance. The 'environmentally responsive work stations', the improved lighting and the improved thermal comfort of the occupants were all significant contributors to a productivity increase of approximately 16%. With an annual payroll of about US$13 million, this represents a theoretical saving of over US$2 million per annum (Romm and Browning, 1994).

20.6 Improving user performance – retrofit design

The Reno Post Office in Nevada, USA, was renovated in 1986 with the sole objective of reducing energy costs. The retrofit design changes cost US$300 000 and were simply based on a lowered ceiling and improved lighting. The resultant energy and maintenance savings came to about US$50 000 per year, which represent a pay-back period of six years. The major benefit, though, arose in the unintended area of worker productivity. A quieter and better lit work environment produced dramatic improvements in work output and quality. Productivity increases stabilized at 6% after one year and the mistake rate dropped to the lowest of any post office in the country.

The Pennsylvania Power & Light Company in the USA experienced problems with their lighting system in a building that housed its drafting engineers. The main problem related to distracting glare and reflections off work surfaces. The lighting system was redesigned from general lighting to task lighting with high-efficiency lamps and ballasts. The cost of the renovation was minimal (US$8362) but the effect was dramatic. Total energy and operating costs fell by 73% resulting in annual savings of US$2035. This represented a pay-back period of 4.1 years and a return on investment of 24%. Based on the time it took staff to complete drawings, productivity increased by 13.2%, which was worth US$42 240 per year. This reduced the payback period from 4.1 years to 69 days and increased the return on investment to 540%. The quality and accuracy of work increased, the health of the workers improved, the sick leave rate dropped by 25% and worker morale skyrocketed. The value of reduced errors was calculated as being worth US$50 000 per year which further increased the return on investment to over 1000% (Romm and Browning, 1994).

20.7 Improving user performance through healthier building design

In certain circumstances a building can actually be a source of ill-health for workers and lead to increased absenteeism and reduced productivity. Increasing evidence has been gathered over the past two decades to show that, as well as more obvious safety problems, subtle problems associated with actual building design and function are having a significant effect on the well-being of inhabitants. This is commonly referred to as the *sick building syndrome* (SBS). In the United States, it has been estimated that SBS costs approximately US$3 billion in lost annual productivity, whilst in Australia it is estimated to cost industry several hundred million dollars each year (Dingle, 1995). The future legal ramifications of SBS for building proprietors in terms of not providing a safe working environment may result in substantial damages payouts and possible financial ruin if not adequately insured.

Dingle found that research has consistently shown that the following design elements appear to increase the prevalence of SBS: wall-to-wall carpets, large areas of upholstery, large amounts of shelving and horizontal surfaces, a sense of crowding, office size, poor lighting, poorly designed workstations, air quality and poor environmental control by individuals. The most significant determinant appears to be HVAC design: '50% of problems are due to poorly designed, operated or maintained heating, ventilation and air conditioning systems' (Dingle, 1995, p. 21). Studies have shown that these problems can be greatly reduced if a building incorporates natural ventilation in all or part of its design (Rowe and Wilke, 1994). In a survey of ten Australian office buildings, ranging in age from five years to over 100 years, Rowe and Wilke found that the 100-year-old building, one of only two naturally ventilated buildings surveyed, easily produced the best results in terms of air quality and lower susceptibility to SBS. The other naturally ventilated building ranked second.

Good maintenance and frequent and effective cleaning are also important. Poor cleaning equipment and practices can lead to elevated levels of dust in the air and a corresponding increase in symptoms. The maintenance and cleaning of HVAC systems is clearly another critical factor. The interaction between the individual worker, the design of their workplace and the demands of their job also have an effect on health. Key design features in this area include lighting, workstation design, interaction with other workers and ergonomic design of office furniture and equipment.

20.8 Improving business performance through flexible and adaptable design

Business process re-engineering, whereby firms continually change their work practices to utilize technological advances most effectively and to meet ever-changing industry demands, is now a worldwide phenomenon. The actual building plays an important part in this process. 'The workplace is changing so dramatically now in terms of job function, organization size and where people work that if you don't have a facility designed with that flexibility you have the potential for wasting space and incurring higher costs when those changes take place' (Patterson, 1996). These changes coupled with rapid technological advances mean that future obsolescence factors become critical design considerations. This necessitates the assessment of not only current but also future end-user needs as well as general changes in the work environment and lifestyles. In this scenario, without the proverbial 'crystal ball', it is difficult to forecast requirements in two to three years' time let alone over the anticipated life of the building. Loose-fit adaptable design is the most logical means of dealing with this dilemma; buildings need to embrace flexible and adaptable design to ensure long-term demand for their use. There are many examples emerging that highlight this need.

In the United States, re-engineering of office space use has led to trends whereby office space requirements are being cut by up to two-thirds as business performance becomes more closely related to physical layout. Focus is shifting from large individual offices to high-performance collaborative team workspace. 'Hot desking' is gaining momentum whereby workers can alternately use the same desk/workspace due to flexible working hours and changed work practices. Offices are now used as an important part of production with emphasis not only placed on worker productivity but also on the productivity of building use. Studies are emerging showing that high-performance office design can improve office efficiency by 30–40% (Ferguson, 1996). Telecommuting is another important consideration. Technological advances will increasingly reduce the need for workers in certain sectors to be in a central workplace. Studies have shown that workers can work more productively from home given the right home working environment. These trends are predicted to impact significantly on office space utilization worldwide. All of these factors have the potential to cut organizations' overhead costs significantly, by reducing upfront and operational costs, and to increase profitability through more efficient work practices.

20.9 Strategies to foster development of occupancy cost analysis

The ability of facility design to influence occupancy costs, and particularly worker productivity, has generally been met with scepticism in the construction community. A major problem is that it is not so difficult to demonstrate retrospectively that 'good' design which focuses on reducing functional use costs makes good economic sense but it is often very difficult to determine how to ensure good design before it has started. It remains a relatively new area that relies largely on anecdotal evidence rather than a sufficient volume of accurate historical data and research. Successful development of this area will need to be underpinned by expert economic appraisal of design influences on occupancy costs and worker productivity. The bottom line of business is profit and there is a need to demonstrate how appropriately targeted building design can improve the performance of the occupants and, in turn, profit levels. Facility owners and occupiers are increasingly becoming aware of the effect of the actual physical work environment (i.e. the building) on worker performance and ultimately business performance and profit levels.

Improvements in worker productivity through good design and the concomitant cost benefits of reduced overhead costs and increased profits have the capacity to far outweigh the impact of other operational costs. As described earlier, the BOMA (1991) study found that, for typical office buildings, staff salaries cost approximately 72 times the cost of energy and the Department of Housing and Construction (1980) study found that staffing costs accounted for 92% of facility life costs. Lippiatt and Weber (1992) found that, on an annualized per square metre basis, employee salaries in office buildings are approximately 13 times building costs. Accordingly, even minor productivity changes can have major cost consequences. Theoretically, a 13% increase in construction costs for productivity-enhancing design can be justified if it increases productivity by 1%. Ward (1987, p. 115) surveyed research on the impact of building design on productivity and found that productivity gains of up to 10% are more than feasible. He highlighted the potential global significance of this by stating that 'given the US annual expenditure of US$1 trillion for white collar salaries a 5% productivity increase translates into US$50 billion per year'.

20.10 Measuring productivity changes

The accurate measurement of occupancy costs and productivity changes remains undeveloped. Lippiatt and Weber (1992, p. 1) contend that 'current literature reveals heightened awareness of the importance of productivity impacts on building design yet little progress has been made toward systematically including productivity impacts in building economic analysis. Without explicit treatment of productivity, it is difficult to justify higher-priced designs on productivity improvements'. They also note that despite the fact that salaries often far outweigh any other life cost element, life cycle costing does not account for the productivity benefits of new technologies and building design. It traditionally judges project alternatives from a strictly building cost perspective:

> Higher priced designs that enhance productivity may make more economic sense. Yet without an economic method for systematically including productivity benefits in building Life Cycle Cost analysis, these designs cannot be justified.

The first step lies in developing appropriate economic evaluation methods. Lippiatt and Weber (1992) have developed two such methods, the Net Benefits

(NB) method and the Multi-Attribute Decision Analysis (MADA) method. The following section describes these methods with hypothetical case examples used to illustrate their practical application. The section is drawn from Lippiatt and Weber (1992, pp. 3–17).

20.10.1 Net Benefits (NB) method

The Net Benefits (NB) method expands the life cycle cost approach to incorporate not only the costs but the benefits of design alternatives, thus measuring economic efficiency rather than simply building costs. In the private sector, economic efficiency is synonymous with maximum profits whilst in the public sector this often relates to maximum net benefits. The method is based on comparing and evaluating design alternatives to a base design solution. The base design will normally represent the design that satisfies the minimum design requirements with the lowest investment cost. NB is calculated as present-value benefits less present-value costs. The following formula summarizes the approach used:

$$NB_{A2:A1} = \sum_{t=0}^{N} \frac{B_t - C_t}{(1+d)_t}$$

where

$NB_{A2:A1}$ = benefits less costs, in present value $, of Project Alternative $A2$ relative to base case Alternative $A1$

B_t = benefits (e.g. productivity gains) for Alternative $A2$ less benefits (if any) for Alternative $A1$ in time period t

C_t = costs for Alternative $A2$ less costs for Alternative $A1$ in time period t

d = discount rate reflecting investor's minimum acceptable rate of return

N = number of years in study period.

NB is calculated for each project Alternative A_i ($i \neq 1$) relative to the base case Alternative $A1$. If NB is positive, project A_i is economic and, if NB is negative, uneconomic. The objective is to select the project with the highest NB, or if no alternative has a positive NB then either to select the base case alternative or no alternative at all. The method requires monetary measures of productivity benefits and, to this end, utilizes a Productivity Impact Function (PIF) system which measures 'cause and effect' relationships between building design and productivity. A set of PIFs is used to map changes in design features to changes in productivity. Because the effect of design features on productivity vary according to the nature of job activity, separate functions can be derived for each job type. Each PIF becomes unique to a specific combination of design feature and job type.

The example in Fig. 20.5 illustrates a hypothetical application of the NB method.

20.10.2 Multi-Attribute Decision Analysis (MADA) method

The second method, Multi-Attribute Decision Analysis (MADA), also incorporates worker productivity in life cycle cost analysis. This technique enables more

Scenario
Two conceptual office designs have been developed to house 1300 managerial, professional/technical and clerical workers. The designs are required to be evaluated in terms of their net benefits with due consideration for productivity impacts.

Design Details

	Base Design	**Alternative Design**
Number of Floors	4	5
Office Floor Area	$9290\,\text{m}^2$	$11\,613\,\text{m}^2$
Individual Worker Floor Area	minimum	25% more than base design
Construction Cost (C_0)	$20 million	$25 million
Operating Costs ($C_{t,t\neq0}$)	not applicable	$20 000 p.a. higher than base design

From a purely building cost perspective, the base design is the best choice. However, the productivity impacts of each design may confer other benefits. The increased floor area of the Alternative Design may lead to better job performance and satisfaction. Contributing factors may include increased status and more room for equipment and private meetings. These factors will affect productivity differently depending on job type.

Productivity Impact Functions (PIFs)
Productivity Impact Functions can be developed to measure the impact of floor area on specific job type. The following table shows the PIFs for each job type. Productivity is expressed as a linear function of floor area and each PIF measures the change in annual productivity attributable to floor area. This is expressed in $/worker.

Job Type	PIF for Floor Area	
Manager	$/yr/person	$= (296 \times \text{m}^2\ \text{office}) - 3161$
Professional/Technical	$/yr/person	$= (487 \times \text{m}^2\ \text{office}) - 3711$
Clerical	$/yr/person	$= (485 \times \text{m}^2\ \text{office}) - 1935$

Productivity Benefits of Alternative Design
In order to calculate the Net Benefits for the Alternative Design, the net productivity benefits (B_t) for this design in each time period need to be computed. The value of B_t is shown in the following table.

Job Type	Annual Salary	Number	Base Design		Alternative Design		Net Productivity Benefits ($/yr)
			m^2/office	Productivity/Person	m^2/office	Productivity/Person	
		1	2	3	4	5	$6 = 1 \times (5-3)$
Manager	$55 000	200	10.68	$0	13.38	$800	$160 000
Prof/Tech	$45 000	775	7.62	$0	9.57	$950	$736 250
Clerical	$25 000	325	3.99	$0	5.02	$500	$162 500

Annual Productivity Benefits for Alternative Design (B_t)	$1 058 750

Column 1 is the number of workers for each job type. Columns 2 and 4 give the office floor area per worker for each design. Columns 3 and 5 give the related productivity/person using the PIFs calculated above. The base design (Column 3) is given no productivity benefits for computation purposes as it meets minimum design requirements. The Alternative Design theoretically confers an annual productivity benefit of nearly $1.1 million. This represents almost 2% of total annual salaries.

Net Benefit for Alternative Design
The Net Benefit of the Alternative Design is calculated using the formula outlined earlier. Assumptions of 25 years for the study period (N) and 10% for the discount rate (d) were made and it was also assumed that wages and energy prices would increase at the general price inflation rate. The results are shown in the table below.

$C_0 Co$	$C_{t,t\neq0}$	$B_{t,t\neq0}$	$B_t - C_t$	$NB_{Alt:Base}$
1	2	3	$4 = 3 - 2$	5
$5 million	$20 000	$1 058 750	$1 038 750	$4.4 million

The Alternative Design has net benefits of $4.4 million and is clearly the most economically efficient choice. The cost of the extra floor area more than pays for itself through the productivity benefits; the 2% increase in productivity more than justifies the 25% additional construction costs.

Fig. 20.5 Net Benefits (NB) method example (Lippiatt and Weber, 1992, pp. 6–9).

than one performance or design attribute to be considered in a decision even if the attributes are not measured in comparable units. It simultaneously accounts for both traditional life cycle cost measures in dollars and non-monetary measures of productivity impact. Productivity impact may be expressed in physical dimensions, such as square metres or decibels of sound energy, or even be based solely on informed judgements. It provides an alternative to developing monetary measures of productivity impact such as with the PIF system when using the NB method. A variety of MADA techniques exist. The Analytic Hierarchy Process (AHP) is one of the more commonly used as it facilitates comparisons between 'apples and oranges' by utilizing a scoring system based on a common denominator. The AHP structures a complex decision into a hierarchy with the decision criteria at the top and the alternatives to be evaluated at the bottom. Decision makers then establish relative weights among the criteria through a sequential process of pairing comparisons. A mathematical technique based on eigenvectors is then used to derive the relative weights from the pairing comparison data. Figure 20.6 illustrates the application of this technique to measure productivity impacts.

20.10.3 Appraisal of NB and MADA techniques

These techniques provide a very useful framework for the quantification of productivity factors. However, Lippiatt and Weber admit that more comprehensive supporting data relating building design to productivity is required to implement them effectively.

Research and data collection

Consequently, further research is required to assist designers by providing sufficiently detailed and reliable data and information. This needs to be combined with general education and awareness programmes to spread the knowledge gained and provide the necessary evidence required to sell this design concept. Implementation will be more quickly facilitated through statistically rigorous data and research rather than *ad hoc* and largely anecdotal information. Nevertheless, in the short term, case studies such as those described herein should be used to raise proprietor awareness of the immense benefits of occupancy cost analysis during the design stage.

20.11 Use of key design features to enhance user performance

Nevertheless, key design features that affect user performance have been identified and can be used now to promote occupancy cost analysis. Although difficult to quantify, the benefits of these key features are clear. The case studies strongly suggest that, in terms of improving user-performance, visual acuity, thermal comfort and air quality are key determinants. Daylighting and natural ventilation are design concepts that can have a tremendous influence in this area. Clearly,

Scenario
A project manager is considering three design alternatives for a building project.
Step 1: Establish Performance Criteria
The manager establishes six relevant performance criteria for this decision:
1. Initial Cost, 2. Operation & Maintenance Cost, 3. Energy Cost, 4. Lighting Effectiveness, 5. Air Quality and 6. Quiet.
1–3 are classed as Life Cycle Cost (LCC) criteria and 4–6 are grouped as Productivity criteria.
Step 2: Establish Weighting for each Criterion
The pairing comparison procedure is used to first determine the importance of the two categories, LCC and Productivity. The project manager decides LCC criteria are twice as important as Productivity criteria. Therefore, weightings of 0.667 are given for LCC and 0.333 for Productivity. The weighting of each pairing of criteria is then made as shown below.

LCC CRITERIA

	Initial Cost	O&M Cost	Energy Cost
Initial Cost	1.000	1.400	1.200
O&M Cost		1.000	0.800
Energy Cost			1.000
Computed Weights	**0.392**	**0.274**	**0.334**

PRODUCTIVITY CRITERIA

	Lighting Eff.	Air Quality	Quiet
Lighting Eff.	1.000	2.000	1.600
Air Quality		1.000	0.800
Quiet			1.000
Computed Weights	**0.392**	**0.274**	**0.334**

Step 3: Rating Design Alternatives With Each Criterion
For the LCC criteria, the project manager uses the LCC $ amounts as the basis for the rating. The AHP method then normalizes the vector of the reciprocals of the LCC $ amounts of all the design alternatives (each reciprocal is divided by the sum of the reciprocals). This makes the ratings inversely proportional to the LCC $ amounts. An example of this rationale is that if a design costs half as much as an alternative it is twice as desirable since two can be built for the price of one. The following table shows the $ amounts for the LCC criteria, the design alternatives and the resultant ratings based on normalizing the reciprocals.

LCC CRITERIA

	Initial Cost		O&M Cost		Energy Cost	
	LCC ($)	Rating	LCC ($)	Rating	LCC ($)	Rating
Design 1	2 000 000	0.336	350 000	0.244	900 000	0.272
Design 2	2 500 000	0.269	238 500	0.358	700 000	0.350
Design 3	1 700 000	0.395	215 000	0.397	650 000	0.377

Proportional data are not available for the productivity criteria so the pairing comparison method is used to rate the design alternatives. The pairing comparison values and resulting computed ratings of the design alternatives are shown below.

LIGHTING EFFECTIVENESS

	Design 1	Design 2	Design 3
Design 1	1.000	1.200	0.800
Design 2		1.000	0.600
Design 3			1.000
Computed Weights	**0.323**	**0.260**	**0.418**

AIR QUALITY

	Design 1	Design 2	Design 3
Design 1	1.000	2.000	1.000
Design 2		1.000	0.500
Design 3			1.000
Computed Weights	**0.400**	**0.200**	**0.400**

QUIET

	Design 1	Design 2	Design 3
Design 1	1.000	1.500	0.750
Design 2		1.000	0.500
Design 3			1.000
Computed Weights	**0.323**	**0.260**	**0.418**

Step 4: Ranking Design Alternatives
The final step involves computing the overall ratings and rating the design alternatives accordingly. This is done by summing the products of the weights times the ratings for each alternative. This is shown in the following table. Design 3 has the highest overall rating and is selected by the project manager.

Rank	Alternative	Rating
1	Design 3	0.400
2	Design 1	0.308
3	Design 2	0.292

Fig. 20.6 Multi-Attribute Decision Analysis (MADA) method example (Lippiatt and Weber, 1992, pp. 9–15).

workers who are comfortable and enjoy their working environment are more likely to work more efficiently and productively. Lippiatt and Weber (1992) found that building design affects productivity through ambient conditions such as noise, air quality, lighting and temperature and through workplace conditions such as enclosure, size and layout. To date, the Buffalo Organization for Social and Technological Innovation (BOSTI), cited in Brill *et al.* (1984), has probably undertaken some of the most significant research and compiled one of the largest databases in this area. BOSTI undertook a five-year research programme to quantify design impacts on worker productivity that involved over 6000 employees in 70 organizations. BOSTI identified six office features that most significantly affect productivity: temperature fluctuation, air quality, glare, noise, enclosure and furniture.

This knowledge can be used now by designers to improve end user performance and provide greater value to the proprietors and users of their buildings. The only problem at this point is establishing the full benefits of these value-adding design features.

20.12 Linking building design with business management

The greatest influence on worker productivity comes from good management and organizational practice. However, ensuring the optimum working environment should be an integral part of this practice and this is where building design has a significant role to play. Worker productivity is intertwined between work environment and management. Design that focuses on a building's functional-use requirements and the needs and wants of the end-users, the occupants, all makes good management sense. Involvement of employees in the design process can lead to more accurately targeted design, increased staff morale and esteem, and genuine desire for the design initiatives to be effective. Staff should also be made aware of management goals and objectives. Good design can provide the catalyst for all of this and, combined with good management practice, can lead to sustainable productivity gains.

20.13 Multi-disciplinary design approach

Owing to the wide variety of variables impinging on this area, research ideally needs to be based on a team approach incorporating input from building owners, architects, engineers, government bodies, services authorities, environmental experts, energy experts, construction cost experts, facilities managers, sociologists and, perhaps most importantly, the end-users. The emergence of facilities/asset management as a discipline in its own right has seen a marked increase in the level of post-occupancy evaluation studies. The incorporation of analyses of the effect of design on worker productivity, currently given only scant attention, could provide the most practical means of building up the necessary data and information.

20.13.1 Focus on end-user functions and requirements

Functional use analysis, an integral component of Occupancy Cost Analysis and described in detail in Chapter 11, also needs further development. Functional use analysis is also integral to the work of designers and many would contend that all the matters raised herein are merely examples of good design. It is the designer's responsibility to examine the functional requirements of the proposed facility, to plan the work environment effectively, minimize circulation and to cater for future needs. However, designers have traditionally spent little time in evaluating the performance of their finished products. The evaluation of successful facility performance lies in the satisfaction of the end-users.

Effective occupancy cost analysis requires more in-depth analysis of the actual functions and requirements of the end-users and evaluation of actual facility performance. As stated earlier, the former should ideally involve a design team approach which includes the end-users if possible. Loose-fit adaptable design has also taken on increasing importance due to obsolescence considerations and the fact that most facilities have a large number of different end-users during their life cycle. With the advent of Facilities Management, post-occupancy evaluation studies are gaining popularity and should provide invaluable assistance to the design team if the data obtained are collected and disseminated in a practical form.

20.13.2 Retrofits

The catalyst for acceptance of occupancy cost analysis lies in the retrofitting (refurbishment) of existing facilities. New facilities represent only a minuscule proportion of the total built environment. As the case studies indicate, the benefits of retrofits can be more readily measured through 'before and after' analyses. If existing facility proprietors and users can be shown that retrofit designs based on accurate occupancy cost analysis can produce considerable financial benefits with short payback periods and increased profit levels, the construction industry may well see a boom in retrofit work.

20.14 Conclusions

Occupancy cost analysis during the pre-design stage has considerable potential to improve value for a proprietor's money quite dramatically. Whilst the value-adding benefits may be difficult to quantify accurately in monetary terms, logic suggests that these benefits are real. The prudent proprietor who advocates such analyses during the design stage for appropriate building types has much to gain and, it could be strongly contended, little to lose. To that end, building proprietors and users need to be better educated in terms of the real costs of building ownership and operation. They need to be shown the true costs and benefits of design alternatives and how achieving optimum value for money may well lie in using a life cost approach during the design stage that incorporates functional use and occupancy cost analysis.

Most buildings are constructed for the purpose of making money and it is time for a shift in design emphasis from the design and construction of an 'empty building' to the design of a 'living facility' where focus is placed on satisfying the needs and improving the performance of actual users with the ultimate objective of increasing profit levels for the firms utilizing and/or owning the facility.

References and bibliography

BOMA (1991) Experience exchange report. Building Owners & Managers Association, USA.

Brill, M. (1984) *Using Office Design to Increase Productivity* (Buffalo: Workplace Design and Productivity Inc).

Department of Housing and Construction (1980) *TI140 AE – Life Cycle Costing – Technical Information*. Technical Bulletin, Sydney.

Dingle, P. (1995) Sick building syndrome defined. *BEST*, November, 19–21.

Green, J., Adams, A., Nelson, S. and Aisbett, K. (1986) *Evaluating Hospital Ward Designs In Use*. School of Health Administration, University of New South Wales.

Ferguson, A. (1996) Time for the office building to lift its productivity too. *Business Review Weekly*, October.

Hatzantonis, S.J. (1992) *Recurrent Costs – A Role for the Quantity Surveyor*. Undergraduate Thesis, University of Technology, Sydney.

Lippiatt, B.C. and Weber, S.F. (1992) *Productivity Impacts in Building Life-Cycle Cost Analysis* (Gaithersburg, USA: US Department of Commerce).

Patterson, M. (1996) Investing success. *Buildings Online The Magazine*, USA.

Peck, M. (1993) Role of design professionals. *Building Owner & Manager*, February.

Romm, J.J. and Browning, W.D. (1994) *Greening the Building and the Bottom Line* (Old Snowmass, USA: Rocky Mountain Institute).

Romm, J.J. and Browning, W.D. (1995) Energy efficient design can lead to productivity gains that far exceed energy savings. *The Construction Specifier*, June, 44–51.

Rowe, D. and Wilke, S. (1994) Sick building syndrome and indoor air quality – perception and reality compared (University of Sydney: Department of Architectural and Design Science).

Service, B. (1998) NSW Commission leads to performance-based contracts. *Chartered Building Professional*, August, Canberra.

Thayer, B.M. (1995) Daylighting and productivity at Lockheed. *Solar Today*, May/June, 26–9.

Ward, R. (1987) Office buildings systems performance and functional-use costs. In *Proceedings of the Fourth International Symposium on Building Economics*, Vol. A, Working Commission W.55 on Building Economics (Copenhagen: International Council for Building Research Studies).

21

Technology and innovation

Marton Marosszeky†

Editorial comment

Innovation and research in construction is generally aimed at improving efficiency in some way – whether the goal is shortened construction time, improved quality of product, reduced rework or rectification work during construction or any of a number of other possible outcomes. What is common to all these goals is that they all provide increased value to clients for the money that they invest in building, usually by reducing costs.

It is unfortunate that the construction industry in some parts of the world, notably the UK and Australia, as discussed in the following chapter, has a history of low levels of investment in research and development (R&D) when compared with other sectors of the economy such as manufacturing. There are a number of major factors which have contributed to this situation:

- the 'one-off' nature of most building projects
- the generally low profit margins achieved by building contractors
- the very tight schedules under which the design and construction of buildings must usually be undertaken and completed
- stringent budgetary constraints typically based solely on initial capital cost rather than on life cycle costs.

All of these factors severely limit the time and/or money that can be invested in innovation or research. They lead instead to the continuing adoption of conventional methods and solutions, whether or not they are necessarily the most appropriate in the specific situations in which they are applied.

In Australia the construction industry accounts for around 7% of GDP, only slightly less than the manufacturing sector (Lenard, 1997). In spite of the major contribution that construction makes to GDP, funding for R&D within the industry is generally very limited. This is an unfortunate situation as improvements in productivity in this sector actually have a much greater effect on GDP than other service industries, so much so that, in Australia, for example, a 10% improvement in productivity in the construction industry would produce a 2.5% increase in GDP (Stoekel and Quirke, 1992).

† Building Research Centre, University of New South Wales, Australia

In the following chapter, Marton Marosszeky, a very active participant in construction-related research, examines the Australian situation in some detail. He looks at a number of areas that are deserving of further investigation and suggests that there are many opportunities for innovation and research which would lead to improved performance in the industry. He also considers the challenges that face the construction industry in our rapidly changing world, which result from the rise of large multinational corporations, global competition and the explosion of new technologies in communication and other areas.

References

Lenard, D. (1997) *Innovation and Industrial Culture in the Australian Construction Industry*. PhD Thesis, University of Newcastle, Australia.

Stoekel, A. and Quirke, D. (1992) *Services: Setting the Agenda for Reform* (Canberra: Department of Industry Technology and Commerce).

21.1 Introduction

This chapter examines the building and construction industry in an international context, although it refers to the Australian industry from time to time regarding specific examples and cases. It also describes the changes confronting the industry community and players, and the challenges that need to be addressed as the industry moves along the path to international competitiveness. The aim of this chapter is to identify how investment in research, development and innovation at both the project and at the enterprise level, can lead to significant competitive advantage.

Within the last 300 years, it has been the past two decades that have seen the most significant transformation in manufacturing processes. Application of IT (information technology), automation, new materials, new equipment, improved organizational structures and participative management have brought unprecedented improvements in productivity. Product quality has improved as client satisfaction has been embraced as the fundamental tenet of business.

Innovation and knowledge are not only integral dynamics promoting the main drivers of change but the key to survival and growth. In recent times some of the best international examples of growth based on innovation have come from Japan. The CEO of Sony Corporation, Akio Morita, as quoted in the CIC (1993) report *Profit From Innovation*, argues that from a corporate perspective, for a company to achieve long-term profit from innovation it is necessary that creativity in technology be accompanied by creativity in strategic management and marketing; he asserts that excellence in technology is not enough.

Fruin (1992) describes the approach of Japanese enterprises as an interactive, dynamic learning system in which companies continually adapt and change, responding to feedback from recent in-house experience, as well as from other players in their own industry and elsewhere. For the construction industry to be truly successful, it needs to emulate this manufacturing culture, adopting similar methods and goals, as they are as relevant to construction as to any other industry.

While construction has changed more slowly than manufacturing, the industry is in transition worldwide. IT and globalization are impacting on all the processes of project procurement. In some countries, significant investments in technology and innovation are changing the way the industry delivers its projects. In Australia, for example, investment in construction innovation is relatively low.

In terms of the construction workforce, roles, relationships, skills, contractual agreements are all changing and need to change even more. The industry's ability to embrace change will determine its success.

The main forces driving innovation in project procurement are:

Market related

- Faster production.
- Reduced overall costs.
- Improvements in quality.
- Improved market position.

Community related

- Sustainability.
- Environmental quality.
- Safety.
- Regulatory compliance.

21.2 Incentives for innovation

In this era of unprecedented international competition, only innovative firms have survived, and only the most innovative have grown. Competitive pressures, throughout all industry, are keener than ever before as a result of the progressive opening of the world economies to market competition on a global scale. This pressure has seen a more rapid change in the nature of manufacturing and retailing worldwide than ever before.

The post-World War II period has seen rapid growth in the international activity of construction industry firms, worldwide. Construction contractors have been winning international contracts because of their specialist experience and knowledge of projects such as hydro and nuclear power plant construction and tunnelling projects. Contractors specializing in lifts, mechanical, hydraulic, fire and electrical services, control systems and facades are now supplying and fixing their systems worldwide. Manufacturers have also moved onto the world stage with operations in the USA, Europe and South East Asia.

This period has also seen many major, long-term clients of the industry, from the manufacturing and retailing sectors, take a more active role in the procurement of their buildings for their own activities, by bringing techniques used in manufacturing into the construction sector. In an endeavour to optimize return on investment and minimize operating costs, they have embraced processes such as value engineering and partnering to get their projects completed efficiently and according to plan.

21.2.1 Demand issues – community needs

Community needs, as expressed through government policy and through direct community actions, reflect concern for environmental sustainability, for worker and community safety and for reliability of the construction industry's products and services.

At a policy level, governments are concerned with promoting industry development, both in the interests of the economic well-being of their region, and as customers for some 30% of the industry's services. They have a responsibility to manage procurement risks and ensure that the public gets value in its investments.

Until 1990, governments in Australia were relatively inactive regarding policy development for the construction industry. By comparison, there have been an increasing number of examples in the recent past, both in Australia (CIDA, CPSC), and overseas (Singapore CIDB) of governments taking an increasingly active role in policy development for the sector (Latham, 1994; NSW DPWS, 1996, 1997, 1998; APCC, 1997).

It has also become evident that the public sector is more interested in procuring services from the private sector rather than the buildings and infrastructure that house and support those services. For example, governments increasingly see their responsibilities as providing education and health services rather than building and maintaining schools and hospitals.

21.2.2 Demand issues – client needs

Gann (1997) makes the point that in the long term the demand for fixed capital assets is moving away from constructed products to equipment and machinery. He notes that the cost of construction has been rising relative to other industries due to the relatively lower rates of productivity growth and because modern buildings are becoming lighter envelopes which house increasingly sophisticated systems that support human activities. It is clients and users who are in a position to benefit from more efficient buildings. In 1964 the US Steel Company conducted an analysis of their costs in the construction and use of their new headquarters building which they built at that time in Pittsburgh. Of the total costs, including their organizational operating costs over 40 years, construction represented only 4%, as did the maintenance and operating costs. The remaining 92% represented salary and operational costs in the day-to-day business of the firms. From this analysis it is evident that the efficiency of organizations in the use of office space is by far the most significant issue in building design.

This observation is fully supported by DEGW's (1996) research, where they report that there is an increasing need for flexibility of space utilization in office buildings. This is shown by the many different space functions, access provisions and servicing levels that tenants require, all of which reflect the endeavours of those trying to achieve a much greater fluidity of work organization than has been typical in the past.

Construction industry clients are articulating a desire for product and process reliability; they want to avoid unexpected surprises during the procurement phase, and subsequently when their building is in use. However, it has to be said that, in many instances, clients either do not have the management skills or the will and understanding of the industry to convert this desire into reality.

There are also signs that in some cases clients are taking an interest in life cycle cost issues, although commonly the desire at policy level to consider these issues is not translated into practice. It seems that during procurement, the trade-offs between capital expenditure and life cycle cost often favour short-term cost reduction.

21.2.3 Supply issues – design

Over the past two decades, primarily as a result of their inability to define and sell the value of their services effectively, the role of the design professionals in the procurement process in Australia has diminished. This has occurred through a combination of price-based market competition and the absence of a clear articulation of the benefits that design professionals can contribute to the process.

It has also been widely recognized that design management is one of the most neglected areas in construction management (Koskela *et al.*, 1997). From the time of the Tavistock report on communication in construction (Higgin and Jessop, 1965), to the present (Coles, 1990), design quality has been identified as a major issue in construction. In his study of construction defects, Josephson (1996) reported that when measured by cost, design-related defects were the biggest category. In a survey of 30 contractors in Sydney, conducted by the author and others in 1996, the most common criticism of design registered was that, in the case of large projects, timeliness, coordination and full resolution of design were all insufficient.

At the same time, while many argue that the design process is under stress, the industry is becoming more industrialized. Consequently, design and planning are becoming increasingly important, yet there are fewer and fewer resources being invested in design. This often occasions rework and rectification during construction, the time when it is most expensive to do so.

There are two underlying reasons for this situation. One is that the value added in design and planning has not been sufficiently recognized by clients and contractors. It is a human foible that intangible, qualitative aspects of products and services are often undervalued. Secondly, the design professions have been caught in a vicious circle in which not only is their input not being valued but their fee base has been reduced, forcing them to reduce the services that they can offer. This has led to a downward spiral in terms of the value added by the design professions in many parts of the process.

Against these organizational issues, communication technologies are permitting organizations to develop designs concurrently at different sites. For example, some organizations are working in sequential shifts between teams in the US and India, others are transferring work between offices within the same organization to even out workloads.

Visualization technology is enabling the full 3D modelling of complex construction and assembly processes during the design stage, thus permitting design and construction optimization. This technology is also making designs more accessible to stakeholders such as the public and future occupants, who traditionally were unable to have a significant input into the processes of the industry.

21.2.4 Supply issues – construction

In spite of the popular observation that, in construction, we still lay brick upon brick, and little has changed, the contrary is the case. In all areas of activity, there has been continual small-scale innovation in the development of new and modified materials, products and equipment. There is no aspect of construction that has remained unchanged. In terms of large-scale changes in process, however, it is true that these have been relatively few and far between.

Gann (1997) notes that traditional local construction industries tend to compete on price rather than on the quality or technical competence, and that generally this has led to an overdeveloped sense of price and an underdeveloped sense of value.

As with manufacturing, all construction has become increasingly equipment oriented, and components are prefabricated wherever possible. This equipment orientation has supported the globalization of the industry in a number of ways; for example, where companies have been successful in the development of new equipment-based construction methods, they usually become internationally competitive and win contracts the world over. Japanese tunnelling technology, for example, has won contracts for Japanese contractors in Europe, the USA and Australia. The most innovative companies, in terms of both technology and management, have become the world leaders in the sector, and today, large-scale construction is a specialization of giant multinational companies. For example, Bechtel specializes in the provision of power generation infrastructure worldwide.

Equipment developments have seen capacity, diversity and flexibility increase. New technology has added numerous features including self-navigation and intelligence to improve safety and efficiency. Construction has become more equipment intensive and more value is being added further up the supply chain through sophisticated product manufacture and prefabrication.

Innovation in process organization, 'lean construction', has become a keen topic of study since the publication of *The Machine that Changed the World* (Womack *et al.*, 1990). The concepts of lean production (a philosophy that incorporates concurrent engineering), total quality management (TQM), just in time (JIT), design for manufacture (DFMA), design for assembly (DFA) and process re-engineering and their application to the management of the construction process have been discussed in some detail by those interested in extending efficiencies in the building industry (Koskela, 1992; Huovila *et al.*, 1994; de la Garcia *et al.*, 1994; Betts and Wood-Harper, 1994). This debate has triggered five international workshops on the subject of lean construction in recent years. Some researchers have already begun to study the potential of certain of these techniques when applied to single aspects of construction (Mohamed *et al.*, 1996; Forsythe, 1997).

In relation to on-site activities, computer and virtual modelling can now provide the basis for more detailed planning and the optimization of construction activities. In the recent past, the management task, which traditionally focused on time and cost, has had to be adapted to include the areas of quality, safety, and environmental planning and control. All these aspects of management have yet to be fully integrated.

21.3 Opportunities for innovation

This section presents some key problem areas where innovation would benefit the industry.

21.3.1 Industry-wide needs

Cooperative as against adversarial relationships between the parties in the process

There is no doubting that the nature of relationships and contractual agreements in any situation promotes certain responses. In many instances the agreements used in the construction sector encourage behaviour that is in conflict with community interests. Under these circumstances it is imperative that the relationship between agreements and outcomes be studied so that agreements which encourage the achievement of community interests are developed and adopted. Some obvious negative examples of this are situations where a focus is placed on short-term costs during construction with the neglect of life cycle costs; another is where behaviour and practice are wasteful of resources and harmful to the environment.

Improved use of IT between the stages in project delivery and project use

Information is costly to develop, whether it relates to design, material quantities, construction planning in relation to materials, systems, costs, quality, safety and/or environmental factors. The transferability of information is being addressed under the STEP Program where the basic agreements for interoperability are currently being developed.

The adoption of performance measurement and benchmarking as drivers of industry development and improvement

There are numerous examples of the use of objective performance comparison based on objective criteria in the manufacturing sector and it has been used in a wide range of settings (Karim *et al.*, 1997). Many firms that are established leaders in their field use this tool on a continuing basis. Interestingly, in many such cases, where comparisons have confirmed the overall leadership of these enterprises, the process has led to the identification of further potential improvements. The introduction of these tools to the construction sector promises similar benefits by providing players with an improved understanding of the relative efficiencies of their different activities.

21.3.2 Areas of concern for the design professions

Identifying and communicating the value created and added by the design professions

Research needs to be conducted to study the efficiency of current industry procurement processes used in the sector. For example, one study could demonstrate the relativity of potential construction cost increases to design fee savings in areas such as structural and mechanical services. Another example is where much of the design resolution is left to the construction phase as against full resolution during design. As design briefs have contracted in scope, the extent of design and estimation tasks undertaken by specialist contractors has increased. This has the impact of slowing the construction phase and of creating duplication of design and estimation among the specialist contractors, probably increasing the overall costs in the system.

The improvement of the quality of design resolution and documentation and design management

Research and application in this area is required to develop a framework for the analysis and coordination of the individual sub-system designs. In relation to the management of the design process, Koskela *et al.* (1997) suggests an approach based on an understanding of the optimal sequence of design tasks and the application of planning techniques to improve process management.

In relation to issues such as documentation error and service quality, performance measurement is likely to be an effective tool to drive process improvement. Some work is already under way in this area at the Building Research Centre (BRC) at the University of New South Wales.

21.3.3 Key needs of contractors

The integration of safety, quality, environment, cost and time in the planning and delivery stages

Over the past seven years the industry has expanded its formal management procedures to include quality, safety and environmental management. While initially, dedicated managers were appointed to these perceived specialist functions, it has become apparent that the best approach is to integrate these responsibilities into the line management functions of the project and construction managers. In this case, it is necessary to develop integrated planning tools that include all of the issues, rather than merely proceeding with the current process of having separate plans for each area. The solution needed is the development of a nested set of management tools that can be viewed either individually or in an integrated manner.

Improved planning and optimization of construction processes on the basis of computer generated models

With this technology the implications of alternative construction strategies can be evaluated and construction systems optimized. An analytical framework needs to be developed to facilitate modelling of construction processes. New knowledge is needed regarding construction reasoning and processes to underpin the development of graphical models. Systems are needed that can model the hierarchical nature of construction operations from elemental tasks to major construction processes.

21.3.4 Technology implementation

Recent developments in areas such as advanced, multi-sensor, field data collection and transputer-based real-time data fusion systems, real-time, millimetre accuracy, 3D positioning systems developments, videogrammetry as well as modelling and developments in control systems have the potential to transform certain construction processes and improve the efficiency of others. Many of these technologies have been developed overseas and often in other industries. Because of the time that it takes to adapt new technologies into any setting, more efficient introduction and implementation systems are needed to implant these technologies into the Australian construction sector. These techniques have the potential for further development in areas such as the construction of models using multi-sensor data and the development of control tools for machines operating in dynamic unstructured environments.

21.3.5 Process redesign

Howell *et al.* (1996) argue that the traditional model of a construction project as purchasing a product, needs to be replaced with the model of construction as a prototyping process. In the latter case, the process is a continuous negotiation between ends and means: this negotiation is completed by representatives of different value sets who must continuously resolve the ends/means question at increasing levels of detail. Such a prototyping process is enhanced by the use of cooperative processes such as partnering, and the intense interaction of concurrent engineering.

Currently, by and large, traditional work packages are executed by the traditional parties through the fragmented processes of the industry, and some *ad hoc* innovations do occur within those traditional trade packages. In order to develop more efficient, restructured processes, it is necessary to involve all the parties in the supply chain in a cooperative, structured approach. This will logically lead to a re-engineering of the supply chain (Marosszeky, 1997) in such a way that a small number of first-tier suppliers will emerge to take responsibility for composite work packages, and thus coordinate the inputs of a group of second-tier suppliers. This structure has the potential to involve the manufacturers in both the design and planning process.

21.4 Challenges to be overcome

There is a tension between an industry organized along Taylorist[1] lines of specialization and fragmentation, and the need to control processes to match tight delivery schedules and the need for a high quality product. In many instances, the opportunity exists for processes to be re-engineered through process redesign and the restructuring of supply chains. To achieve and maintain international competitiveness, the industry has to embrace change, invest in learning and innovation, as well as overcome some of its conservative traditions.

One of the main impediments to innovation in the sector is the high degree of fragmentation in the industry. For example, in Australia, the average number of employees of the 148 000 organizations in the sector is three (NSW DPWS, 1998). Consequently there are few models of larger-scale innovation. Gann (1997) estimates that of the 200 000 private construction contractors in the UK construction industry only 1% have the critical mass of five or more qualified professionals who may be capable of systematic technical development.

Larger organizations have both the financial and technical capacity to innovate. However, most members of the industry are inhibited by their perceptions of risk and reward under the existing structure. The argument is often put that there is little incentive to take additional risk as any one party is only adding approximately 15% of the value in the supply chain. Decision makers will be liberated from this conservative viewpoint once they realize that the substantial benefits to be had from process redesign are at the disposal of the innovator in the re-engineering process and that the rewards are likely to be considerable relative to current profit margins. Further, it will be found that the most innovative company in a sector will be able to attract increasing market share because of its competitiveness on price and service provision.

There are two notable examples from Switzerland where relatively small to medium architectural practices have taken a key role in building system development involving process redesign (Marosszeky, 1980). In both instances the architects conceived the design and joined with manufacturers to supply and install complete building systems. In one case a single manufacturer delivered the entire building, in the other a team of specialist manufacturers worked together in a consortium to bid for projects.

21.5 The payback on innovation

The CIC report, *Profit from Innovation* (CIC, 1993), provides several case studies of companies that have benefited from innovation. These range from construction companies that have innovated successfully in techniques and equipment and have therefore gained a competitive advantage in certain types of work, to consultants who have developed new software-based design products that have helped them to improve their market position.

Generally it is very difficult to calculate the payback on investments in R&D, as all the potential benefits are hard to identify and in some instances very difficult to quantify. A relevant measure is that only those companies that invest in innovation really thrive. However, in the case of two research and development projects conducted at the BRC at the University of New South Wales, a notional calculation was possible to give some indication of payback.

The first was a project to evaluate concrete repair material properties before use. This project was a pre-competitive project, one with public outcomes. The objective was to reduce the risk of failure in all structural concrete repairs. Currently many repairs to concrete structures fail through shrinkage, and need to be redone within about five years. There are no existing international standard tests for the objective comparison of the shrinkage characteristics of concrete repair materials and this project developed such a test.

Concrete infrastructure repair expenditure in New South Wales is estimated at $8 million annually. This is based on the estimate of senior staff at the state Roads and Traffic Authority for its own repair costs in 1996. It is assumed that 50% of this expenditure is on structural repairs. Based on experience to the present it is reasonable to assume that 25% of structural repairs fail within five years due to the premature cracking of the repair material, thereby limiting a potential 25-year life. On this basis the value of repairs which will fail prematurely is $1 million per annum or one eighth of the total cost.

The BRC has now developed an objective test for the assessment of these materials making it possible to select the best materials. The project cost approximately $150 000. For NSW alone, by enabling the selection of materials that will not fail, the payback is the order of 35 times the investment over five years.

A second case study examined the payback on a project, which developed the re-engineering of the timber floor in cottages on sloping sites (Forsythe, 1997). This was a commercial project, which typifies the savings that can be achieved through process re-engineering. BRC researchers studied waste, process flow and innovation in techniques for this work package. The project resulted in a 20% saving in cost and in time. The investment was $250 000.

In Australia, approximately 150 000 houses are built each year, of which approximately 30% have timber floors. If just 10% are on sloping sites, based on these figures, and once all houses on sloping sites use the new techniques, the potential annual cost saving represents a payback of more than 20 times the original cost.

21.6 Who should pay for innovation?

This section should be prefaced by the observation that, based on international comparisons, the Australian construction sector has traditionally invested in innovation at a very low level. OECD reports between 1989 and 1993 show that Japan and Sweden are the highest investors in construction R&D, investing approximately 3% of construction generated GDP while Australia ranks well down with investment levels at about one third of this rate. Innovation projects fall into the categories of competitive research (work with proprietary outcomes) and pre-competitive research, where the outcomes are public or shared by a group of competing stakeholders. In this latter case the benefits accrue to the stakeholders initially and to the entire industry eventually.

In Australia, government support for construction industry research has been considerable. This has been directed, in the main, to the CSIRO Division of

Building, Construction and Engineering (DBCE). The construction research community in universities, however, enjoys very little government support. In contrast to this, the equivalent to the CSIRO DBCE in the UK, the Building Research Establishment, has been privatized and substantial funds are available for construction research through the Science and Engineering Research Council and through European funding sources. In the area of applied construction research, the public purse has traditionally supported innovation in two areas, management of technology and industry development. The first category includes research into issues such as regulations, occupational health and safety, and environmental impacts. The second category, industry development, includes projects that develop management tools such as benchmarking to improve industry efficiency, support the uptake of improved management practices and IT, or develop information about home or overseas markets to enhance competition and to enable Australian companies to enter overseas markets.

Private sector investment is rightly directed at the improvement of competitiveness. It is important to recognize, however, that there are areas of pre-competitive research and development that improve the efficiency of the stakeholders in the research and eventually flow through to the entire industry. Industry leaders need to be proactive in identifying such opportunities, as they cannot afford to wait for governments to provide the research.

Of concern is the fact that, over the past decade, Australian government funding mechanisms have significantly reduced their support for research in relation to the regulatory system. The health of this system is critical to the construction industry. Because of the fragmented nature of this industry, risk management in relation to the failure of materials and products is essentially provided by the technical knowledge that underpins the regulatory system.

Gann (1997) asserts that, in the UK, there is a clear market failure in the development of technical know-how with respect to SMEs[2] in construction. There appears to be a similar problem in the Australian construction sector. Government policies have been directed for some time at the large organizations of the industry, on the basis that change at that level will trickle down to the smaller organizations which make up the bulk of the sector. While this is no doubt effective for those smaller organizations working on major projects, a very significant part of the industry is not exposed to these influences. The BRANZ[3] model in New Zealand is one that specifically sets out to support the SMEs in the sector and it achieves its goals very successfully. BRANZ support is partly based on a nominal industry levy to enable it to provide services to this sector.

Private sector research and development in the Australian construction industry is generally carried out by the material and product manufacturers and is aimed at the development of improved products and the improvement of manufacturing processes. Generally, the rest of the sector is a low investor in R&D. This paper has suggested areas where investment would benefit different industry groups.

Perhaps the main challenge for the private sector in relation to its research and development needs is that it should take the initiative. While recognizing that innovation is essential to long-term success, it is also important to acknowledge that some innovation is needed in the cooperative development of factors that affect market efficiency. There are many examples of other industries managing to leverage their investment in R&D with matching investment from government. This, however, requires a greater level of industry cohesion than has been evident in the past.

21.7 Conclusions

This chapter has considered the needs for research and innovation as a key driver of process improvement in construction. It has looked at impediments and challenges to the conduct of research and has suggested mechanisms for funding based on international examples.

While it has presented an international perspective, it is written on the basis of detailed experience in, and knowledge of, the Australian construction sector. In this sense, the Australian construction sector has been used as a case study and many of the examples are drawn from it. In spite of this, the structure of the sector has fundamental similarities around the world, regulations have developed gradually along similar lines and consequently the general observations and conclusions are in some measure universal.

The differences between competitive research and pre-competitive research are explained and the case is put for industry investment in certain areas of pre-competitive work on the basis of the benefits that will accrue to the stakeholders. Examples are given of the payback on research drawn from some analyses developed by the author at the BRC, University of New South Wales.

This chapter identifies some of those areas of research and innovation in management and technology that will benefit industry development and promote its efficiency. Finally, the need for industry to take control and develop its own research agenda is articulated. The industry needs to leverage its own investment in R&D with matching government funds. There are mechanisms for this within current government policy in most developed countries.

In this regard it is noted that investment in research, development and innovation in the UK, Europe and Japan is significantly higher than in Australia. Those countries who are significant investors in innovation will reap the benefits in terms of their competitiveness in the international marketplace.

Endnotes

1. Taylor was an early advocate of the specialization of tasks in manufacturing to improve efficiency based on time and motion studies. This movement led to over-fragmentation in production processes in which responsibility was difficult to attribute and in which coordination became the major management task. There has been a more recent swing back to teamwork and cooperation.
2. Small and medium enterprises (SMEs), which are now recognized as being the main driver of employment growth in the economy.
3. The Building Research Association of New Zealand is funded in part through a government levy on construction; these funds are directed at supporting the information needs of SMEs.

References and bibliography

APCC (1997) *Construct Australia: Building a Better Construction Industry in Australia* (Australian Procurement and Construction Council).

Betts, M. and Wood-Harper, T. (1994) Re-engineering construction: a new management research agenda. *Construction Management and Economics*, **12**, 551–6.

CIC (1993) *Profit from Innovation* (London: Construction Industry Council).

Coles, E.J. (1990) *Design Management; A Study of Practice in the Building Industry*. Occasional Paper No. 40. The Chartered Institute of Building.

DEGW (1996) *From Envisioning to Implementation*. Property Council of Australia Annual Congress, September.

de la Garcia, J.M., Alcantara, P., Kapoor, M. and Ramsh, P.S. (1994) Value of concurrent engineering for A/E/C industry. *Journal of Management in Engineering*, May–June, 46–55.

Fruin, W.M. (1992) *The Japanese Enterprise System* (Oxford. Clarendon Press).

Gann, D. (1997) Should governments fund construction research? *Building Research and Information*, **25** (5), 257–67.

Higgin, G. and Jessop, N. (1965) *Communications in the Building Industry* (London: Tavistock Publications).

Huovila, P., Koskela, L. and Lautanala, M. (1994) Fast or concurrent the art of getting construction improved. *2nd International Workshop on Lean Construction*, Santiago, Chile, September, 1–9.

Karim, K., Marosszeky, M., de Valence, G. and Miller, R.M.A. (1997) *Benchmarking Construction* (Sydney: Building Research Centre). ISBN 0 7334 1612 8, A2.

Koskela, L. (1992) *Application of the New Production Philosophy to Construction*. Technical Report 72. Centre for Integrated Facility Engineering, Department of Civil Engineering, Stanford University, California.

Koskela, L., Ballard, G. and Tanhuanpaa, V.P. (1997) Towards lean design management. In: *Proceedings of the Fifth Annual Conference of the International Group for Lean Construction*, Department of Civil Engineering, Griffith University, Gold Coast, Queensland.

Latham Report (1994) *Constructing the Team: Joint Review of Procurement and Contractual Relations in the UK Construction Industry* (London: HMSO).

Marosszeky, M., Jaselkis, E., Smithies, T. and McBride, A. (1997) Reengineering the construction process using 'first tier' suppliers. *Construction Process Reengineering*. Griffith University, Gold Coast, Queensland.

NSW DPWS (1996) *The Construction Industry in New South Wales; Opportunities and Challenges*. Discussion Paper (Sydney: New South Wales Department of Public Works and Services).

NSW DPWS (1997) *A Perspective of the Construction Industry in NSW in 2005*. Discussion Paper (Sydney: New South Wales Department of Public Works and Services).

NSW DPWS (1998) *Information Technology in Construction: Making IT Happen*. Discussion Paper (Sydney: New South Wales Department of Public Works and Services).

Womack, J.P., Jones, D.T. and Roos, D. (1990) *The Machine that Changed the World* (New York: Simon and Schuster).

22

Dispute resolution

John Twyford†

Editorial comment

Disputes are costly – usually the only ones to benefit from disputes are the legal advisers who are called in to assist with resolving them. The adversarial nature of the relationship between contractors and clients is something of a tradition, but surely an unwanted one, as it all too often permeates the day-to-day dealings between those involved in project delivery and leads to unproductive and hostile working relationships between the parties.

Disputes that do arise can add substantially to the final cost of a project. Direct effects include increased costs arising from the engagement of legal personnel, arbitration, court proceedings, preparation of documents and other evidence, time lost by staff who are required to give evidence personally and so on.

Indirect costs include reduced quality of work, either due to acceleration of work to meet tightened schedules, or loss of motivation as a result of poor workplace relations. Major disputes, which can substantially delay completion of a project, may mean missing an important opening date (e.g. a retailer who hopes to start trading in time for Christmas) or loss of prospective tenants who require space to be available for occupation by a specified time.

Regardless of how disputes arise, their speedy settlement can help to ameliorate their effects. An essential part of the good pre-planning of any construction project is the setting up of a framework or mechanism for the handling of disputes, such that when disputes do arise they can be resolved as quickly as possible.

It is best, of course, if disputes can be avoided completely, but there are a number of factors that are characteristic of the construction industry that make dispute avoidance difficult: the large number of people involved (the client, designers, contractor, subcontractors, consultants), the large amounts of money involved, the competitive nature of the industry, and the reliance on low bid tendering as the means of selecting contractors.

The impact of disputes on the ultimate value gained by the client is not restricted to simple financial loss through increased contract price or the expense of litigation; disruptions to schedules can lead to hastily completed work, lack of cooperation or a lack of interest in quality outcomes. The end result for the client is less return on investment, in terms of both increased costs and reduced quality.

† University of Technology, Sydney, Australia

In this chapter, John Twyford looks at the various avenues that are available for resolving disputes once they do arise, and how some standard forms of contract set in place agreed mechanisms for dispute resolution.

22.1 Introduction

There is a need for the parties to a proposed contract to discuss and agree on an appropriate regime of dispute resolution before the contract is entered. This may be an unpalatable task and inconsistent with the euphoria that normally accompanies the successful conclusion of a negotiation. Even so, the matter is of prime importance as the method of dispute resolution agreed upon could determine whether or not the cost and time parameters originally contemplated by the parties are met. If a dispute does arise the cost to the parties can be considerable in terms of the legal costs to manage the dispute, lost time (for example the productive time lost in executives instructing lawyers), the unbudgeted moneys a party might be ordered to pay by the tribunal hearing the dispute, and the cost to the principal in not having the project completed on time. In addition to these moneys, an unsuccessful party is frequently ordered to pay the legal costs of the successful party. For these reasons it is important to look at the causes of dispute and ways in which particular disputes might be resolved. There is no one method that is universally efficacious and often it is appropriate to employ a number of procedures, each specific to the matter in dispute, the stage the project has reached, the sums of money at stake and the effect of a failure to resolve the dispute on the overall project.[1]

22.2 Causes of disputes

The classical building dispute involves a contractor suing for moneys alleged to be due under a contract 'and being met by a claim for abatement of the price or cross-claims founded on an allegation that the performance of the contract has been defective'.[2] This is to oversimplify the matter as the underlying causes are to be found in the structure of the relationship. There would appear to be a number of factors contributing to the frequency of construction industry disputes, especially in Australia. Dorter and Sharkey (1990) list no fewer than 38 matters that can be made the subject of a claim or dispute. These include the following.

22.2.1 Calling of tenders prior to the completion of the design

It is understandable that a principal would wish to get a project under way for a number of reasons; however, a premature commencement leaves the contract open ended. It is almost certain that a contractor faced with this situation will claim for delay costs, variations to make the uncompleted design workable, and adjustment of prime cost items that could not be priced at the time of tender. Whilst the principal might have been prepared from some additional costs arising from these circumstances, the true cost comes as a shock.

22.2.2 Lack of legal precision in the standard documentation in use in the construction industry

Certainly during the 1970s and early 1980s in Australia the standard documents used in the industry left leeway for contractors to make claims for delays. The position of contractors was enhanced by a series of decisions of the courts.[3] Since then there have been attempts to shift the balance but the large number of standard documents produced has meant that the opportunity for judicial interpretation is limited.

The fact that most contracts in the construction industry are let by means of competitive tender has engendered an adversarial attitude in contractors towards their clients. There exists the distinct prospect for a contractor to increase his or her profit at the expense of the principal by making claims, and equally for the principal to control the costs by rejecting claims (sometimes unjustifiably).

Frequently, the contract document used by the parties will misallocate the risk inherent in the construction process. This can be the accidental result of the use of standard industry documents or the direct consequence of a party using its stronger bargaining position to dictate the terms of a transaction. In a tight market a principal can require that the contractor take a large measure of the design risk whilst it is clear that this is a matter within the principal's control. Equally, in Australia, contractors have succeeded in shifting the risk for labour relations onto the principal. Again, it is the contractor who can best control this risk.

22.2.3 Low return on investment

The Australian construction industry is undercapitalized and, whilst the return on invested capital is good, the return on turnover is low. There are virtually no barriers to entry to the industry and this, together with the low level of capitalization, has encouraged risk-taking behaviour. Contractors frequently bid low to secure a contract hoping to realize a profit by making claims. Often the claims will be based on deficiencies that the contractor has seen in the contract documentation at tender time but withheld, knowing that a claim can be made at a later date. This is hardly the basis of a sound business transaction. Contractors who are prepared to do this are aided and abetted by a well organized 'claims industry'.

22.2.4 Poor contract documentation

Inaccurate contract documentation has been a fertile source of claims. Inaccuracies in contract design documentation lead to the need for variations that are priced at premium rates and result in the principal paying delay costs. In Australia, claims arising out of bills of quantity have been epidemic and it is now rare for a principal to supply a bill as a tender document. Where bills are supplied the accuracy is carefully disclaimed.

22.3 Dispute resolution processes

If the parties make no provision for dispute resolution in their contract then any dispute will need to be litigated in the courts. Proceedings in the courts are conducted on an adversarial basis and produce a final answer that binds both parties. Equally, although a creature of the agreement of the parties, commercial arbitration is adversarial in nature and produces a result that binds both parties. Both litigation and arbitration are said to be curial methods of dispute resolution on the basis that the word *curia* refers to a court and the giving of a final decision. Although there is a significant place for curial dispute resolution, dissatisfaction with the process has led to the evolution of the so-called *alternative dispute resolution* (ADR). ADR is a consensual process and is directed towards bringing the parties to the point where the dispute is settled on some mutually agreeable basis. The outcome of the ADR process is therefore unenforceable until the result is embodied in some form of binding contract such as a deed of settlement. The discussion continues with an examination of the various techniques of dispute resolution, the way in which each is implemented and the strengths and weaknesses of each.

22.4 Curial dispute resolution

22.4.1 Litigation in courts

If the parties make no provision in their contract for dispute resolution any disputes will need to be heard by a court. The court will apply the ordinary legal principles to the resolution of the dispute. This will mean that the party alleging failure on the part of his or her contractual partner will need to demonstrate a breach of contract, a breach of a duty of care, an instance of unjust enrichment or some threatened breach of his or her legal rights before the court will act. The decision will be made by a judge in the chosen jurisdiction after the parties have crystallized the issues in a formal pleading process. In most instances the remedy awarded will be a requirement for the party deemed at fault to pay compensation. The compensation thus payable is properly called damages and the damages are assessed in accordance with the principles described in Chapter 17.

In certain circumstances the court will grant an injunction (an order to the defendant to desist in illegal conduct) or a decree for specific performance (an order directing the defendant to carry out the terms of a contract). Most jurisdictions have a tiered judicial system graduated according to the sum of money claimed. In NSW there are three levels, with the smaller claims being heard by a magistrate in the Local Court, the intermediate claims (presently to a monetary limit of A$750 000) in the District Court and the major claims in the Supreme Court. A party who commences proceedings in the wrong court is penalized as to the recovery of his or her legal costs. In most Australian jurisdictions the Supreme Courts have created specialist lists to deal with building and engineering matters. This involves rules that require the parties to narrow the issues in dispute as closely as possible, provide written evidence and submit to a rigorous timetable for the prosecution of the litigation. The purpose of these rules is to enable the litigation to be concluded as soon as possible to enable the parties to get on with their commercial relationship. To assist in the speedy resolution of issues, judges have been appointed to the jurisdiction who have skills in dealing with construction industry disputes. A further innovation in most jurisdictions permits judges to refer the technical issues in the proceedings for resolution by an expert referee.[4] The referee then furnishes a report to the judge who brings down a judgment embodying the technical findings of the report. In this way the litigants have the best of both worlds: the technical skill of an expert and the legal knowledge of a judge.[5]

In several Australian jurisdictions there exists, alongside the court system, a system of tribunals to deal with domestic building disputes. These tribunals are consumer oriented. The court systems have the disadvantage of being formalistic, slow and expensive. The parties to court proceedings are locked into a process that has little scope for lateral thinking. Settlement often comes at the doorstep of the court at a time when little in the way of expended legal costs can be salvaged. After litigation the business relationship between the parties is usually severed. The advantages are that the court gives an authoritative legal answer which is less likely to be the subject of a successful appeal. As indicated above, the use by the court of a referee provides both the best technical and the best legal answer. The procedure is expensive, time consuming and suited only to major disputes.

22.4.2 Commercial arbitration

The English Courts recognized arbitration as a viable method of dispute resolution as early as 1609[6] and the first statutory recognition came 89 years later.[7] It is now the preferred method of dispute resolution in the building and construction industry in the Common Law world. Arbitration is a consensual process and depends upon the parties to a dispute making an agreement to refer a dispute to arbitration. Provided the agreement meets certain requirements and the parties act within the confines of the law, arbitration is capable of producing a solution to a dispute that is recognized by the State and enforceable in the court system. What follows is a brief description of the important stages in the arbitration process.

Arbitration agreement
Since the consent of the parties is essential, there must exist an agreement to refer the dispute to arbitration before the process can begin. This agreement can be given at the time the contract is entered (here the agreement usually takes the form of a clause in the contract) or after the dispute has arisen. There is in Australia a requirement that arbitration agreements be in writing.[8] Care should be taken in drawing an arbitration agreement as the agreement is the source of the arbitrator's power. The draftsperson should say if the arbitrator's powers are to be confined to contractual disputes or if they will extend to other legal remedies that the parties might have in tort or under statute.[9] It is possible for parties to a contract (or for that matter in most other legal disputes) to agree after the dispute has arisen to refer the matter to arbitration.

Initiation of the arbitration process

The parties need to be in dispute before the process can begin. This usually arises from the demand of one party failing to be met by the other. Claims arising out of the valuation of variations, extensions of time and the consequences of failures to grant extensions of time, delay costs, adjustments of the contract sum and inter-pretation of contractual provisions can all give rise to disputes. Standard forms of construction contract have an arbitration agreement which provides machinery for crystallizing the existence of a dispute. This requires a party making a claim to give notice of his or her demands to the other party. If the demands are not met within a limited time the parties are deemed to be in dispute and the party making the claim can take steps to have an arbitrator appointed. There is an increasing tendency in drafting these agreements to require the parties to engage in some form of ADR before arbitration begins and the parties' attitudes harden. The making of a claim does not of itself give rise to a dispute. The High Court of Australia has held that the failure to pay an unchallenged progress or final certificate is not a dispute.[10]

Identification of the arbitrator

The standard forms of contract contain provisions allowing the parties to identify the organization that they desire to nominate the arbitrator. If the parties do not make a nomination there is a default position whereby the nomination is made by the president of an interested organization. The original practice of vesting this gift in a trade association such as the Master Builders Association of NSW has fallen into disuse because of the perceptions of bias. In Australia, this role is frequently performed by the Institution of Arbitrators Australia; in Singapore, by the Singapore International Arbitration Centre and, in Malaysia, by the Regional Centre for Arbitration. This development is to be welcomed as, in addition to the perceived neutrality of these organizations, they seek to raise the standard of arbitration and educate arbitrators generally. Even so, parties entering a contract should think about this matter carefully. The level of experience and skill of the arbitrators on the lists maintained by these organizations (at least in Australia) varies. There is a lot to be said for the parties to a contract identifying an arbitrator at the time the contract is entered. Once a dispute has arisen, the party seeking the nomination of an arbitrator needs to request the president of the nominating body to make a nomination. The request should be accompanied by documentary evidence verifying the president's right to make such nomination. Usually a copy of the relevant contract clause and the correspondence evidencing the dispute will suffice.

Preliminary conference

Once nominated, the proposed arbitrator will call the parties to a preliminary conference. At the conference the arbitrator will see that the parties are agreeable to him or her hearing the matter in terms of the proposed arbitrators' impartiality, availability and the parties' willingness to meet the fees requested. Once these matters are resolved, the arbitrator will set a timetable for the parties to prepare pleadings (written statements of the basis of their claims and defences), exchange particulars (precise details of the arithmetical and factual aspects of the claims and defences) and the discovery and inspection of documents. The purpose of these steps is to ensure that each of the parties is aware of the other's position, to encourage settlement and reduce the area of disputation. The date for the hearing is set at this meeting and the parties are under considerable pressure to meet any agreed timetable.

The hearing

In most arbitrations the hearing will proceed along adversarial lines, much the same as proceedings in court. The party making the claim (plaintiff) will be required to demonstrate his or her entitlement by adducing evidence. The witnesses are examined in chief and then cross-examined. There is, however, an increasing trend for the evidence in chief to be given in written form and the witnesses then subjected to cross-examination. The plaintiff has the obligation of showing that his or her assertions are correct. Evidence is evaluated according to the civil standard (on the balance of probabilities) and not beyond a reasonable doubt as required by the criminal law. Generally the formal rules of evidence do not apply to arbitrations unless the parties decide otherwise.[11] The arbitrator must exercise extreme care in deciding what evidence to admit and what to reject as this is a function that can easily be characterized as favouring one side or the other and a breach of the rules of natural justice.[12] With the consent of the parties, an arbitrator is entitled to use an inquisitorial method of conducting the proceedings. In Australia and New Zealand[13] it is arguable that consent might not be necessary; however, an arbitrator embarking on this course of action would need to be very mindful of the rules of natural justice. It is possible for an arbitrator to conduct the arbitration on the documents alone without oral evidence. This is the preferred procedure, for example, for resolution of London Commodity Exchange and Maritime disputes.[14]

The references in this discussion have been to *an* arbitrator. Whilst the use of the singular implies the use of one arbitrator it should be understood that it is perfectly competent for the parties to appoint more than one arbitrator. In particular, two arbitrators are often used with each party nominating an arbitrator of their choice. In these circumstances it is necessary to provide for a referee in case the arbitrators fail to agree. To avoid this problem the modern practice is to use one arbitrator only.

The award

At the conclusion of the hearing the arbitrator publishes an award to the parties thereby completing his or her work. The award must be in writing, signed by the arbitrator and include a statement of the arbitrator's reasons for making the award. In terms of content, the award must answer the issues posed by the parties in their pleadings. The parties may by agreement dispense with the need of the arbitrator to give reasons.[15] Unless the dispute is minor, parties are advised against dispensing with reasons as this makes it almost impossible for a court to review the award should one of the parties be dissatisfied. If a party fails to comply with the terms of an award (usually the payment of money), the successful party can apply to the court for leave to enforce the award as though it were an order of the court.[16] In practice this means seizing the defendant's property in satisfaction of the debt or initiating insolvency proceedings. In all jurisdictions an arbitrator has the power to order the unsuccessful party to pay the unsuccessful party's costs and the costs of the arbitration.

Judicial review of awards

Where an arbitrator has been guilty of misconduct he or she may be removed by the court or any award made set aside.[17] Misconduct does not connote any moral culpability but rather is a technical word describing a failure by the arbitrator to comply with the basic legal requirements of arbitration or a failure to observe the rules of natural justice. There are limited rights to appeal to the court on questions of law arising during the proceedings or out of the award. The question of law must be one of significance to the commercial community and not merely of concern to the parties.[18]

How effective is arbitration?

The main advantage of arbitration is the fact that the tribunal has a knowledge of the industry where the dispute arises. This to some extent reduces the need to rely on expert witnesses and thereby reduces costs. Parties, however, seem not to make as much use as is desirable of the arbitrator's skills, preferring to adduce expert witnesses on technical matters. Much of the advantage is lost and arbitration would benefit as a dispute resolution mechanism if proceedings became more inquisitorial. The English commodity trade arbitrations provide an excellent example of how this can work. Both the award and the proceedings are private and the fact of the proceedings and the result are known only to the parties and this is seen as an advantage by the commercial community. Arbitration is as expensive as proceedings in court and perhaps more so. The arbitrator has to be paid, rooms hired and transcript writers provided. These are costs normally borne by the State where the courts are utilized.

International arbitration

Where the contracting parties are nationals of different countries it is unlikely that either will wish to submit to the judicial system of the country of the other party in the event of a dispute. For this reason arbitration is the preferred method of dispute resolution for these transactions. The arbitration agreement identifies a neutral international body to appoint the arbitrator and to provide the infrastructure for the arbitration. There are several highly respected bodies that carry out this function, including the International Chamber of Commerce, with its headquarters in Paris, and the London Court of Arbitration. The arbitrations are conducted according to the rules of these bodies and a great deal of flexibility is permitted to the parties in the manner of prosecuting its case. The parties may choose the laws of which country governs the transaction, the law that determines the manner of conducting the arbitration or choose to have the issues resolved without reference to a legal system according to principles of general justice and fairness (amiable composition). The New York Convention (1927) provides that signatories to the convention recognize foreign arbitral awards and facilitate the enforcement of those awards within their legal systems. Australia, New Zealand, Singapore and Malaysia have all signed the convention. Subsequently, the United

Nations Commission on International Trade Law (UNCITRAL) has prepared a model law for the conduct of international arbitrations. To date Australia, New Zealand, Singapore and Malaysia have adopted the model law while England has not.

22.5 Alternative dispute resolution (ADR)

Where both parties wish to resolve a dispute, the chances of success using ADR are high, but they are unfortunately much less so when only one party is cooperative. Like arbitration, ADR is a consensual process and very often an arbitration agreement in a contract will be preceded by a clause requiring the parties to meet at least once to explore the chances of reaching a non-curial resolution to the dispute. Some care needs to be taken in working under these clauses because they are susceptible to abuse by parties who wish to delay the obligation to pay money. The various techniques of ADR are described below and each is basically a form of structured negotiation. Care needs to be taken in the use of these terms as they are used interchangeably. In particular, mediation often shades into conciliation. Any session of ADR will be presided over by a facilitator whose role will be to attempt to get the parties to settle the dispute. It is emphasized that it is the parties who resolve the dispute. The chairperson is sometimes referred to as a *third party neutral* and acts primarily as an agent of reality; that is, the parties are encouraged to look at their claims in a dispassionate light and drop ambit claims. There has been some discussion of the legal and ethical position of a mediator who is faced with a party who wishes to settle on terms that are unjust or clearly disadvantageous to that party. The conventional wisdom in ADR is that all settlements are good. The situation just described goes against that convention.

22.5.1 Mediation

A mediator opens by explaining that the object of the process is to encourage the parties to arrive at their own settlement. Each party is encouraged openly to articulate his or her expectations. The mediator will ask the parties to discuss the issues in each other's presence and then continue the discussion separately. He or she encourages the parties to examine the causes of the dispute, their separate and joint interests and promotes good communication. It is at the separate meetings that the mediator has the opportunity to encourage the parties to be realistic about their expectations. Beyond what has been mentioned above, the mediator remains non-interventionist during the proceedings to allow the parties to arrive at their own resolution of the dispute. If the parties do arrive at a solution it is wise for the mediator to advise them to have their legal advisers prepare the necessary documents to give effect to the agreement. This poses the slight risk that the settlement will fail; however, mediators do not have the immunity from claims for negligence that arbitrators enjoy and the fact that the parties have been independently advised on the proposed settlement is a measure of protection. The mediator will usually prepare a minute of the intended settlement for the parties. It is permissible for the parties to be legally represented in a mediation.

22.5.2 Conciliation

Conciliation is a natural extension of the mediation process. It embraces all of the steps outlined above and then involves the mediator/conciliator shifting to a more interventionist role. There is a considerable advantage in a conciliator possessing technical skills in the area of the dispute. In addition to the mediation function, the conciliator may, in an effort to bring about a settlement:

- assess the rights and obligations of the parties
- value the work performed
- give an opinion as to the outcome of formal proceedings
- make an assessment of likely damages payable.

22.5.3 Case presentation or mini trial

This is an American idea that is suitable for substantial disputes involving corporate disputants. Both parties present an outline of their evidence and contentions to a panel comprising the chief executives of each of the parties, and a chairperson who has experience in mediation. Frequently a retired judge has been used in this role. The proceedings are conducted by the executive staff of each of the corporations participating, which forces both parties to appreciate the best points of the other side's case. A provisional result is then given by the panel. If the provisional result is not accepted, the work done in preparation is not wasted as much of it can be used in later proceedings. The scheme has, on occasions, achieved spectacular results. One reason for this no doubt is the fact that the senior executives within a corporation are appraised of the way middle-management are dealing with a dispute and have the authority to compromise a matter if it is in the overall interests of the corporation to do so.

22.5.4 Independent expert appraisal

The neutral third party investigates the facts and provides an objective and impartial assessment. Usually this assessment will relate to the value, measurement or quality of some aspect of a project. The report of an expert can be used in settlement negotiations or the parties can agree to make it binding. If the latter course is followed, it is important to realize that using an expert's report in this way must be distinguished from arbitration. On occasions the distinction is subtle. Generally, arbitration is a quasi-judicial proceeding carrying with it the obligation on the arbitrator to act fairly and hear both parties. An arbitrator's award can be enforced in the same manner as a judgment of the court. The report of an expert is produced pursuant to a contractual arrangement between the parties to obtain and abide by the report. If a party fails to comply with an expert's appraisal the other party must sue for breach of contract. This distinction must be borne in mind when drawing dispute resolution clauses as the modern trend is to combine both arbitration and expert appraisal.[19]

22.6 Conclusions

Disputes are so common as to make it wise for the parties to make provision for their resolution at the outset. There are a number of strategies available and a wise contracting party will choose a combination of these strategies. Disputes involving large sums of money and complex legal issues seem, in one way and another, to come before the courts. If this is the likely situation, the contract should reflect this and allow a party in dispute to choose either arbitration or proceedings in court.[20] All standard contracts published in Australia still rely on arbitration as the prime method of dispute resolution.[21] In addition, all require some resort to ADR[22] prior to commencing litigation. One contract uses expert appraisal for resolving minor disputes during the execution of the works.[23] These provisions make a great deal of sense as 'minor' disputes have the potential to delay the works with serious consequences and cumulatively may trigger litigation that lasts for years after the project is completed. For this reason it is wise to draft time bar clauses to ensure that these matters are dealt with as the work progresses.

Endnotes

1. As Mahoney J.A. pointed out in *Ferris v. Plaister* (1994) 34 NSWLR 475 at pp. 494–5: 'there is no magic wand for settlement of disputes.'
2. Per Lord Wilkinson-Brown in *Linden Gardens Trust Ltd v. Lenesta Sludge Disposal Ltd* [1994] 1 AC 85 at p. 105.
3. These decisions culminated with the decision *Taylor Woodrow International Ltd v. The Minister of Health* 19 SASR 1 where a clause in a contract permitting the contractor to recover 'loss and expense' for most delays effectively shifted the risk for the performance of the subcontractors on to the principal.
4. In NSW, the Supreme Court maintains a list of suitable experts. The persons on the list are usually members of the Institute of Arbitrators Australia and are members of the architectural, building, engineering or quantity surveying professions.
5. In NSW the power of the court to refer a matter to a referee is provided for in Part 72 of the Supreme Court Rules.
6. *Vynior's Case* (1609) 8 Co. Rep, 81b.
7. Arbitration Act 1698, 9 & 10 William III c.15.
8. *Commercial Arbitration Act 1984* (NSW) s4; it is noted that this statute is reproduced in all Australian State and Territorial jurisdictions and is referred to as the *Uniform Arbitration Acts* (Aust), which title has been used throughout this chapter. This is not so in New Zealand where the agreement may be oral or in writing, *Arbitration Act 1996* First Schedule s7. In Singapore and Malaysia, failure to use a written arbitration agreement takes the arbitration outside the scope of the *Singapore Arbitration Act 1985* s2 and the *Malaysian Arbitration Act 1952* s2 thereby depriving the successful party of the machinery of the Act to enforce the award.

9. For example, in Australia it has been held that, given the appropriate words in the arbitration agreement, an arbitrator can grant relief under the *Trade Practices Act 1974* (Federal), *IBM Australia Ltd v. National Distribution Service Ltd* (1991) 22 NSWLR 466. In Australia, if it were desired that the arbitrator have inquisitorial powers, it would be wise to make mention of this in the arbitration agreement although it is arguable that s19(3) of the *Uniform Arbitration Acts* gives such a right.

10. *Plucis v. Fryer* (1967) 41 ALJR 192.

11. *Uniform Arbitration Acts* (Aust) s19(3), *Arbitration Act 1996* (New Zealand) Schedule 1, article 19. The position is less clear in Singapore where s2(1) of the Evidence Act 1990 provides that the Act does not apply to arbitration although arbitrators as a matter of practice have some regard to the rules. The position is similar in Malaysia.

12. The rules of natural justice are said to be encapsulated in the Latin maxims *Nemo judex in causa sua* (a person cannot be a judge in his own cause) and *Audi alteram partem* (both parties must be give an opportunity to be heard). The classic statement of the rule is that of Lord Hewart in *The King v. Sussex Justices* [1924] 1 KB 256 at p. 259: '[J]ustice should not only be done, but it should manifestly and undoubtedly be seen to be done.' See also the remarks of Cole J. in *Xuereb v. Viola* (1989) 18 NSWLR 453 where at p. 472 his Honour pointed out that it was necessary for a party to have 'a fair opportunity to place their evidence and arguments'. For a Malaysian example see *Tan Kooi Neoh v. Chuah Tye Imm* [1958] MLJ 123.

13. *Uniform Arbitration Acts* (Aust) s14, Arbitration Act 1996 (New Zealand) Schedule 2, article 3.

14. These arbitrations are said to enjoy a very high level of satisfaction with the litigants.

15. *Uniform Arbitration Acts* (Aust) s29, *Arbitration Act 1996* (New Zealand) Schedule 1, article 31.

16. *Uniform Arbitration Acts* (Aust) s33, *Arbitration Act 1996* (New Zealand) Schedule 1, article 35, Singapore Arbitration Act s20 and Malaysian Arbitration Act 1952 s27.

17. *Uniform Arbitration Acts* (Aust) s42 & s44, *Arbitration Act 1996* (New Zealand) Schedule 1, article 34, Singapore Arbitration Act s17 and Malaysian Arbitration Act 1952 s24.

18. *Uniform Arbitration Acts* (Aust) s38 & s39, *Arbitration Act 1996* (New Zealand) Schedule 5, article 5; the law of New Zealand does not require that the point be of general significance but one substantially affecting the rights of one or more of the parties, and Singapore Arbitration Act s28 & s30.

19. The distinction is well explained by Davenport (1995, p. 216).

20. This policy has been adopted in the JCC documents published by Master Builders Australia, The Royal Australian Institute of Architects and the Property Council clause 13.03, the CIC1 published by the RAIA clause 7.2 and AS2124 published by Standards Australia clause 47.2.

21. JCC clause 13.04, CIC1 clause 7.3, AS2124 clause 47.3, PC1 published by the Property Council clause 15.11, AS4000 published by Standards Australia

clause 42.3 and C21 published by NSW Department of Public Works and Services clause 63.
22. JCC clause 13.02, CIC1 clause M7, AS2124 clause 47.2 alternative 1, PC1 clause 15.12, AS4000 clause 42.2 and C21 clause 60.
23. PC1 clause 15.2.

References and bibliography

Davenport, P. (1995) *Construction Claims* (Sydney: The Federation Press).

Dorter, J.B. and Sharkey, J.J.A. (1990) *Building and Construction Contracts in Australia Law and Practice* (Sydney: The Law Book Company).

Jacobs, M.S. (1990) *Commercial Arbitration Law Practice* (Sydney: The Law Book Company).

Robinson, N.M., Lavers, A.P., Tan, K.H. and Chan, R. (1996) *Construction Law in Singapore and Malaysia*. Second edition (Butterworths Asia).

23

The way forward

Gerard de Valence and Rick Best†

23.1 Introduction

The theme that has been pursued throughout this book is the search for value in buildings. In following this theme the authors have introduced a broad range of concepts and techniques that can be considered and applied during the earliest stages of the procurement process. There is also a wealth of background information which underpins the whole process of project initiation and planning.

23.2 Winds of change

The salient points that have emerged from the various discussions, and which are, in many cases, common to the discussions of different authors are as follows:

1. The perceived importance of a greater integration of the activities of those who are participants in the complex process which begins with a 'notion to build' and ends with post-occupancy evaluation.
2. A move, by clients, towards a single point of responsibility for building projects, coupled with the use of a variety of alternative delivery systems.
3. A broadening of the concept of value for money, particularly as more professional clients move towards the concept of life-cost planning, rather than concentrating almost exclusively on initial capital cost.
4. The increasing use of a variety of techniques for systematic project analysis as a basis for decision-making.
5. A change in the nature of buildings, with increased complexity resulting from an increase in services (in buildings such as hospitals and hotels), or greater reliance on passive engineering (solar power and heating, natural ventilation, daylighting).
6. A growing awareness of the effects of indoor environments on the health and performance of building occupants.

† University of Technology, Sydney, Australia

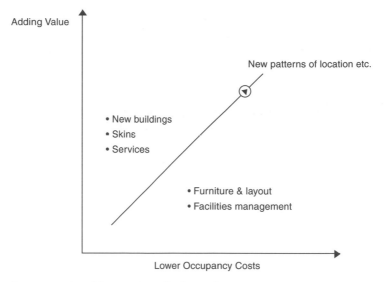

Fig. 23.1 Client expectations (after Francis Duffy of DEGW).

7. The increasing emphasis on the environmental performance of buildings, internally and externally, i.e. in terms of the quality of the indoor environment, and the effects and impacts which buildings have on their surroundings at local, regional, national and global levels.

Figure 23.1 shows one approach to these issues. Clients are seeking two main outcomes from the building industry; first, they want lower occupancy costs and better utilization of the space they use and pay for. Secondly, they want industry partners who will be able to add value to their businesses and projects. This should not be seen as an overall change in client objectives, rather it demonstrates that clients are appreciating the principles of value for money and are seeking this through a balance between the diamond of scope, time, cost and quality, and the broader objective of adding value to their business.

To deliver on this broader objective, the way that projects are conceived, initiated and delivered needs to be re-thought and re-engineered, with the focus moving from the product (buildings and structures) to the process (managing the flow of information, materials and resources). In doing so the traditional approaches taken by many of the consultants, contractors and suppliers in the building industry will be challenged.

23.2.1 Integration and teamwork

The importance of greater coordination and teamwork throughout the procurement process is a recurring theme. This is not to say that the concept of a team approach is new; the project management approach to construction which has become so dominant in many places in the past two decades reflects the need for a highly structured and professional approach to the management of the team of people who are involved in the delivery of building projects. The emergence of the project manager, a highly skilled individual whose purpose is the efficient management of the team, has been a response to the increased complexity of both the products and the process of the design and construction of buildings. With that increased complexity have come greater demands for coordination at all stages of the process.

The outstanding success of projects such as the ING (formerly NMB) Bank Headquarters is evidence of the benefits that a true team approach can produce (see Chapters 18 and 21); however, there is still a feeling that this mode of work amounts to 'design by committee', and therefore pushes compromise to a point where the end result satisfies no one. There are those in the industry, however, who believe that this approach is the correct one; for example, British architect Tom Jestico:

> The problem we all face is that the construction process militates against teamwork. All too often we are thrown together and told to produce a building when the consultants have never met each other.
>
> I prefer the synthesis of teamwork – the idea of buying into a project, writing down the mission and then refining it. Bunn (1998, p. 25)

Advances in information technology and the emergence of integrated communications networks and common protocols have facilitated the establishment of the shared electronic database. This, together with the greater processing power of current computers and improved software, provides the foundation for previously unimagined cooperative working among consultants, and much finer control of all aspects of the process by clients and their representatives.

23.2.2 Alternative delivery systems

Project management independent of the design team was an initial move towards single-point responsibility in project delivery; more recently there has been a strong move to more innovative methods of procurement, generally characterized by the existence of an entity that carries the responsibility for most or all of the aspects of procurement. This 'turnkey' approach is not new, but in recent years it has been used much more widely as clients believe that their projects will be delivered more quickly and more cheaply when this method is adopted. Generally known as design and build (D&B) or design and construct (D&C), this method is considered by many to be both more convenient and more efficient, with design and construction combined under a single management umbrella, and better opportunities for the use of concurrent engineering techniques.

The impact of the D&B approach on the quality (and hence the value) of buildings produced is problematic; those who question the worth of D&B suggest that buildings procured under D&B contracts are often of lower quality than would be the case had a more traditional approach been adopted. Common criticisms include problems which result from design and construction proceeding concurrently, and cost-cutting during construction through the deletion of components or reductions in the quality of components due to unexpected overruns in the earlier stages of the process.

Clients who are looking for maximum value for money need to clarify their expectations of quality and value as early as possible, and then choose a delivery system that will best satisfy those expectations. One of the outcomes might include increased emphasis on build and maintain contracts, or warranty and insurance schemes for buildings, or other measures (such as those suggested in the Latham Report, 1994) as clients seek value from ongoing performance rather than just construction. Clients will increasingly use alternative delivery systems because the advantages tend to outweigh the disadvantages from the client's perspective. The use of D&B, BOT and BOOT on projects in rapidly industrializing countries has raised familiarity with alternative procurement methods, and has often involved complex consortia or joint venture structures. At the same time clients in the developed countries are increasingly using non-traditional forms of procurement on a wider range of projects as they redefine their expectations of the industry.

23.2.3 Value and time

Life cycle costing is a technique that evaluates the total cost of ownership, including capital and running costs, over the life of a building. The important elements in life cycle costing are energy use, repair and replacement of fittings and finishes, structural renewal, and replacement of engineering services components. Total operating costs over the life of a building may be as much as ten times the original capital costs, as buildings typically have an economic life of 50 years or more, while the useful life of many services components varies from just a few years up to around 30 years. Consequently, building services will need to be almost totally replaced over the life of a building, and many components will be replaced many times.

Actual construction costs reflect the compromise between minimizing life costs and minimizing project costs as design decisions which reduce life costs often result in higher capital costs. Life cycle costing can lead to better financial performance for clients in the long run as a result of more inclusive investment analysis. Occupants benefit from better buildings, while the community will often benefit from the improved environmental performance of buildings.

23.2.4 Problem definition and analysis

The value a client ultimately gains from a building is largely determined during the earliest stages of the procurement process. There are now a number of well-established techniques available to assist decision-makers to arrive at balanced conclusions about financial strategies and project selection, and to assist designers in problem definition and analysis, briefing and conceptual planning. Some of these techniques can be applied in different ways at different stages of the process, others may be merely included as part of a strategic management framework at the beginning of the procurement process but not actually implemented until later.

A number of these techniques have been described here; what is common to them all is that they seek, through a structured and systematic analytical approach, to break projects and project components down into manageable pieces. These pieces can then be examined, individually and collectively, in order to increase the understanding of the project parameters by the various parties involved in the decision to build, and the subsequent design and construction process.

Improved understanding of client expectations, more and better project information, and clearer problem definition all help in the quest for better value for money.

23.2.5 Building services and passive engineering

There is good reason to believe that clients in the years ahead will increasingly see building services as the area that offers the greatest scope for adding value to their buildings. There are two factors at work here: first, the quality of building services is a major determinant of the life cost of the buildings which house them, and secondly, the return on a building (however measured) depends in large part on the quality of the services.

Over the past few decades the share of total building costs accounted for by building services, for many categories of building, has shown a steady increase. This is due to the increased sophistication of space conditioning systems, greater use of building management and control systems, more comprehensive security and communications systems, and increased costs associated with code compliance. The result is that many buildings are significantly more complex now than their counterparts of a decade or two ago, and cost more to own and operate. As a result of these changes, the average building now has between a quarter and a third of its capital cost in services, while in services-intensive buildings, such as hospitals and premium office buildings, the proportion may be as high as 50%.

By contrast, there is a new generation of buildings in which services costs represent a decreasing proportion of total capital cost and which also offer greatly reduced life costs. These buildings rely on passive engineering techniques such as natural ventilation and solar engineering; energy use and running costs are substantially reduced. Those who construct these buildings may pay a premium in terms of increased initial outlay in order to gain long-term saving, but there are many who believe that these buildings should not be more expensive (e.g. Fedrizzi, 1995; Jestico, quoted in Bunn, 1998) in the first instance.

What is clear, however, is that the proportion of total cost attributable to services will generally decrease while the cost of the building structure will increase. This requires cost professionals to reconsider their conventional view of the apportionment of budgets to various building elements. Value for money for the client will depend heavily on the correct appraisal of alternatives with sufficient weight being given to all the functions of components, and a thorough appreciation of the designers' intentions.

23.2.6 Indoor environmental quality

Several authors have emphasized the critical relationship that exists between indoor environments and the health and productivity of the people who live and work in them. The most important point to note here that it is often the case that buildings which provide the best conditions are often also those which cost less to operate as they rely more heavily on passive engineering.

Building owners and tenants are finding that those who actually occupy their buildings have strong feelings about things such as being able to open windows or control the lighting levels in their work area. Having control over their surroundings is important to many people and when such control is given to them it has demonstrable effects on attitude, attendance, efficiency and output.

23.2.7 Environmental performance

There is increasing pressure on those who procure, design, construct and operate buildings to continually improve the performance of buildings from an ecological or environmental standpoint. Some of this pressure comes from legislation as energy standards, building codes, waste management regulations and the like are being brought in to combat ongoing environmental degradation resulting from greenhouse gas emissions, resource depletion, water, air and soil pollution, deforestation, loss of biodiversity and so on.

It has now become imperative for those who plan to build to ensure that any development proposal satisfies the ever more rigorous environmental planning and environmental impact assessment requirements that are now in place in many parts of the world. Failure to address those concerns at the earliest possible stage leaves developers exposed to the possibility of having to make major amendments later in the process or risk outright rejection of proposals that do not meet those requirements.

Further pressure is being applied by individuals and organizations who are concerned about the environment, and the general public is beginning to question the environmental performance of many manufactured goods, including buildings. A 'green' corporate image has become a desirable commodity, as organizations can display a caring image to the public and this has become a useful marketing tool. Clearly the value which those who build ultimately gain can be greatly affected by the way in which these procedures are handled.

23.3 Conclusions

Both the process of building procurement and the products of that process are very complex. Large amounts of money are invested each year in the design, construction, renovation and demolition of buildings in all parts of the world. Any investor, whether they invest in buildings, stocks, financial institutions, gems, or anything else does so in the hope of increasing the value of what they own. Those who invest in buildings naturally hope to get the best value for their money; however, the traditional processes of project appraisal and selection, design, contracting and operation have been demonstrated to have significant shortcomings.

The pressure that has been placed on contractors to prepare detailed tender submissions in short tender periods has been noted, and also that clients often under-specify or poorly document their requirements, where a detailed brief would reduce the frequency of variations to the work during construction. Further, contractors have had reason to be frustrated knowing that the client, when assessing tenders, is likely to focus only on price and not value. This practice alone can, in part, be blamed for the industry's attitude towards providing greater value in its services to clients. In addition, the traditional practice of low bid tender assessment by clients can result in the contractor being forced into an adversarial position from the start of a project with little or no option other than to pursue profit, or minimize loss, through aggressive claims and variations practices.

References and bibliography

Bunn, R. (1998) Rodric Bunn talks to Tom Jestico. *Building Services,* September, 25.

Fedrizzi, S.R. (1995) Going green: the advent of better buildings. *ASHRAE Journal,* December, 35–8.

Latham Report (1994) *Constructing the Team: Joint Review of Procurement and Contractual Relations in the UK Construction Industry* (London: HMSO).

Index

Page numbers in **bold** refer to figures; page numbers in *italics* refer to tables.